Plant Proteomics

METHODS IN MOLECULAR BIOLOGY™

John M. Walker, Series Editor

386. Peptide Characterization and Application Protocols, edited by *Gregg B. Fields, 2007*

385. Microchip-Based Assay Systems: *Methods and Applications,* edited by *Pierre N. Floriano, 2007*

384. Capillary Electrophoresis: *Methods and Protocols,* edited by *Philippe Schmitt-Kopplin, 2007*

383. Cancer Genomics and Proteomics: *Methods and Protocols,* edited by *Paul B. Fisher, 2007*

382. Microarrays, Second Edition: *Volume 2, Applications and Data Analysis,* edited by *Jang B. Rampal, 2007*

381. Microarrays, Second Edition: *Volume 1, Synthesis Methods,* edited by *Jang B. Rampal, 2007*

380. Immunological Tolerance: *Methods and Protocols,* edited by *Paul J. Fairchild, 2007*

379. Glycovirology Protocols, edited by *Richard J. Sugrue, 2007*

378. Monoclonal Antibodies: *Methods and Protocols,* edited by *Maher Albitar, 2007*

377. Microarray Data Analysis: *Methods and Applications,* edited by *Michael J. Korenberg, 2007*

376. Linkage Disequilibrium and Association Mapping: *Analysis and Application,* edited by *Andrew R. Collins, 2007*

375. In Vitro Transcription and Translation Protocols: *Second Edition,* edited by *Guido Grandi, 2007*

374. Biological Applications of Quantum Dots, edited by *Marcel Bruchez and Charles Z. Hotz, 2007*

373. Pyrosequencing® Protocols, edited by *Sharon Marsh, 2007*

372. Mitochondrial Genomics and Proteomics Protocols, edited by *Dario Leister and Johannes Herrmann, 2007*

371. Biological Aging: *Methods and Protocols,* edited by *Trygve O. Tollefsbol, 2007*

370. Adhesion Protein Protocols, *Second Edition,* edited by *Amanda S. Coutts, 2007*

369. Electron Microscopy*: Methods and Protocols, Second Edition,* edited by *John Kuo, 2007*

368. Cryopreservation and Freeze-Drying Protocols, *Second Edition,* edited by *John G. Day and Glyn Stacey, 2007*

367. Mass Spectrometry Data Analysis in Proteomics, edited by *Rune Matthiesen, 2007*

366. Cardiac Gene Expression: *Methods and Protocols,* edited by *Jun Zhang and Gregg Rokosh, 2007*

365. Protein Phosphatase Protocols: edited by *Greg Moorhead, 2007*

364. Macromolecular Crystallography Protocols: *Volume 2, Structure Determination,* edited by *Sylvie Doublié, 2007*

363. Macromolecular Crystallography Protocols: *Volume 1, Preparation and Crystallization of Macromolecules,* edited by *Sylvie Doublié, 2007*

362. Circadian Rhythms: *Methods and Protocols,* edited by *Ezio Rosato, 2007*

361. Target Discovery and Validation Reviews and Protocols: *Emerging Molecular Targets and Treatment Options, Volume 2,* edited by *Mouldy Sioud, 2007*

360. Target Discovery and Validation Reviews and Protocols: *Emerging Strategies for Targets and Biomarker Discovery, Volume 1,* edited by *Mouldy Sioud, 2007*

359. Quantitative Proteomics by Mass Spectrometry, edited by *Salvatore Sechi, 2007*

358. Metabolomics: *Methods and Protocols,* edited by *Wolfram Weckwerth, 2007*

357. Cardiovascular Proteomics: *Methods and Protocols,* edited by *Fernando Vivanco, 2006*

356. High-Content Screening: *A Powerful Approach to Systems Cell Biology and Drug Discovery,* edited by *D. Lansing Taylor, Jeffrey Haskins, and Ken Guiliano, and 2007*

355. Plant Proteomics: *Methods and Protocols,* edited by *Hervé Thiellement, Michel Zivy, Catherine Damerval, and Valerie Mechin, 2007*

354. Plant–Pathogen Interactions: *Methods and Protocols,* edited by *Pamela C. Ronald, 2006*

353. Protocols for Nucleic Acid Analysis by Nonradioactive Probes, *Second Edition,* edited by *Elena Hilario and John Mackay, 2006*

352. Protein Engineering Protocols, edited by *Kristian Müller and Katja Arndt, 2006*

351. *C. elegans:* *Methods and Applications,* edited by *Kevin Strange, 2006*

350. Protein Folding Protocols, edited by *Yawen Bai and Ruth Nussinov 2007*

349. YAC Protocols, *Second Edition,* edited by *Alasdair MacKenzie, 2006*

348. Nuclear Transfer Protocols: *Cell Reprogramming and Transgenesis,* edited by *Paul J. Verma and Alan Trounson, 2006*

347. Glycobiology Protocols, edited by *Inka Brockhausen, 2006*

346. *Dictyostelium discoideum* Protocols, edited by *Ludwig Eichinger and Francisco Rivero, 2006*

345. Diagnostic Bacteriology Protocols, *Second Edition,* edited by *Louise O'Connor, 2006*

344. Agrobacterium Protocols, *Second Edition: Volume 2,* edited by *Kan Wang, 2006*

343. Agrobacterium Protocols, *Second Edition: Volume 1,* edited by *Kan Wang, 2006*

342. MicroRNA Protocols, edited by *Shao-Yao Ying, 2006*

341. Cell–Cell Interactions: *Methods and Protocols,* edited by *Sean P. Colgan, 2006*

340. Protein Design: *Methods and Applications,* edited by *Raphael Guerois and Manuela López de la Paz, 2006*

339. Microchip Capillary Electrophoresis: *Methods and Protocols,* edited by *Charles S. Henry, 2006*

METHODS IN MOLECULAR BIOLOGY™

Plant Proteomics

Methods and Protocols

Edited by

Hervé Thiellement

Michel Zivy

Catherine Damerval

UMR de Génétique Végétale, La Ferme du Moulon, Gif-sur-Yvette, France

Valérie Méchin

UMR de Chimie Biologique, Thiverval Grignon, France

HUMANA PRESS ✻ TOTOWA, NEW JERSEY

999 Riverview Drive, Suite 208
Totowa, New Jersey 07512

www.humanapress.com

This publication is printed on acid-free paper. ∞
ANSI Z39.48-1984 (American Standards Institute)

Permanence of Paper for Printed Library Materials.

Cover illustration: Figure 1 of Chapter 3, "Protein Extraction from Cereal Seeds," by G. Branlard and E. Bancel.

Cover design by Patricia F. Cleary.

For additional copies, pricing for bulk purchases, and/or information about other Humana titles, contact Humana at the above address or at any of the following numbers: Tel.: 973-256-1699; Fax: 973-256-8341; E-mail: orders@humanapr.com; or visit our Website: www.humanapress.com

Printed in the United States of America. 10 9 8 7 6 5 4 3 2 1

eISBN 1-59745-227-0; 13-digit ISBN 978-1-58829-635-1

Library of Congress Cataloging in Publication Data

Plant proteomics : methods and protocols / edited by Hervé Thiellement ... [et al.].

p. cm. —(Methods in molecular biology ; 355)

Includes bibliographical references and index.

ISBN 1-58829-635-0 (alk. paper)

1. Plant proteins. 2. Plant proteomics. I. Thiellement, Hervé. II. Methods in molecular biology ; v. 355

QK898.P8P53 2006

772'.62—dc22 2006041207

Preface

The aim of *Plant Proteomics: Methods and Protocols* is to present up-to-date methods and protocols used by recognized scientists in the world of plant proteomics. If this world was a very small one twenty-five years ago when the first papers were published, it has since experienced exponential growth, and in most countries around the world there are laboratories working on plant proteomics.

Two-dimensional gel electrophoresis is still the basic method used, but it has been improved greatly with IPG in the first dimension (Chapter 13) and with new detection methods with fluorochromes (Chapters 14 and 15). Significant progress has been achieved in protein extraction, which is particularly difficult with plant tissues containing phenols, proteases, and other secondary metabolites that interfere with proteins. Standard procedures have been optimized (Chapters 1 and 2) for peculiar tissues (Chapters 3, 4, and 5) and cellular compartments (Chapters 6 to 10). These methods rely on improvements made in the solubilization of proteins from membranes (Chapters 11 and 12). Mass spectrometry was a revolution that permitted the high throughput identification of proteins separated by 2D gels (Chapters 19 and 20) but also from blue native 1D gels (Chapters 27 and 28) despite the fact that Edman sequencing can still be useful (Chapter 18). Associated with other techniques such as 2DLC or LC of intact proteins, mass spectrometry also permits the identification of polypeptides from complexes (Chapters 21 and 22). A rapidly expending area of plant proteomics concerns the analysis of post translational modifications and protein–protein relationships (Chapters 24 to 29). Finally, proteomics produces different kinds of data (qualitative and quantitative data, spectrometric data) that must be organized in databases to be shared by the community (Chapter 23) and statistic tools are necessary to analyze the abundance of data efficiently (Chapters 16,17)

The authors contribute by explaining in full detail their experimental methodology, either in the wet or in the dry lab, allowing a novice to successfully undertake the described method. Experts in proteomics will also appreciate the chapters dealing with approaches with which they are not familiar.

Hervé Thiellement
Michel Zivy
Catherine Damerval
Valérie Méchin

Contents

Contributors

AJITH ANAND • *The Samuel Roberts Noble Foundation, Ardmore, Oklahoma*
EMMANUELLE BANCEL • *INRA Station d'Amélioration et Santé des Plantes, Clermont Ferrand, France*
MURIEL BARDOR • *CNRS UMR 6037, IFRMP 23, GDR 2590, Université de Rouen, Mont Saint-Aignan, France*
GÉRARD BRANLARD • *INRA Station d'Amélioration et Santé des Plantes, Clermont Ferrand, France*
HANS-PETER BRAUN • *Abteilung Angewandte Genetik, Naturwissenschaftliche Fakultät, Universität Hannover, Hannover, Germany*
LINDA BRECI • *Department of Chemistry, The University of Arizona, Tucson, Arizona*
DELPHINE CENTENO • *Laboratoire de Protéomique, INRA, Montpellier, France*
JAMILA CHAÏB • *INRA Avignon, UR de Génétique et Amélioration des Fruits et Légumes, Montfavet, France*
FRANÇOIS CHEVALIER • *Laboratoire de Protéomique, INRA, Montpellier, France*
MIREILLE CHEVALLET • *Laboratoire d'Immunologie, CEA, Grenoble, France*
CATHERINE DAMERVAL • *UMR de Génétique Végétale, La Ferme du Moulon, Gif-sur-Yvette, France*
FRÉDÉRIC DELOM • *Plant Biochemistry & Molecular Physiology, INRA, Montpellier, France*
KÅRE ENGKILDE • *BioCentrum-DTU, Biochemistry and Nutrition Group, Technical University of Denmark, Lyngby, Denmark*
HOLGER EUBEL • *ARC Centre of Excellence in Plant Energy Biology, The University of Western Australia, Perth, Australia*
MIREILLE FAUROBERT • *INRA Avignon, UR de Génétique et Amélioration des Fruits et Légumes, Montfavet, France*
LOÏC FAYE • *CNRS UMR 6037, IFRMP 23, GDR 2590, Université de Rouen, Mont Saint-Aignan, France*
TANJA FEILNER • *Department of Biochemistry, University College Cork, Cork, Ireland*
ANNE-CATHERINE FITCHETTE • *CNRS UMR 6037, IFRMP 23, GDR 2590, Université de Rouen, Mont Saint-Aignan, France*
LISA GIACOMELLI • *Department of Plant Biology, Cornell University, Ithaca, New York*
RÉMY GIBRAT • *Plant Biochemistry & Molecular Physiology, INRA, Montpellier, France*

FERNANDO GONZÁLEZ-CAMACHO • *Centro de Investigaciones Biológicas (CSIC), Madrid, Spain*
ANGELIKA GÖRG • *Technische Universität München, Fachgebiet Proteomik, Freising-Weihenstephan, Germany*
PAUL A. HAYNES • *Bio5 Institute for Collaborative Bioresearch and Department of Biochemistry and Molecular Biophysics1, The University of Arizona, Tucson, Arizona*
EMILY HATTRUP • *Department of Biochemistry and Molecular Biophysics, The University of Arizona, Tucson, Arizona*
JOSHUA L. HEAZLEWOOD • *ARC Centre of Excellence in Plant Energy Biology, The University of Western Australia, Perth, Australia*
JESCO HEINEMEYER • *Abteilung Angewandte Genetik, Naturwissenschaftliche Fakultät, Universität Hannover, Hannover, Germany*
FATIMAH HICKMAN • *Bio5 Institute for Collaborative Bioresearch, The University of Arizona, Tucson, Arizona*
SUSANNE JACOBSEN • *BioCentrum-DTU, Biochemistry and Nutrition Group, Technical University of Denmark, Lyngby, Denmark*
JOHANN JOETS • *UMR de Génétique Végétale, La Ferme du Moulon, Gif-sur-Yvette, France*
MATTHEW KEELER • *Department of Biochemistry and Molecular Biophysics, The University of Arizona, Tucson, Arizona*
JULIA KEHR • *Max-Planck-Institute of Molecular Plant Physiology, Potsdam, Germany*
BIRGIT KERSTEN • *RZPD German Resource Center for Genome Research GmbH, Berlin, Germany*
SETSUKO KOMATSU • *Laboratory of Gene Regulation, Department of Molecular Genetics, National Institute of Agrobiological Sciences, Tsukuba, Japan*
OLIVIER LANGELLA • *UMR de Génétique Végétale, La Ferme du Moulon, Gif-sur-Yvette, France*
CÉLINE LALANNE • *INRA, UMR BIOGECO, Equipe de Génétique, 69 route d'Arcachon, 33612 Cesta Cédex, France*
ZHENTIAN LEI • *The Samuel Roberts Noble Foundation, Ardmore, Oklahoma*
JESSICA LETARTE • *Bio5 Institute for Collaborative Bioresearch2, The University of Arizona, Tucson, Arizona*
DAGMAR LEWEJOHANN • *Abteilung Angewandte Genetik, Naturwissenschaftliche Fakultät, Universität Hannover, Hannover, Germany*
SYLVIE LUCHE • *Laboratoire d'Immunologie, CEA, Grenoble, France*
VALÉRIE MÉCHIN • *UMR de Chimie Biologique, Thiverval Grignon, France*
FRANCISCO JAVIER MEDINA • *Centro de Investigaciones Biológicas (CSIC), Madrid, Spain*

A. Harvey Millar • *ARC Centre of Excellence in Plant Energy Biology, The University of Western Australia, Perth, Australia*
Kirankumar S. Mysore • *The Samuel Roberts Noble Foundation, Ardmore, Oklahoma*
Luc Negroni • *Faculté de Médecine, Plate-forme Protéomique, Nice, France*
Esther Pelpoir • *INRA Avignon, UR de Génétique et Amélioration des Fruits et Légumes, Montfavet, France*
Jean-Benoit Peltier • *Department of Plant Biology, Cornell University, Ithaca, New York*
Christophe Plomion • *NRA, UMR BIOGECO, Equipe de Génétique, Cestas, France*
Thierry Rabilloud • *Laboratoire d'Immunologie, CEA, Grenoble, France*
Martijn Rep • *Plant Pathology, Swammerdam Institute for Life Sciences, University of Amsterdam, Amsterdam, The Netherlands*
Valérie Rofidal • *Laboratoire de Protéomique, INRA, Montpellier, France*
Michel Rossignol • *Laboratoire de Protéomique, INRA, Montpellier, France*
Véronique Santoni • *Biochimie et Physiologie Moléculaire des Plantes, INRA, Montpellier, France*
Nicolas Sommerer • *Laboratoire de Protéomique, INRA, Montpellier, France*
Ib Søndergaard • *BioCentrum-DTU, Biochemistry and Nutrition Group, Technical University of Denmark, Lyngby, Denmark*
Lloyd W. Sumner • *The Samuel Roberts Noble Foundation, Ardmore, Oklahoma*
Wojciech Szponarski • *Plant Biochemistry & Molecular Physiology, INRA, Montpellier, France*
Hervé Thiellement • *UMR de Génétique Végétale, La Ferme du Moulon, Gif-sur-Yvette, France*
Olivia Tran Dinh • *CNRS UMR 6037, IFRMP 23, GDR 2590, Université de Rouen, Mont Saint-Aignan, France*
Maria V. Turkina • *Division of Cell Biology, Linköping University, Linköping, Sweden.*
Alexander V. Vener • *Division of Cell Biology, Linköping University, Linköping, Sweden.*
Bonnie S. Watson • *The Samuel Roberts Noble Foundation, Ardmore, Oklahoma*
Walter Weiss • *Technische Universität München, Fachgebiet Proteomik, Am Forum 2, Freising-Weihenstephan, Germany*
Klaas J. van Wijk • *Department of Plant Biology, Cornell University, Ithaca, New York*
Michel Zivy • *UMR de Génétique Végétale, La Ferme du Moulon, Gif-sur-Yvette, France*

1

Total Protein Extraction with TCA-Acetone

Valérie Méchin, Catherine Damerval, and Michel Zivy

Summary

We describe a procedure allowing extraction of total proteins that performs efficiently with a large variety of plant tissues, based on simultaneous precipitation and denaturation with TCA and 2ME in cold acetone. We also describe protein solubilization prior to IEF, either in classical rod gels or in IPGs, using two different solutions. The procedure is easy to carry out. The major caveats are (1) keep samples at low temperature during extraction, and then (2) manage protein samples at about 22 to 25°C to avoid urea precipitation.

Key Words: protein extraction; protein solubilization; TCA; acetone; total proteins; chaotropes; detergents; reducing agents; plant proteomic.

1. Introduction

In the context of proteomic studies, comparison of 2D gels requires well-resolved proteins; streaking and smearing must be avoided, as well as artifacts caused by proteolysis. Protein patterns must be reproducible from gel to gel. Sample preparation is thus a crucial step prior to electrophoresis. The main difficulties with plant tissues are low cellular protein content, the presence of proteases and interfering compounds such as phenolics, pigments, lipids, nucleic acids, and others *(1)*. After extracting as many different proteins as possible, one has to solubilize them in a solution compatible with isoelectric focusing (IEF).

After working with a large variety of plant tissues, we developed a method in which proteins are denatured and precipitated in a mixture of 2-mercapto-ethanol (2ME) and trichloracetic acid (TCA) in cold acetone. It was derived from the method of Wu and Wang *(2)*, who showed that TCA precipitation efficiently inhibits protease activity in plant tissues. Proteins were then solubilized in the urea-K_2CO_3-sodium dodecyl sulfate (SDS) (UKS) solution *(3)*. This procedure gives highly reproducible gels with a good spot resolution in large

From: *Methods in Molecular Biology, vol. 335: Plant Proteomics: Methods and Protocols*
Edited by: H. Thiellement, M. Zivy, C. Damerval, and V. Méchin © Humana Press Inc., Totowa, NJ

pH and Mr ranges, and it has become very popular under the name of the TCA/acetone method ***(4,5)***. The UKS solubilization solution was developed for the first dimension in IEF rod gels: the pH gradient was generated by carrier ampholytes upon voltage application. The replacement of these classical gels by immobilized pH gradient gels (IPGs) significantly improved the reproducibility of the first-dimension separation. At the end of the 1990s, the development of commercially available large-size IPGs led to their wide adoption as IEF media. However, running protein samples on IPGs prompted modifications in solubilization procedures, because of high salt content and the presence of ionic detergent (SDS) in UKS that are not compatible with the high voltages required to perform separation in such gels ***(6)***.

We describe the different steps of our procedure, from protein precipitation and denaturation to resolubilization in a solution suitable for subsequent IEF in IPGs.

2. Materials

1. Precipitation solution: 10% TCA (w/v), 0.07% 2ME (v/v) in cold acetone. This solution must be freshly prepared and stored at –20°C until use (*see* **Note 1**). **Caution:** All three components are toxic, and the solution must be prepared under a hood on.
2. Rinsing solution: 0.07% 2ME (v/v) in cold acetone. This solution can be stored at –20°C for about 1 mo.
3. R2D2 solubilization solution: 5 *M* urea, 2 *M* thiourea, 2% 3[3-cholaminopropyl diethylammonio]-1-propane sulfonate (CHAPS; w/v), 2% *N*-decyl-*N*,*N*-dimethyl-3-ammonio-1-propane sulfate (SB3-10) (w/v), 20 m*M* dithiothreitol (DTT), 5m*M* phosphine, 0.5% Pharmalyte 4-6.5 (v/v), 0.25% Pharmalyte 3-10 (v/v) in double-distilled (dd)H_2O. The solution is slightly heated (below 30°; *see* **Note 2**) to aid urea solubilization. It is aliquoted and stored at –80°C for months (*see* **Note 3**).
4. UKS solubilization solution: 9.5 *M* urea, 5 m*M* K_2CO_3, 1.25% SDS, 5% DTT, 6% Triton X-100, 2% ampholines 3.5 to 9.5 in ddH_2O. K_2CO_3 is prepared as a 2.8 % (w/v) stock solution and SDS as a 10% (w/v) filtered stock solution, Triton X-100 is provided as a 20% solution. Then 3 mL ddH_2O are added to the other components until the urea is solubilized (by heating below 30°C; *see* **Note 2**). Solutions of 40 mL are usually prepared and then aliquoted and stored at –80°C for months.
5. Determination of protein concentration: The protein content of the samples (either in UKS or R2D2) is evaluated using the 2-D Quant Kit from Amersham Biosciences. The method involves a precipitation of proteins from the sample, followed by specific binding of copper ions to proteins as described by the manufacturers instructions (ref. 80-6483-56). The procedure is compatible with common sample preparation reagents.
6. IPG strip rehydration solution (for UKS only):
   a. Solution A: 7 *M* urea, 2 *M* thiourea, 1.4% CHAPS (w/v), 16 m*M* DTT, 5 m*M* phosphine, 0.3% Pharmalyte 3-10 (v/v) in ddH_2O.

b. Solution B: 7 *M* urea, 2 *M* thiourea, 0.5% CHAPS (w/v), 10 m*M* DTT, 5 m*M* phosphine, in ddH_2O.

3. Methods

3.1. Protein Precipitation and Denaturation

1. Grind the plant tissues in a mortar and pestle in liquid nitrogen to obtain a fine powder (*see* **Note 4**).
2. Transfer about 200 µL of powder to a weighed 2-mL Eppendorf tube. Cover with 1.8 mL of the cold TCA-2ME-acetone solution (*see* **Note 5**), mix, and then store at –20°C for 1 h. TCA and acetone denature and precipitate the proteins. The solution inactivates the phenoloxidases and oxidases, preventing phenol oxidation into quinones, which would result in protein binding into insoluble complexes. It has also been shown to inactivate proteases *(3)*, as well as phenol extraction *(7,8)*. Acetone alone allows solubilization of the pigments, lipids, and terpenoids possibly present in the tissue. 2ME prevents the formation of disulfide bonds during precipitation.
3. Centrifuge for 10 min at 10,000*g* (in a refrigerated centrifuge, below 4°C).

3.2. Rinsing with 2ME-Acetone Solution

1. Discard the supernatant and resuspend the pellet in 1.8 mL of cold rinsing solution (*see* **Note 5**). Store at –20°C for 1 h. This step eliminates the acidity caused by TCA that would impede protein recovery.
2. Centrifuge for 15 min at 10,000*g* (in a refrigerated centrifuge, below 4°C).
3. Discard the supernatant. This step is repeated twice (*see* **Note 6**).
4. Dry the pellet under vacuum for about 1 h to eliminate the acetone fully (alternatively, dry for 20–30 min in a SpeedVac without heating).
5. Weigh the pellet (*see* **Note 7**).

3.3. Protein Solubilization

1. The amount of UKS or R2D2 buffers for protein solubilization depends on the plant tissue; for instance, we use 60 µL/mg dry powder for leaf tissue (maize, rape) and 50 µL/mg dry powder for maize kernels.-
2. Resolubilization is achieved by vortexing for 1 min. At this stage the sample still contains cellular debris.
3. Centrifuge for 15 min at 10,000*g* (25°C), and collect the supernatant in a 1.5-mL Eppendorf tube.
4. Centrifuge again (15 min, 25°C), and transfer the supernatant to a new Eppendorf tube. Samples containing solubilized proteins can then be stored at –50°C or –80°C for months.

Solutions used to resuspend and solubilize proteins generally contain chaotropes, detergents, and reducing agents *(9)*. Chaotropes unfold proteins by breaking noncovalent bonds. The most commonly used chaotropes are urea and thiourea. They are often used in combination, which improves protein solubilization. Detergents are needed to improve protein solubilization in the presence of chaotropic agents. CHAPS, a zwitterionic detergent, has solubilizing properties similar to those of Triton X-100 (a nonionic detergent) but has a more powerful effect in preventing protein-protein interaction. SB3-10 is even more efficient than CHAPS in solubilizing protein, but is poorly soluble at high concentration in urea. Thiol reducing agents (DTT is the most commonly used) and phosphines prevent disulfide bond formation. Phosphine in the form of tris carboxyethyl phosphine (TCEP)-HCl (which is nonvolatile, stable, and soluble in aqueous solutions, contrary to phosphine in the form of TBP) is known to be more selective and efficient than DTT, since it keeps its reducing power at acidic pH and at pH above 7.5.

The R2D2 solution, which was specifically designed for solubilization of proteins prior to IEF in IPGs ***(6)***, associates urea with thiourea (*see* **Note 2**), CHAPS with SB3-10, and DTT with phosphine.

The UKS solution was developed earlier, before the finding that the addition of thiourea and zwitterions could increase protein solubilization. Thus, it contains high-molarity urea (close to saturation; *see* **Note 2**) instead of a mixture of urea and thiourea and Triton X-100 instead of CHAPS and SB3-10. The specificity of UKS relies mainly on the presence of the ionic detergent SDS. SDS greatly helps the solubilization of proteins and, probably because of its ability to denaturate proteins, it limits the activity of proteases that may occur even in the presence of urea at saturation ***(10)***. The presence of SDS in the sample is in principle incompatible with IEF because it is a ionic detergent, but it is very well tolerated at this concentration, probably because of the simultaneous presence of Triton X-100, with which it can form micelles that migrate to the cathodic end of the IEF gel. UKS also contains K_2CO_3, which alkalizes the buffer and therefore limits protein-protein and protein–nucleic acid interactions, as well as protease activity.

R2D2 is easier to use than UKS, because it is used for protein solubilization and IPG strip rehydration; a different rehydration solution must be prepared for UKS-solubilized samples, to compensate for the excess of SDS and other compounds. As few proteins seem to be more soluble in one or the other of these two solutions, it is recommended to stick with one. A priori UKS should be preferred only in samples in which high levels of protease activity are suspected.

In both UKS and R2D2 solutions, adding ampholytes improves the resolution in the IEF dimension.

3.4. Preparation of Samples for IEF

1. Protein samples must be centrifuged once again before use and the pellet discarded.

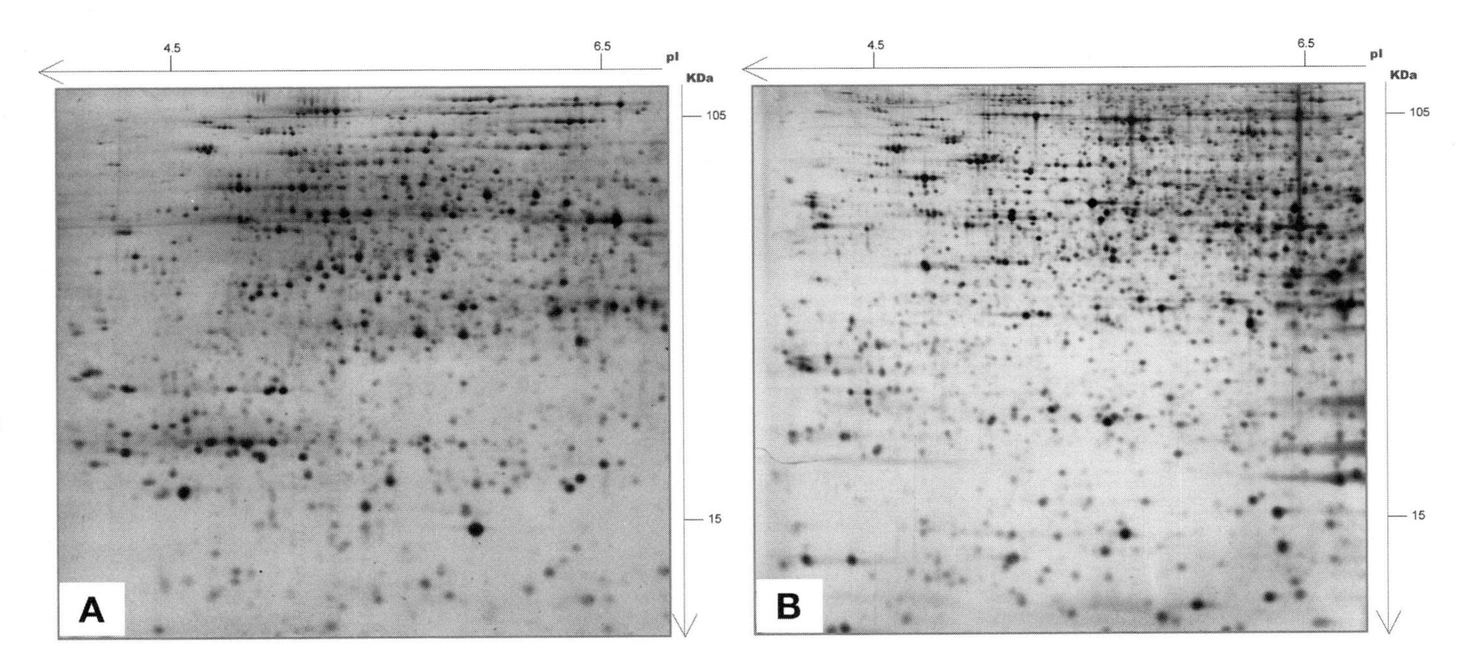

Fig. 1. 2-D gels obtained using R2D2 to solubilize proteins extracted following the TCA/acetone procedure. **(A,B)** Maize leaf and 14 DAP maize endosperm protein gels, respectively. **(C,D)** Rape stem and rape root protein gels, respectively. First 50 µg **(A,B)** or 250 µg of proteins **(C,D)** are loaded onto IPG (pH 4–7 linear) using the rehydration loading method. Passive rehydration is allowed for 1 h. Then active rehydration is performed at 22°C for 12 h at 50 V. IEF is achieved by 0.5 h at 200 V, 0.5 h at 500 V, 1 h at 1000

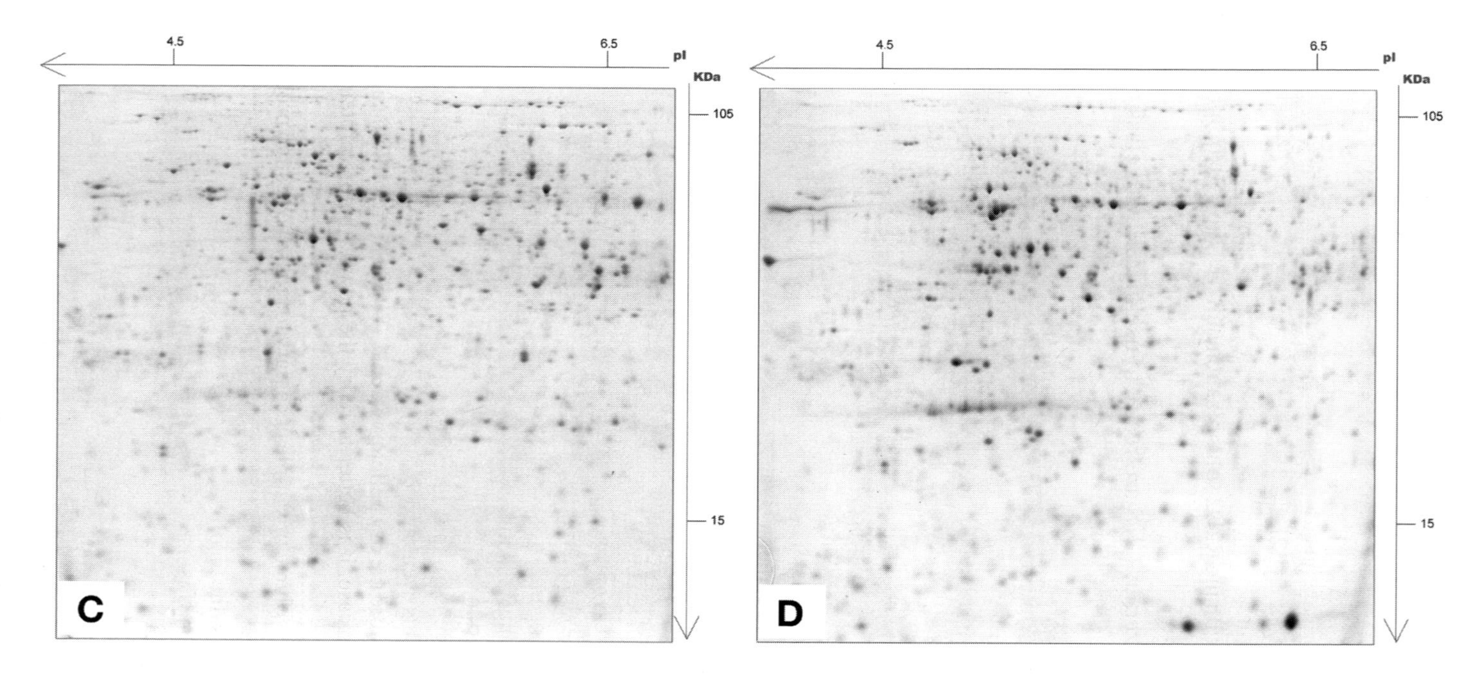

Fig. 1. (*continued*) V, and 10,000 V for the duration to reach 84,000 Vh. After IEF, IPG strips are equilibrated and then sealed at the top of the 1-mm-thick second-dimensional gel with the help of 1% low melting agarose. We use continuous 11% T, 2.67% C home-made gels with piperazine diacrylamide (PDA) as the crosslinking agent. After SDS-polyacrylamide gel electrophoresis (PAGE), proteins are visualized by silver staining **(A,B)** or colloidal Coomassie Blue G250 staining **(C,D)**.

2. About 50 µg of total proteins are loaded per IEF gel (24 cm long, either classical rod gels or IPGs) for analytical gels revealed with silver nitrate (**Fig. 1A** and **B**), or 150 to 500 µg for gels revealed with colloidal Coomassie blue (**Fig. 1C** and **D**).
3. Samples solubilized in R2D2 are ready to use in IEF with IPGs. One has just to complement the necessary volume of solubilized sample to reach the desired protein amount with the R2D2 solubilization buffer, up to 450 µL for active rehydration *(6)*.

Samples solubilized in UKS are ready to use in IEF with classical rod gels and ampholytes *(3)*. For IEF in IPGs, they must be complemented up to 450 µL with another solution (A or B; *see* **Note 8**), which brings in thiourea, CHAPS, and phosphine, thus improving resolution by lowering the proportion of salt and SDS in the sample.

4. Notes

1. The precipitation and rinsing solutions must be cold when used. Always keep an acetone bottle at 4°C to prepare these solutions.
2. Both UKS and R2D2 solutions contain high molarity of urea and must not be heated above 30°C (which may be tempting to solubilize urea), because isocyanate ions are produced that would result in protein carbamylation.
3. SB3-10 is poorly soluble in urea (this is why the urea concentration has been limited to 5 *M*), and it is necessary to be patient when melting a frozen R2D2 aliquot. An aliquot can be efficiently used when the solution is perfectly limpid (i.e., when all R2D2 constituants are solubilized).
4. A fine powder must be obtained for efficient protein extraction. This may require precrushing of hard material (e.g., mature maize grains). It is also possible to use an automatic cryogenic crusher with a metallic ball (6 mm diameter).
5. Up to pellet drying, it is important to work at a low temperature (below 4°C) to limit protease action.
6. Several rinsing steps can profitably be applied with highly pigmented samples, so that a white pellet is eventually recovered. It is possible to extend rinsing overnight.
7. It is possible to store the dry pellet powder at –80°C before protein solubilization. However, in this case, it is better to redry the powder before protein resolubilization.
8. According to the volume of sample necessary to reach the desired protein amount, solution A or B is used. When 20 to 90 µL are sufficient, solution A is used to complement the solution of UKS-solubilized proteins. When the required volume is higher, solution B is used for complementation. We thus obtain suitable ionic conditions for IEF in IPGs.

References

1. Damerval, C., Zivy, M., Granier, F., and de Vienne, D. (1988) Two-dimensional electrophoresis in plant biology, in *Advances in electrophoresis* (Chrambach, A., Dunn, M., and Radola, B., eds.), VCH, Weinheim, New York, pp. 265–340.

2. Wu, F. and Wang, M. (1984) Extraction of proteins for sodium dodecyl sulfate polyacrylamide gel electrophoresis from protease-rich plant tissues. *Anal. Biochem.* **139,** 100–103.
3. Damerval, C., de Vienne, D., Zivy, M., and Thiellement, H. (1986) Technical improvements in two-dimensional electrophoresis increase the level of genetic variation detected in wheat seedling proteins. *Electrophoresis* **7,** 52–54.
4. Saravanan, R. S. and Rose, J. K. (2004) A critical evaluation of sample extraction techniques for enhanced proteomic analysis of recalcitrant plant tissues. *Proteomics* **4,** 2522–2532.
5. Carpentier, S. C., Witters, E., Laukens, K., Deckers, P., Swennen, R., and Panis, B. (2005) Preparation of protein extracts from recalcitrant plant tissues: an evaluation of different methods for two-dimensional gel electrophoresis analysis. *Proteomics* **5,** 2497–2507.
6. Méchin, V., Consoli, L., Le Guilloux, M., and Damerval, C. (2003) An efficient solubilization buffer for plant proteins focused in immobilized pH gradients. *Proteomics* **3,** 1299–1302.
7. Hurkman, W. and Tanaka, C. (1986) Solubilization of plant membrane proteins for analysis by two-dimensional gel electrophoresis. *Plant Physiol.* **81,** 802–806.
8. Granier, F. (1988) Extraction of plant proteins for two-dimensional electrophoresis. *Electrophoresis* **9,** 712–718.
9. Herbert, B. (1999) Advances in protein solubilisation for two-dimensional electrophoresis. *Electrophoresis* **20,** 660–663.
10. Colas des Francs, C., Thiellement, H., and de Vienne, D. (1985) Analysis of leaf proteins by two-dimensional gel electrophoresis: protease action as exemplified by ribulose bisphosphate carboxylase/oxygenase degradation and procedure to avoid proteolysis during extraction. *Plant Physiol.* **78,** 178–182.

2

Phenol Extraction of Proteins for Proteomic Studies of Recalcitrant Plant Tissues

Mireille Faurobert, Esther Pelpoir, and Jamila Chaïb

Summary

Phenol extraction of proteins is an alternative method to classical TCA-acetone extraction. It allows efficient protein recovery and removes nonprotein components in the case of plant tissues rich in polysaccharides, lipids, and phenolic compounds. We present here a tried and tested protocol adapted for two dimensional electrophoresis (2-DE) and further proteomic studies. After phenol extraction, proteins are precipitated with ammonium acetate in methanol. The pelleted proteins are then resuspended in isoelectric focusing buffer, and the protein concentration is measured with a modified Bradford assay prior to electrophoresis.

The important points for successful use of this protocol are (1) keeping samples at very low temperature during the first step and (2) careful recovery of the phenolic phase after the centrifugations, which are major features of this protocol.

Key Words: extraction method; proteins; phenol; plant proteomic; membrane proteins; two-dimensional gel electrophoresis; glycoproteins.

1. Introduction

Plant protein extraction is the first step in proteomic studies. Plant tissues contain relatively low levels of proteins whose extraction is often rendered difficult by the presence of other compounds, such as cell wall and storage polysaccharides, lipids, and phenolic compounds. The solubility of plant proteins is closely associated with their intracellular localization, and proteins are classically extracted by either aqueous buffer, detergents, or direct precipitation *(1)*. Besides the most commonly used trichloroacetic acid (TCA)/acetone precipitation method *(2)*, phenol extraction followed by methanol/ammonium acetate precipitation was reported by Hurkman and Tanaka *(3)* in 1986 for proteomic studies. The authors emphasized the efficiency of the method in removing

From: *Methods in Molecular Biology, vol. 335: Plant Proteomics: Methods and Protocols*
Edited by: H. Thiellement, M. Zivy, C. Damerval, and V. Méchin © Humana Press Inc., Totowa, NJ

nucleic acids, which interact with proteins and give poor resolution and high background in two-dimensional electrophoresis (2-DE).

Phenol extraction was first developed to purify (deproteinize) carbohydrates and then nucleic acids. For molecular biologists, phenol extraction is now the standard and preferred way to remove proteins from nucleic acid solutions.

Phenol is the simplest aromatic alcohol; it contains a polar [OH] group bound to an aromatic ring. It exhibits weak acidic properties and is corrosive and poisonous. Phenol is partially miscible with water: when saturated with water the aqueous layer contains about 7% phenol and the organic layer about 28% water. It interacts with proteins mainly via hydrogen bonding and causes proteins to become denatured and soluble in the organic phase. Then, contrary to widespread belief, proteins are not in the interface but in the phenol phase.

The phenol extraction method is mainly reported for recalcitrant plant tissues or organs such as wood ***(4)*** potato and rapeseed seedlings ***(5)***; potato, apple, and banana leaves ***(6)***; olive leaf ***(7)***; and tomato, avocado and banana fruits ***(8)***.

Comparison of TCA/acetone and phenol extraction protocols led Carpentier et al. ***(6)*** and Saravanan and Rose ***(8)*** to the observation that the two methods were efficient in extracting proteins from recalcitrant tissues, but phenol extraction was most efficient in removing interfering substances and resulted in the highest quality gels with less background and less vertical streaking. The two methods minimize the protein degradation often encountered during sample preparation, owing to endogenous proteolytic activity. It was also pointed out that the phenol method yielded a greater number of glycoproteins ***(8)***.

The phenol extraction procedure has a high clean-up capacity. It also acts as a dissociating agent decreasing molecular interaction between proteins and other materials ***(6)***. The major drawbacks of the protocol are that it is time consuming (at least 6 h) and that phenol and methanol are toxic.

2. Materials

1. Phenol: Tris-HCl saturated, pH 6.6/7.9 (Amresco-Interchim, Biotechnology Grade).
2. Extraction buffer: Prepare a solution of 500 m*M* Tris-HCl, 50 m*M* EDTA, 700 m*M* sucrose, 100 m*M* KCl and adjust pH to 8.0 with HCl. This solution can be stored for a week at 4°C.
   Add just before extraction 2% β-mercaptoethanol and 1 m*M* phenylmethylsulfonyl fluoride (PMSF; *see* **Note 1**).
3. Precipitation solution: 0.1 *M* ammonium acetate in cold methanol. This solution is stored at –20°C.
4. Isoelectric focusing buffer: 9 *M* urea, 4% CHAPS, 0.5% Triton X-100, 20 m*M* DTT, 1.2% Pharmalytes pH 3 to 10 (*see* **Note 2**). Triton X-100 is provided as a 10% solution. This solution can be aliquoted and stored at –20°C for months.

5. Determination of protein concentration: protein concentration is evaluated according to a modified Bradford assay using the dye reagent from Bio-Rad (*see* **Note 3**).

3. Methods

3.1. Protein Extraction

1. Fresh plant tissue is frozen in liquid nitrogen after harvest and ground to a fine powder within precooled steel cylinders of an automatic cryogenic crusher (*see* **Note 4**).
2. Then 1 g of ground tissue is suspended in 3 mL of extraction buffer in a 15-mL Falcon tube, vortexed, and incubated by shaking for 10 min on ice (*see* **Note 5**).
3. Afterward, an equal volume of Tris-buffered phenol is added, and the solution is incubated on a shaker for 10 min at room temperature (*see* **Note 6**).
4. To separate insoluble material (in the pellet), for aqueous and organic phases, the sample is centrifuged for 10 min at 5500*g* and 4°C. The phenolic phase, which is on the top of the tube (*see* **Note 7**), is recovered carefully to avoid contact with the interphase and poured into a new tube.
5. This phenol phase is then back-extracted with 3 mL of extraction buffer. The sample is shaken for 3 min again and vortexed. Centrifugation for phase separation is repeated for 10 min at 4°C and 5500*g*.
6. The phenol phase still on the top of the tube is carefully recovered and poured into a new tube; 4 vol of precipitation solution are added. The tube is shaken by inverting, and the sample is incubated for at least 4 h or overnight at –20°C.
7. Proteins are finally pelleted by centrifugation (10 min, 5500*g* at 4°C).
8. After centrifugation, the pellet is washed three times with cooled precipitation solution and finally with cooled acetone. After each washing step, the sample is centrifuged for 5 min at 5500*g* and 4°C.
9. Finally, the pellet is dried under vacuum (*see* **Note 8**).

The proteins are first extracted in a Tris buffer containing several protecting agents. EDTA inhibits metalloproteases and polyphenol oxidases by chelating metal ions. PMSF irreversibly inhibits serine proteases. β-Mercaptoethanol is a reducing agent that prevents protein oxidation. Moreover, as a precaution against protease activity, the temperature must be kept below 4°C, and samples should be placed on ice during the first step of the extraction process (*see* **Note 5**). The extraction period should also be minimized. The presence of KCl is related to its "salting in" effect, improving the solubility and then the extraction of proteins.

An alternative method to classical phenol extraction has been proposed by Wang et al. *(7)*. Extraction is carried out in the presence of sodium dodecyl sulfate (SDS) and is termed phenol/SDS extraction. It maximized protein yields in olive leaf tissue, displayed a good 2-DE resolution, and gave more spots with increased intensity than phenol alone. However, the addition of SDS did not improve extraction in the case of banana, apple, and potato leaves *(6)*.

With sucrose, the Tris buffer is heavier than Tris-buffered phenol. So, during the phase separation the phenol phase is "pushed" on top, which facilitates recovery of the phenol phase (*see* **Note 7**).The upper phenol phase contains cytosolic and membrane proteins.

Buffering the phenol with Tris to pH 8.0 (*see* **Note 6**) ensures that nucleic acids are partitioned to the buffer phase and not to the phenol-rich phase *(6)*.

Proteins are usually precipitated by addition of salts or water-miscible organic solvents. Here a combination of both is used. Four volumes of methanol efficiently precipitate most proteins. However, methanol poorly precipitates proteins from acidic solutions. An organic base or a buffer (ammonium acetate) solves this problem.

3.2. Protein Solubilization and Quantification

1. The final pellet is resuspended in IEF buffer. In our conditions, starting with 1 g of fresh tomato fruit tissue, 200 μL of IEF buffer are needed.
2. The sample is incubated for at least 1 h (sometimes more) at room temperature under agitation. Do not heat samples; this would lead to carbamylation of proteins.
3. For quantification, several dilutions of ovalbumin standard are made in IEF buffer (8 dilutions from 0 to 60 μg/μL). Then 10 μL of 0.1 *N* HCl are added to every samples. The final volume is adjusted to 100 μL with water, either for standard curve samples or for tissue sample.
4. Then 3.5 mL of diluted dye reagent are added, and the optical density is read at 595 nm.

To estimate the protein concentration in plant samples, the Bradford assay *(9)* is more appropriate than the Lowry *(10)* and biuret methods, which are based on the quantification of phenolic compounds *(1)*. However, direct quantification in sample solubilization buffers is not possible owing to interference with IEF buffer components. We therefore use the modified procedure of Ramagli and Rodriguez *(11)*, which is based on acidification of the sample buffer. It allows direct quantitation of protein solubilized in sample buffers containing urea, carrier ampholytes, nonionic detergents, and thiol compounds.

4. Notes

1. **Caution:** β-mercaptoethanol and PMSF are toxic compounds. PMSF can be prepared as a stock solution 200 m*M* in isopropanol, aliquoted, and stored at –20°C.
2. Don't add too much water to solubilize CHAPS and urea powders; for a 25 mL final volume, add only about 10 mL of water. When preparing this solution, avoid heating above 30°C, to prevent protein carbamylation. Solubilization may take time.

3. The diluted dye reagent is prepared according to the standard macroassay procedure as described in the Bio-Rad instruction manual.
4. It is very important to obtain a fine powder; the finer it is, the more efficient are the protein extraction and the removal of contaminants. The powder should also be homogenous for accurate sample comparison.
5. At this step it is important to work at low temperature to limit protease activity.
6. Tris-buffered phenol is prepared according to the manufacturer's recommendations and is stored at 4°C. In the bottle, the phenol phase is below the Tris phase. Pipet the whole required volume at once to avoid bottle manipulation and ambiguous separation of the two phases.
7. The trick here is to use sucrose in the extraction buffer to invert the phases.
8. It is possible to delay pellet resolubilization by storing well-dried protein pellets at –80°C. Be careful to prevent rehydration of the pellet by placing it in a vacuum chamber while warming up.

References

1. Michaud, D. and Asselin, A. (1995) Application to plant proteins of gel electrophoretic methods. *J. Chromatogr. A* **698,** 263–279.
2. Damerval, C., Zivy, M., Granier, F., and de Vienne, D. (1988) Two-dimensional electrophoresis in plant biology. *Adv. Electrophoresis* **2,** 263–340.
3. Hurkman, W. J. and Tanaka, C. K. (1986) Solubilisation of plant membrane proteins for analysis by two-dimensional gel electrophoresis. *Plant Physiol.* **81,** 802–806.
4. Mijnsbrugge, K. V., Meyermans, H., Van Montagu, M., Bauw, G., and Boerjan, W. (2000) Wood formation in poplar: identification, characterization, and seasonal variation of xylem proteins. *Planta* **210,** 589–598.
5. Mihr, C. and Braun, H. P. (2003) Proteomics in plant biology, in *Handbook of Proteomics* (Conn, P., ed.), Humana Press, Totowa, NJ, pp. 409–416.
6. Carpentier, S. B., Witters, E., Laukens, K., Deckers, P., Swennen, R., and Panis, B. (2005) Preparation of protein extracts from recalcitrant plant tissues: an evaluation of different methods for two-dimensional gel electrophoresis analysis. *Proteomics* **5,** 2497–2507.
7. Wang, W., Scali, M., Vignani, R., et al. (2003) Protein extraction for two-dimensional electrophoresis from olive leaf, a plant tissue containing high levels of interfering compounds. *Electrophoresis* **24,** 2369–2375.
8. Saravanan, R. S. and Rose, J. K. C. (2004) A critical evaluation of sample extraction techniques for enhanced proteomic analysis of recalcitrant plant tissues. *Proteomics* **4,** 2522–2532.
9. Bradford, M. M. (1976) A rapid and sensitive method for the quantitation of microgram quantities of protein utilizing the principle of protein-dye binding. *Anal. Biochem.* **72,** 248–254.
10. Lowry, H., Rosebrough, J., Farr, A. L., and Randall, R. J. (1951) Protein measurement with the folin phenol reagent. *J. Biol. Chem.* **193,** 265–275.

11. Ramagli, L. S. and Rodriguez, L. V. (1985) Quantification of microgram amounts of protein in two dimensional polyacrylamide gel electrophoresis sample buffer. *Electrophoresis* **6,** 559–563.

3

Protein Extraction from Cereal Seeds

Gérard Branlard and Emmanuelle Bancel

Summary

Seeds may contain different components such as starch and complex carbohydrates that can seriously reduce protein extraction. The proteins in cereal seeds are usually classified in four groups according to their solubility criteria: albumins, globulins, prolamins, and glutelins. They can be specifically extracted. A general procedure for extracting the proteins present in green seeds or immature cereal kernels is given. Then several procedures mostly adapted to cereal seeds are reported for: (1) the whole storage proteins (mostly prolamins and glutelins); (2) the albumins–globulins extracted using salt buffer; (3) the amphiphilic proteins extracted using a phase partitioning process; and (4) the proteins strongly attached to or within the starch granules of the seed endosperm. These procedures have been used for 2-D electrophoresis and proteomic analyses.

Key Words: Seeds; albumins; globulins; amphiphilic proteins; storage proteins; starch.

1. Introduction

Extracting the proteins from seeds is generally performed without much difficulty. However, to be successful, this first step of proteomic analysis requires special care because the seed is not a homogeneous tissue. Seed composition and tissue can vary considerably between species. Soon after fertilization, cell division results in different tissues such as the endosperm, embryo, scutellum, cotyledon, aleurone layer, and envelopes, each of which has a different biochemical composition. Like most plant tissues, seeds may contain different components that can seriously reduce protein extraction or modify the protein diversity revealed using two-dimensional electrophoresis (2-DE). The peripheric layers of the seeds are often composed of hemicellulose, lignans, polyphenols, and arabinoxylans. In many species, the cell wall material is composed of arabinoxylans and arabinogalactans. The seed is often a reservoir of components, and some tissues can be rich in polysaccharides and lipids as

From: *Methods in Molecular Biology, vol. 335: Plant Proteomics: Methods and Protocols*
Edited by: H. Thiellement, M. Zivy, C. Damerval, and V. Méchin © Humana Press Inc., Totowa, NJ

well as many secondary metabolites. These tissues are generally poor in protein but often rich in proteases *(1)*. To extract the seed proteins satisfactorily, it is necessary to take certain criteria into consideration in solubilizing protein for electrophoretic analyses *(2)* and to avoid sources of variation that could seriously interfere with 2-DE separation such as vertical streaking, smearing, reduction in the number of spots revealed, and so on. For cereal kernels, the main storage components are starch and complex carbohydrates (Fig. 1), which may prevent easy extraction of the proteins.

The many proteins present in the seed are usually classified into four groups according to their solubility in specific solvents used successively to extract the proteins *(3)*. In finely ground seed, albumins are water-soluble proteins, whereas globulins have to be extracted subsequently with an NaCl solution. Both these proteins are soluble, in contrast to the storage proteins found in cereal kernels. Albumins and globulins each represent about 10% of the total proteins of cereal seeds and more than 85% of the total proteins of leguminous seeds *(4)*. Storage proteins were initially classified as prolamins and glutelins. They are both hydrophobic and specifically accumulate in small vacuoles or protein bodies. Prolamins such as gliadins, hordeins, zeins, avenins, and rizines (in wheat, barley, maize, oat, and rice, respectively), are monomeric, and they are typically soluble in 70% ethanol. Glutelins (glutenins in wheat) are generally polymerized and extracted with an acidic solvent including a reducing agent. As the two classes of storage proteins (prolamins and glutenins) are rich in both prolin and glutamine, they were included in the prolamin class *(5)* and other specific solvents are currently used to extract storage proteins *(6–8)*.

Depending on the species, it is possible that not all seed proteins will be fully extracted. Several criteria should be taken into consideration when a method of extraction is chosen. Because the proteins are not the major component in seed, they are often aggregated or linked to other components such as cell wall material or starch granules, or they may be not readily soluble owing to the coarse fragmentation of the seed resulting from milling or grounding. The latter phenomenon is crucial in the case of cereals in which genetic factors may influence the kernel hardness of the endosperm and hence the granulometry of the flour. The regularity of the granulometry of the crushed or milled seeds will influence the reproducibility of the protein extraction. The water content of the seed also influences the granulometry, and all material should be kept at constant moisture before sampling. Furthermore, even when seeds are harvested from a pure homozygous genotype, they may not be of identical age depending on their position on the spike, the pod, the capitulum, and so forth. For example 2 or 3 d may be required for the fertilization of the ovaries of all the flowers located between the medium and distal part of the

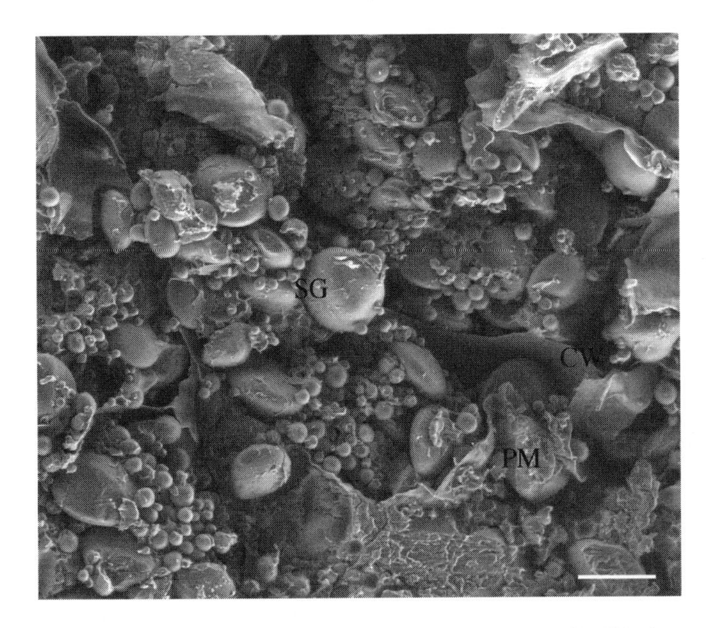

Fig. 1. Environmental scanning electron microscopy (F.E.I. Co, Eindoven, Holland) of the mature wheat kernel showing the starch granules (SG) surrounded to varying degrees by a protein matrix (PM); the cell wall (CW) is also indicated. The texture of the cereal endosperm may make complete extraction of the proteins present in the protein matrix difficult. Scale bar = 20 μm. (Photo courtesy of Gaillard-Martinie B. INRA.)

wheat spike. The resulting kernels are often of different size, and the quantitative composition of the proteins present in some of their tissues may be different.

Several classical procedures used to extract seed proteins are described in the following section. They can be divided into two types: (1) those considered as general procedures in which the major proteins present in seed can be revealed using appropriate extracting solutions, and (2) procedures better suited to cereal proteins, in which four classes can be revealed, plus one specific to starch granules.

2. Materials

All chemicals should be of analytical grade.

2.1. General Procedures for Extracting the Whole Seed Proteins

1. Extracting solution: 10% (v/v) trichloroacetic acid (TCA) diluted in glacial acetone containing 0.07% (v/v) of 2-mercaptoethanol and 0.4% (v/v) of plant cocktail inhibitor proteases (Sigma, Steinheim, Germany, cat. no. P9599). This solution, called TCA/acetone extracting solution, should be stored at –20°C.

2. Washing solution: the pelleted proteins are washed in glacial acetone solution with the same ingredients as above but without TCA, and stored at –20°C.
3. Solubilization solution: 7 *M* urea, 2 *M* thiourea, 4% (w/v) 3[3-cholaminopropyl diethylammonio]-1-propane sulfonate (CHAPS), 2% (w/v) dithiothreitol (DTT), 0.4% (v/v) protease inhibitor cocktail (Sigma), 1% (v/v) carrier ampholytes 3 to 10 and 1% (v/v) of carrier ampholytes 6 to 9.5 (or 4–7 or 6–11 depending on the IPG strip gradient used in the isoelectric focusing (IEF) step (*see* **Note 1** and **2**).
4. Alkylation solution: 4-vinylpyridine (Aldrich, cat. no. V3, 40-4)
5. Glycerol solution: 60% (v/v) glycerol diluted in water with bromophenol blue traces.
6. Protein determination: use the Bradford assay (Sigma, cat. no. B6916) or a 2D Quant kit (Amersham, cat. no. 80 6483 56) with bovine serum albumin (BSA) as the protein standard.

2.2. Specific Protein Extraction

2.2.1. Albumins and Globulins

1. Sodium-phosphate buffer: 50 m*M* sodium phosphate buffer, pH 7.8, 0.1 *M* NaCl containing 0.4% (v/v) protease inhibitor cocktail (Sigma). This extracting buffer can be stored at 4°C for up to 4 wk before use.
2. Dialysis: the dialysis is performed using the slide-A-lyzer® dialysis cassette 7K MWCO, 0.1 to 0.5 mL capacity (Pierce, www.piercenet.com).

2.2.2. Amphiphilic Proteins

1. Tris-HCl buffer: 0.1 *M* Tris, 5 m*M* EDTA, 0.25 *M* NaCl brought to pH 7.8 with HCl. This Tris-HCl buffer can be stored at 4°C for several weeks.
2. Triton X-114 solution: add 2% (v/v) Triton X-114 and 0.4% (v/v) of protease inhibitor cocktail (Sigma) to the above Tris-HCl buffer. This Triton solution should be prepared just before use.
3. Precipitation solution: 1 vol diethylether and 3 vol ethanol prepared under the fume hood before use.

2.2.3. Starch Granule Proteins

Deionized water should be used throughout the procedure.

1. Density gradient: 90% (v/v) cesium chloride (CsCl; Prolabo).
2. Washing solution: 4% (w/v) CHAPS and 2% (w/v) DTT.

3. Methods

3.1. General Procedure for Whole Seed Proteins

The purpose of the general procedure is to extract proteins from the finely crushed whole seed without considering which specific protein families could be revealed. The procedure proposed by Damerval et al. ***(9)*** and Granier ***(10)***, which is presented in Chapter 1, can be used for all seeds. It is particularly suitable for the extraction of proteins from any green mature seed but also for the early stages of cereal kernel formation either for envelope tissues, embryo or the developing endosperm, which can be expelled through the cut end by pressing the kernel. Specific tissues such as the aleurone layer, embryo, and endosperm can be isolated from the developing kernel and the mature kernel. In contrast to the embryo, which is easily isolated, the aleurone layer requires time and experience to be able to dissect under the microscope sufficient material for protein extraction and separation using 2-DE.

Given possible variation between the seeds harvested on one individual plant, it is advisable to collect the green seeds from representative or known parts of the ear, capitulum, pod, etc.

Depending on the procedure used, precipitation before solubilization (*see* **Subheading 3.1.1.**) or direct solubilization (*see* **Subheading 3.1.2.**), some differences may appear in the quantity of proteins extracted or in changes in the protein profile revealed on the 2-DE gels. In any case, once a procedure has been adopted, the same procedure should be used throughout the experiment.

3.1.1. Precipitation before Solubilization

After collection, the green material from the ear, pod, capitulum, etc., should immediately be weighed. The green seed material is frozen in liquid nitrogen before being crushed with a mortar and pestle.

1. The powder (100 mg) is immediately added with 1 mL of TCA/acetone glacial extracting solution (*see* **Note 3–5**). The proteins are left to precipitate overnight and then centrifuged at 15,000*g*, at 4°C, for 15 min.
2. The pellet is washed twice with 1 mL of the washing solution centrifuged at 15,000*g*, at 4°C for 15 min, and the supernatant is discarded (*see* **Note 6**).
3. The pellet obtained from 100 mg is air-dried to remove residual acetone and then stored at –80°C (*see* **Note 7**). This procedure allows the proteins to be concentrated before resolubilization with 1250 µL of the solubilization solution or with an IEF buffer containing specific solvents such as chaotropes and detergents.

3.1.2 Direct Resolubilization

Seed proteins can also be extracted directly from crushed kernels without previous TCA/ acetone extraction and precipitation. Detergents and chaotropic agents are often more efficient in extracting the proteins directly using the native tissue than using aggregated and precipitated forms, and this procedure also prevents protein loss caused by incomplete precipitation and/or resolubilization. The procedure described here does not include organic precipitation using glacial acetone, which is useful to concentrate proteins and remove salts and other organic compounds. This acetone precipitation procedure was proposed only recently ***(11)***. Another extracting procedure was recently reported in which proteins soluble in KCl were extracted from developing wheat kernels ***(12,13)***. In barley the proteins were first extracted using 5 m*M* Tris-HCl, pH 7.5, 1 m*M* $CaCl_2$ ***(14,15)***. Thiourea/urea lysis buffer containing protease inhibitor cocktail, Dnase I, Rnase , Triton X-100, and DTT was used for *Arabidopsis thaliana* seeds ***(16)***. The procedure described here is mainly oriented to cereal endosperm proteins that are mainly composed of storage proteins (prolamins and glutelins). It does not eliminate some other proteins; in particular many albumins and globulins will also be extracted together with storage proteins.

1. The material should be weighed immediately after collection and then crushed (*see* **Note 8** and **9**) or milled (*see* **Note 10**).
2. Add 500 µL of the solubilization solution to 40 mg of wholemeal flour in an Eppendorf tube, mix on a vortex, and incubate at room temperature for 1 1/2 h.
3. After sonication (20 W, 20 s), the slurry is let to rest 30 min and centrifuged (9000*g*, 5 min, 20°C). Then 250 µL of the supernatant is collected. The main content of the supernatant is the storage protein fraction.
4. The protein concentration in the extract can then be estimated using either the Bradford ***(17)*** assay (Sigma) or the 2D Quant kit (Amersham).
5. Add to 250 µL of the supernatant 20 µL of carrier ampholytes (according to the IPG strip gradient used in the IEF step) and 4 µL 4-vinylpyridine (*see* **Note 11**). After a contact of 15 min and vortexing, 254 µL of the glycerol solution is added (*see* **Note 12**). The protein extract can be stored at –80°C until use or analyzed directly in 2-DE.

3.2. Specific Protein Extraction

3.2.1 Albumins and Globulins

The simultaneous extraction of the two proteins classes (i.e., water-soluble albumins and salt-soluble globulins) has been optimized for the IEF. Thus, after being extracted with the sodium-phosphate buffer as described elsewhere ***(18)***, the protein extract is desalted and the proteins are then precipitated and solubilized in a buffer compatible with the IEF. Albumins and globulins of wheat kernel were recently revealed on 2-DE using this procedure ***(19)***. Whole grains (including envelopes and with or without the embryo depending on the objective) are crushed or milled following the procedure described in **Subheading 3.1.2.** above and in **Notes 8–10**. Extraction is performed as follows:

1. 100 mg of wholemeal flour is added to 1 mL of sodium phosphate buffer, and continuously stirred at 4°C for 2 h. After centrifugation (8000*g*, 10 min, 4°C), 500 µL of supernatant is collected (*see* **Note 13**).
2. Supernatant is dialyzed against cold water at 4°C for 24 h according to the manufacturer's instructions (*see* **Note 14**).
3. The desalted sample is left to precipitate with 1 mL of acetone at –20°C overnight and then centrifuged (8000*g*, 4°C, 5 min).

4. The supernatant is removed and the pelleted proteins are rinsed three times with glacial acetone and centrifuged (8000*g*, 4°C, 5 min). The pellet is finally air-dried to remove residual acetone.
5. The recovered pellet is solubilized in 250 µL of solubilization buffer (*see* **Subheading 2.1.**) with occasional vortexing. After 1 1/2 h at room temperature, the solubilized pelled is sonicated (20 W, 20 s), and the slurry is left to rest 30 min and centrifuged (9000*g*, 5 min, 20°C). The supernatant is collected.
6. The supernatant containing the protein extract is collected and supplemented with 20 µL carrier ampholytes and 4 µL of 4-vinylpyridine. The solution is vortexed and left to rest for 15 min.
7. Then 254 µL of the glycerol solution is added. The solution can be stored at –80°C until use or directly analyzed in 2-DE (*see* **Note 12**).

3.2.2. Amphiphilic Proteins

These proteins are composed of two types of amino acid sequences, one that is rich in hydrophilic amino acids (mostly Lys, Arg, and His) and the other in hydrophobic amino acids (such as Val, Leu, and Ile). Most, if not all, of the membrane proteins are amphiphilic. The extraction procedure described here was based on the sequential procedure described elsewhere (20), with some modifications for 2-DE of amphiphilic proteins of wheat seed ***(21,22)***.

1. Kernels should be crushed as described (*see* **Notes 8–10**).
2. The proteins are extracted by stirring 250 mg of flour in 7.5 mL of Triton X-114 solution at 4°C for 1 h and centrifuged at 10,000*g*, at 4°C, for 30 min.
3. The supernatant is heated at 37°C for 45 min for the phase partitioning and then centrifuged at 5000*g*, at 22°C for 10 min. The upper light phase should be removed with care (*see* **Note 15**).
4. Five volumes (approximately 10 mL) of the precipitation solution are added to the lower detergent-rich phase containing the amphiphilic proteins. The proteins are left to precipitate at –20°C overnight.
5. After centrifugation at 2000*g*, at –8°C, for 10 min, the pellet is washed three times with 10 mL of diethylether-ethanol solution and centrifuged at 2000*g*, at –8°C, for 10 min.
6. Finally the pellet is washed with 5 mL of diethylether, centrifuged, and dried under vacuum at room temperature. The dry pellet can be stored at –80°C until further use. For isofocussing, the solubilization of the dry pellet is obtained by adding 250 µL of the solubilization solution (*see* **Subheading 2.1.**).

3.2.3. Starch Granule Proteins

Starch is the major component (70–75% dry base) of cereal kernel endosperm. The starch granules, which come from amyloplasts, develop during grain formation and in the mature endosperm. A multimodal distribution of starch granule size is generally observed in cereal seeds (**Fig. 1**). The procedure below describes the extraction of the proteins strongly attached to and/or present within the starch granules of the cereal kernel. This procedure was developed from previous studies on wheat granule binding starch synthase ***(23)*** with some modifications ***(24)***.

1. One or two half-kernels are crushed with a mortar and pestle at room temperature, and the floury material is manually separated from the envelope tissue (*see* **Note 16**).
2. Soak 40 to 60 mg of flour in 1 mL of deionized water at 4°C overnight. The slurry is centrifuged at 10,000*g*, 4°C, for 2 min.
3. Then 300 μL of cold water is added to the pellet and the mixture is vortexed.
4. The slurry is layered with 1 mL of 90% (v/v) CsCl, left to rest at 4°C for 5 min, and then centrifuged at 14,000*g*, at 4 °C, for 5 min. The supernatant is carefully pipeted off.
5. Then 1 mL of the washing solution containing CHAPS and DTT is added to the pellet. The mixture is vortexed and then heated at 37°C for 15 min with vortexing every 5 min. The slurry is centrifuged at 14,000*g*, at 4°C, for 3 min, then the supernatant is gently pipeted off.
6. The starch granule pellet is centrifuged (using the same parameters) and washed three times with water as follows: 1 mL of water is added to the pellet, and the slurry is vortexed, left to rest for 5 min at ambient temperature, and centrifuged at 14,000*g*, at 4°C, for 3 min. The supernatant is carefully pipeted off each time.
7. The pellet is then washed one last time with 1 mL of glacial acetone for 5 min, vortexed, and centrifuged as in **step 6**. The starch granule pellet can be air-dried and weighed and stored at –20°C if needed.
8. The starch granule pellet is resuspended using the washing solution at a 1:5 (w/v) ratio. Usually 40 mg of dried starch granules are suspended by vortexing with 200 μL of a freshly made solution of 4% (w/v) CHAPS and 2% (w/v) DTT at room temperature (*see* **Note 16**). The suspension is left to rest for 5 min and then heated at 100°C for 5 min, which causes the starch granules to release their content.
9. The slurry is cooled on ice for 10 min and then centrifuged at 17,000*g*, at 4°C, for 15 min. The supernatant containing the proteins strongly attached or present within the starch granules is collected and stored at –80°C or used directly for 2-DE. For isofocusing, add 150 μL of the solubilization solution (*see* **Subheading 2.1.**) to 100 μL of supernatant.

4. Notes

1. The most commonly used reducing agent is DTT, but the choice of the reducing agent is primarily sample specific, and tributyl phosphine (TBP) can also be used ***(25)***.
2. The solution should be prepared before use or can also be aliquoted and stored at –80°C.
3. TCA, which strongly influences folding and precipitation of proteins, may prevent their complete resolubilization. TCA can thus be omitted from the extraction solution.

4. DTT at 20 m*M* can be used instead of 0.07% 2-mercaptoethanol as the reducing agent.
5. The use of plant protease inhibitor is recommended in all cases, since resolubilization, which for some material requires successive vortexing and sonication (20 w, 20 s), is generally performed at positive temperatures.
6. Although the TCA/acetone procedure allows the proteins to be concentrated, it should be noted that some low-molecular-weight proteins may require centrifugation with higher parameters (for example at 30,000*g*, at 4°C, for 30 min) to be pelleted.
7. The protein pellet can be air-dried then stored at –80°C or directly resolubilized without storage.
8. When green material or immature kernels are used, the green seeds can be frozen (using nitrogen) and then finely crushed under liquid nitrogen using a mortar and pestle.
9. Material can also be lyophilized and stored at –20°C until analysis.
10. When the sample is made of mature cereal kernels, the grains are often milled using for example, a Cyclotec 14920 mill (Tecator, Höganäs, Sweden) with a 0.75-mm mesh sieve. The wholemeal flour must then be extracted immediately.
11. Alkylation of the free SH can be performed with iodo-acetamide instead of 4-vinyl pyridine. Alkylation is usually performed only after the IEF step during the equilibration step of the IPG gel strips. The subunits of storage proteins are usually better separated when the alkylation is performed after isofocusing.
12. The glycerol solution was first adopted for cup loading and maintained when proteins were added to IPG strips by passive rehydration. This solution can be removed for some IEF separations, like albumin and globulin separation.
13. Albumins or water-soluble proteins can be directly extracted in chilled distilled water as performed for *Arabidopsis thaliana* seeds *(16)*.
14. Since acetone precipitation allows partial removal of salts, the dialysis step can be left out, particularly when small quantities of protein extract have to be loaded on IEF. When large quantities (>150 μg) of protein extract have to be loaded, dialysis can reduce horizontal streaking caused by interfering substances in the extracting solution.
15. The phase partitioning enables the amphiphilic proteins to be collected in the lower dense phase. The light phase should be carefully removed with a pipete, as many proteins have been observed at the frontier between the light and dense phase.
16. Starch granules can be isolated from one, two, or more embryo-less kernels. Alternatively, the seeds can be cut in half and the remaining embryo part used for seedling purposes. The amount of starch granules to be used to extract the proteins present within or strongly attached to the granules can be higher than 40 mg. This amount was sufficient to detect on silver-stained 2-DE gels the majority of the wheat starch granule proteins. A similar approach was recently used on barley ***(26)***.

References

1. Tsugita, A. and Kamo, M. (1999) 2D Electrophoresis of plant proteins. *Methods Mol. Biol.* **112,** 95–97.

2. Rabilloud, T. (1996) Solubilization of proteins for electrophoretic analyses. *Electrophoresis* **17**, 813–829.
3. Osborne, T. B. (1907) The proteins of the wheat kernel. *Publication of the Carnegie Institution, Washington,*Washington, DC.
4. Gueguen, J. and Lemarie, J. (1996) Composition, structure et propriétés physicochimiques des proteins de légumineuses et d'oléagineux, in *Protéines Végétales* (Godon, B. E., ed.), Lavoisier, Paris, pp. 80–119.
5. Shewry, P. R., Tatham, A. S., Forde, J., Kreis, M., and Miflin, B. J. (1986) The classification and nomenclature of wheat gluten proteins: a reassessment. *J. Cereal Sci.* **4,** 97–106.
6. Marion, D., Nicolas, Y., Popineau, Y., Branlard, G., and Landry, J. (1994) A new improved sequential extraction procedure of wheat proteins, in *Wheat Kernel Proteins-Molecular and Functional Aspects.* Universita Degli Studi della Tuscia Consiglio Nationale delle Ricerche, pp. 197–199.
7. Fu, B. X. and Kovacs, M. I. P. (1999) Rapid single-step procedure for isolating total glutenin proteins of wheat flour. *J. Cereal Sci.* **29,** 113–116
8. Landry, J., Delaye, S., and Damerval, C (2000) Improved method for isolating and quantitating α-amino nitrogen as nonprotein, true protein, salt-soluble proteins, zeins, and true glutelins in maize endosperm. *Cereal Chem.* **77,** 620–626.
9. Damerval, C., de Vienne, D., Zivy, M., and Thiellement, H. (1986) Technical improvements in two-dimensional electrophoresis increase the level of genetic variation detected in wheat-seedling proteins. *Electrophoresis* **7,** 52–54.
10. Granier, F. (1988) Extraction of plant proteins for two-dimensional electrophoresis. *Electrophoresis* **9,** 712–718.
11. Islam, N., Tsujimoto, H., and Hirano, H. (2003) Proteome analysis of diploid, tetraploid and hexaploid wheat: towards understanding genome interaction in protein expression. *Proteomics* **3,** 549–557.
12. Vensel, W. H., Tanaka, C. K., Cai, N., Wong, J. H., Buchanan, B. B., and Hurkman, W. J. (2005) Developmental changes in the metabolic protein profiles of wheat endosperm. *Proteomics* **5,** 1594–1611
13. Wong, J. H., Cai, N., Balmer, Y., et al. (2004) Thioredoxin targets of developing wheat seeds identified by complementary proteomic approaches. *Phytochemistry* **65,** 1629–1640.
14. Østergaard, O., Melchior, S., Roepstorff, P., and Svensson, B. (2002) Initial proteome analysis of mature barley seeds and malt. *Proteomics* **2,** 733–739.
15. Østerggard, O., Finnie, C., Laugesen, S., Roepstorff, P., and Svensson, B. (2004) Proteome analysis of barley seeds: identification of major proteins from two-dimensional gels (pI 4-7). *Proteomics* **4,** 2437–2447.
16. Gallardo, K., Job, C., Groot, S. P. C., et al. (2001) Proteomic analysis of arabidopsis seed germination and priming. *Plant Physiol.* **126,** 835–848.
17. Bradford, M.M. (1976) A rapid and sensitive method for the quantitation of microgram quantities of protein utilizing the principle of protein-dye binding. *Anal. Biochem.* **72,** 248–254.

18. Nicolas, Y., Martinant, J. P., Denery-Papini, S., and Popineau, Y. (1998) Analysis of wheat storage proteins by exhaustive sequential extraction followed by RP-HPLC and nitrogen determination. *J. Sci. Food Agric.* **77,** 96–102.
19. Majoul, T., Bancel, E., Triboi, E., Ben Hamida, J., and Branlard, G. (2004) Proteomic analysis of the effect of heat stress on hexaploid wheat grain: characterization of heat-responsive proteins from non-prolamins fraction. *Proteomics* **4,** 505–513.
20. Blochet, J. E., Kaboulou, A., Compoint, J. P., and Marion, D. (1991) Amphiphilic proteins from wheat flour: specific extraction, structure and lipid binding properties, in *Gluten Proteins 1990* (Bushuk, W. and Tkachuk, R., eds.), American Association of Cereal Chemistry, St. Paul, MN, pp. 314–325.
21. Branlard, G., Amiour, N., Igrejas, G., et al. (2003) Diversity of puroindolines as revealed by two-dimensional electrophoresis. *Proteomics* **3,**168–174
22. Amiour, N., Merlino, M., Leroy, P., and Branlard, G. (2002) Proteomic analysis of amphiphilic proteins of hexaploid wheat. *Proteomics* **2,** 632–641.
23. Zhao, X.C. and Sharp, P.J. (1996) An improved 1-D SDS-PAGE method for the identification of three bread wheat "waxy" proteins. *J. Cereal Sci.* **23,** 191–193.
24. Marcoz-Ragot, C., Gateau, I., Koenig, J., Delaire, V., and Branlard, G. (2000) Allelic variants of granule-bound starch synthase proteins in European bread wheat varieties. *Plant Breeding* 119, 305–309.
25. Skylas, D. J., Mackintosh J. A., Cordwell, S. T., et al. (2000) Proteome approach to the characterisation of protein composition in the developing and mature wheat-grain endosperm. *J. Cereal Sci.* **32,** 169–188.
26. Borel, M., Larsson, H., Falk, A., and Jansson, C. (2004) The barley starch granule proteome-internalized granule polypeptides of the mature endosperm. *Plant Sci.* **166,** 617–626.

4

Protein Extraction from Xylem and Phloem Sap

Julia Kehr and Martijn Rep

Summary

It is well known that phloem and xylem vessels transport small nutrient molecules over long distances in higher plants. The finding that proteins also occur in both transport fluids was unexpected, and the function of most of these proteins is not yet well understood. This chapter outlines how proteins can be obtained and purified from xylem and phloem saps to perform subsequent proteomic analyses.

Key words: Xylem sap; phloem sap; protein precipitation; proteomics.

1. Introduction

Higher plants contain vascular bundles that permit nutrient distribution as well as communication between even the most distant plant parts. These bundles contain two types of transport units, the xylem and the phloem. The main function of the xylem is to transport water and nutrients to the aerial tissues; the phloem allocates organic assimilates, like sugars and amino acids, from the site of production to all other plant parts. Proteins have also been found in xylem and phloem saps of different plants.

Although xylem elements are dead cells, xylem sap of healthy and pathogen-infected plants contains proteins at low concentrations that nevertheless appear to have specific functions, for instance, in the maintenance, reorganization, or reinforcement of cell walls or in defence against pathogens (***1–5*** and references therein).

The transporting tubes that build the phloem are called sieve elements (SEs). SEs are highly specialized cells lacking nuclei and ribosomes and are thus not equipped for transcription and translation. The transport fluid in these tubes is normally designated as phloem sap. In this sap, more than a hundred proteins

From: *Methods in Molecular Biology, vol. 335: Plant Proteomics: Methods and Protocols*
Edited by: H. Thiellement, M. Zivy, C. Damerval, and V. Méchin © Humana Press Inc., Totowa, NJ

have been detected in different species ***(6)***, and a reasonable number of these proteins has been identified and characterized ***(7–11)***. They are likely to be imported from the tightly associated companion cells (CCs) through specialized plasmodesmata ***(12)***. In contrast to xylem fluid, protein concentrations in phloem sap can reach quite high levels (up to 60 mg/mL in cucurbits) ***(13)***. It is believed that these proteins could play a role in phloem maintenance and defence, but some may also be crucial for long-distance communication ***(7,10,11,14)***.

This chapter describes sampling procedures for xylem and phloem sap and outlines how the proteins can be extracted and purified so that they are suitable for downstream proteomic analyses.

2. Materials

2.1. Collection of Xylem Sap

1. Razor blade.
2. Screw-capped tubes.
3. Containers (e.g., pots) and ice to collect xylem sap.

2.2. Concentration of Xylem Sap Proteins

1. Amicon Centriprep YM-3 and/or Amicon Centricon Plus-20 filter units for concentration and partial purification of xylem sap proteins.
2. 12.5% TCA in 100% acetone supplemented with 0.0875% β-mercaptoethanol added just prior to use (35 μL β -mercaptoethanol per 40 mL) or 80% acetone for protein precipitation.

2.3. Collection of Phloem Sap

1. Sterile razor blade or hypodermic needle.
2. Filter paper.
3. Pipet, reaction tubes, and ice to collect phloem sap.

2.4. Purification of Phloem Sap Proteins

1. 2 *M* HCl.
2. 2 *M* NaOH (analytical grade).
3. Acetone/methanol/dithiothreitol (DTT) (90% [v/v] /10 % [v/v] /10 m*M*). 100% acetone (analytical grade)

3. Methods

3.1. Collection of Xylem Sap

Xylem sap is a part of the extracellular space of plants. It has a unique composition, including specific proteins, the presence and concentration of which can vary depending on the condition of the plant. Because pure xylem sap is

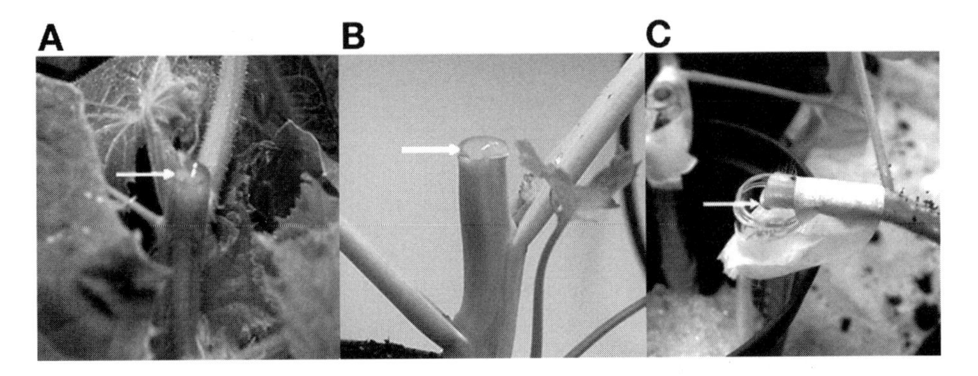

Fig. 1. Root pressure exudation of xylem sap from the cut stems of a cucumber **(A)**, an oilseed rape **(B)**, and a tomato **(C)** plant. Stems are cut approximately 10 cm above soil level, and then the sap exuding from the root side after washing can be collected.

easy to obtain from most plants, the tools of proteomics can be readily applied to identify xylem sap proteins. For preparation of protein samples for separation on one-dimensional electrophoresis (1-DE) or 2-DE gels, issues to take into account are the low protein concentration and the presence of oligosaccharides (and glycosylated proteins). There are basically two steps in the application of a xylem sap protein sample on a 1-DE or 2-DE gel: collection of xylem sap and concentration of xylem sap proteins.

1. Xylem sap can be obtained by cutting the stems with a razor blade and collecting the sap exuded spontaneously (driven by root pressure) from the remaining stem on the root side (**Fig. 1**). The stems can be cut at any height, but, in general, the closer the cut is to the base of the stem, the higher the yield of sap. However, enough of the stem (around 10 cm) should remain to connect to a tube on ice (*see* **Note 1**).
2. Sap collection
   a. The remaining stem segment is placed horizontally, and a tube is taped to the stem segment. The tube is surrounded by ice in a container, and sap is collected for up to 6 h (**Fig. 1C**; the ice container needs to be refilled several times during sap collection; *see* **Note 2**).
   b. Alternatively, the exuding xylem sap can be collected repeatedly every few minutes with a pipet into a reaction tube placed on ice *(4,5)*.
3. In any case, it is important to wash the cut surface before sample collection to remove the content from cut cells and the phloem sap that exudes directly after cutting (*see* **Note 3**).
4. Before concentrating proteins, it is advisable to remove any particulate matter like soil particles, microbial cells, or tissue remnants by centrifugation.

3.2. Concentration of Xylem Sap Proteins

Xylem sap contains proteins but also carbohydrates (*see* **Note 4**) and other compounds like amino acids, salts, and (in pathogen-infected plants) polyphenols ***(15–17)***.

For 1-DE analysis, simple precipitation of proteins suffices, even when using ethanol (4 parts of 95% EtOH/5% 0.1 *M* NaAc to 1 part of protein sample), which also precipitates poly- and oligosaccharides.

To prepare samples for isoelectric focusing, the sap should be first partially purified and concentrated using Amicon filter units and then precipitated with trichloracetic acid (TCA)/acetone.

3.2.1. Protein Concentration and Partial Purification with Amicon Filter Units

Either Amicon Centriprep YM-3 (at most 15 ml, cutoff: 3 kD) or Amicon Centricon Plus-20 (larger volumes, cutoff: 10 kD) filter units can be used to concentrate and partially purify xylem sap proteins.

1. After centrifugation to remove particulate matter, add at most 15 mL (Centriprep YM-3) or 19 mL (Centricon Plus-20) xylem sap to the sample container.
2. Spin Centriprep YM-3 for approx 75 min at 4°C at 3000*g* until equilibrium is reached (between fluid levels inside and outside the filtrate collector).
3. Spin Centricon Plus-20 for 15 min at 4°C at 4000*g*. Decant filtrate. To the Centricon Plus-20, another 19 mL xylem sap can now be added to sample filter cup.
4. Spin for another 15 min.
5. Decant filtrate again and repeat spin and decant steps if further concentration is desired.
6. Collect concentrate (at least 500 μL) directly from Centriprep YM-3.
7. Collect concentrate (at least 200 μL) from Centricon Plus-20 by spinning the inverted unit for 5 min at 1000*g*.

3.2.2. Protein Precipitation with TCA/Acetone

1. Add 4 parts of either 12.5% TCA in 100% acetone with 0.0875% β-mercapto-ethanol to 1 part of protein sample and mix.
2. Incubate for at least 45 min at –20°C (or 1–4 h at 0°C for 80% acetone), and then centrifuge tubes for 30 min in a centrifuge at maximum speed (refrigeration not required).
3. Discard supernatant and wash the pellet three times with ice-cold 100% acetone.
4. Remove remaining liquid with a pipet, and dry the pellet at room temperature. The pellet can be solubilized in rehydration buffer (2 *M* thiourea, 7 *M* urea, 4% 3[3-cholaminopropyl diethylammonio]-propane sulfonate (CHAPS), 10 m*M* DTT) *(5)* for isoelectric focusing or store dry at –20°C.

3.3. Collection of Phloem Sap

Phloem sap is much more difficult to obtain from most plant species than xylem sap, and the major challenge of phloem sap proteomics is therefore to obtain enough material to perform proteomic studies. However, several different methods to obtain phloem samples can be applied, their feasibility largely depending on the plant species of interest.

In a few species (e.g., *Ricinus*, Cucurbitaceae, *Yucca*), phloem sap can be collected after applying small incisions ***(8)*** or excising whole organs ***(18)***. Plants not suitable for this procedure can be sampled by EDTA-facilitated exudation ***(19–21)***. In many plant species, the aphid stylet technique ***(22)*** can also yield phloem sap of high purity but only in small amounts.

The established plant model species, like thale cress (*Arabidopsis thaliana*) or rice (*Oryza sativa*), hardly allow the collection of sufficient amounts of SE exudate for proteomic analysis. In *Arabidopsis*, the sample amounts obtainable are not sufficient for proteomic approaches ***(10)***, whereas in rice, only some highly abundant phloem sap proteins have been identified from planthopper stylet exudate ***(6,23)***. Most current information about the identity of phloem polypeptides comes from cucurbits ***(8)*** *Ricinus* ***(10)***, and recently also from oilseed rape ***(7)***, in which sap collection by exudation is relatively easy. Two variants of the exudation technique are described below.

1. Variant 1.
   a. Cut the main stem or petiole with a razor blade (**Fig. 2**).
   b. Alternatively, hypodermic needles can be used to puncture the plant (*see* **Note 5**).
2. Variant 2.
   a. Proceed with the shoot side and dry the cut surface with a filter paper (*see* **Note 6**).
   b. Remove the first droplet exuding from the incision with filter paper.
   Collect the subsequently exuding sap with a pipet in a reaction tube that is kept on ice (*see* **Note 7**).

3.4. Purification of Phloem Sap Proteins

1. When separating phloem proteins in 1-DE or 2-DE gels, the high concentrations of sugars and other organic materials present in this transport fluid have to be taken into account. In addition, plants possess mechanisms to respond to wounding events that include the polymerization of proteins ***(24)***.
2. For 1-DE, phloem samples can be directly expelled into 1-DE sample buffer ***(25)***, heated, and applied to a gel at room temperature.
3. When performing subsequent 2-DE, proteins should be purified by precipitation before the samples can be applied to the isoelectric focusing gels or strips.

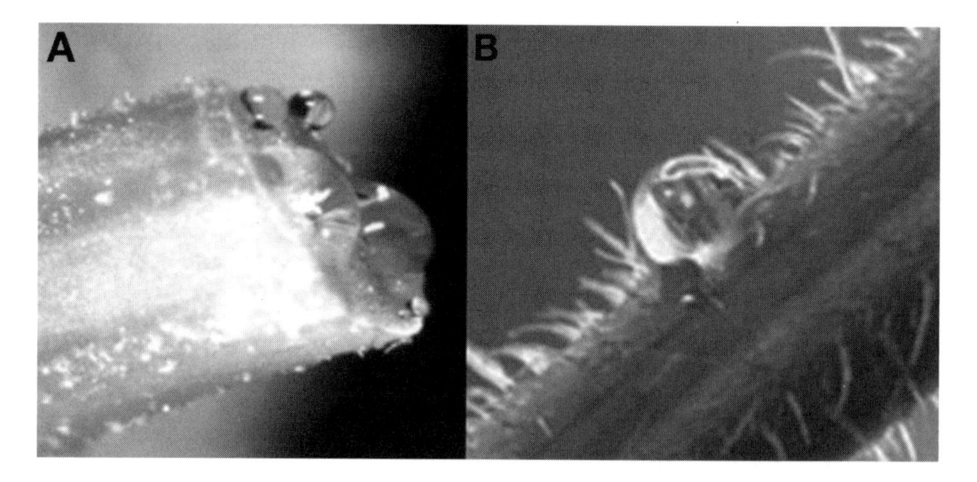

Fig. 2. Phloem sampling from a cucumber plant using the exudation technique. Stems are cut with a razor blade **(A)** or severed with a hypodermic needle **(B)**, and the sap exuding (after removing the first droplets with a filter paper) can be collected on ice. Note that in **(A)** the shoot side of the cut stem is used for phloem sap collection whereas in the root side xylem sap can be obtained (*see* **Fig. 1**).

3.4.1. Depletion of the Major Phloem Proteins 1 (PP1) and 2 (PP2)

The two major phloem proteins, phloem filament protein PP1 and phloem lectin PP2, occur in all dicots and can make up more than 50% of total phloem proteins in certain species (e.g., *Cucurbitaceae*). Under oxidative conditions, these two proteins form insoluble polymers linked by disulfide bridges ***(26,27)*** that can result in streaking and bad resolution on 1-DE and 2-DE gels. To remove these proteins, acidification followed by neutralization may be used ***(28)***. All the following steps can be performed at room temperature.

1. Adjust sample to pH 2.0 with 2 *M* HCl.
2. Neutralize to pH 7.5 with 2 *M* NaOH.
3. Centrifuge at 15,000*g* for 15 min.
4. Collect supernatant and discard the copious white pellet that contains PP1 and PP2 (*see* **Note 8**).

3.4.2. Precipitation of Total Phloem Sap Proteins by Acetone/Methanol/DTT

To remove carbohydrates and other disturbing components, proteins can be precipitated by treatment with acetone/methanol/DTT *(7)*.

1. Add 3 volumes of ice-cold sample precipitation solution.
2. Precipitate at –20°C overnight (*see* **Note 9**).
3. Centrifuge for 5 min at 6800*g* and 4ºC (*see* **Note 10**) and discard supernatant.

4. Wash precipitate twice with 100% cold acetone, and centrifuge at 6800*g* and 4°C for 5 min.
5. Discard supernatant and air-dry pellet for 5–10 min at room temperature (*see* **Note 11**).
6. Dissolve the pellet in buffer for either 1-DE or for 2-DE (*see* **Note 12**).

4. Notes

1. Since xylem sap contains proteases, care has to be taken that samples are kept cool all the time; protease inhibitors can be used to reduce the risk of protease digest.
2. Sap yields after 6 h of bleeding of tomato plants vary widely. From healthy, 6- to 7-wk-old plants, more than 10 mL can be obtained from an individual plant, but occasionally a plant will yield nothing or very little. The conditions of plants will obviously influence sap yield, but we have not found specific conditions that reproducibly affect yield. Diseased plants, such as those infected with the vascular wilt fungus *Fusarium oxysporum*, have much lower sap yields.
3. Protein or sugar concentration can be used as a measure for phloem contamination; low concentrations of both substances indicate a high purity of xylem sap samples.
4. Poly- and oligosaccharides react with common reagents to measure protein concentration, like Bradford and bicinchoninic acid (BCA; Sigma). Their presence in xylem sap leads to an overestimation of protein content.
5. Due to turgor pressure that is highly positive in SEs and negative in xylem elements of living plants, only phloem sap will exude from such incisions.
6. It is important to keep in mind that phloem sap is obtained from the shoot side, whereas xylem sap can be obtained from the other side (the root containing part) of the divided stem.
7. Since phloem sap usually contains protease inhibitors at high concentrations, the addition of chemical protease inhibitors is normally not required.
8. It is important to check the protein composition (e.g. by analysing the protein composition of the pellet by 1-DE) after this precipitation step, because usually not exclusively PP1 and PP2, but also other proteins can precipitate during the procedure.
9. Do not precipitate for extended periods since proteins might become insoluble.
10. Centrifugation at higher speed leads to solubilization problems.
11. Do not dry by vacuum centrifugation since this results in poor solubility.
12. If the pellet does not dissolve, incubation for longer times, sonication, or incubation at higher temperatures might be useful. If some insoluble particles remain, centrifuge samples before applying the samples to gels.

References

1. Rep, M., Dekker, H. L., Vossen, J. H., et al. (2003). A tomato xylem sap protein represents a new family of small cysteine-rich proteins with structural similarity to lipid transfer proteins. *FEBS Lett.* **534,** 82–86.

2. Rep, M., Dekker, H. L., Vossen, J. H., et al.. (2002). Mass spectrometric identification of isoforms of PR proteins in xylem sap of fungus-infected tomato. *Plant Physiol.* **130,** 904–917.
3. Sakuta, C. and Satoh, S. (2000). Vascular tissue-specific gene expression of xylem sap glycine-rich proteins in root and their localization in the walls of metaxylem vessels in cucumber. *Plant Cell Physiol.* **41,** 627–638.
4. Buhtz, A., Kolasa, A., Arlt, K., Walz, C., Kehr, J. (2004). Xylem sap protein composition is conserved among different plant species. *Planta* **219**, 610–618.
5. Kehr, J., Buhtz, A. and Giavalisco, P. (2005). Analysis of xylem sap proteins from *Brassica napus*. *BMC Plant Biol.* **5,** 11.
6. Fukuda, A., Okada, Y., Suzui, N., Fujiwara, T., Yoneyama, T., and Hayashi, H. (2004). Cloning and characterization of the gene for a phloem-specific glutathione S-transferase from rice leaves. *Physiol. Plant.* **120,** 595–602.
7. Giavalisco, P., Kapitza, K., Kolasa, A., Buhtz, A., and Kehr, J. (2006). Towards the proteome of *Brassica napus* phloem sap. *Proteomics* **6,** 896–909.
8. Walz, C., Giavalisco, P., Schad, M., Juenger, M., Klose, J., and Kehr, J. (2004). Proteomics of curcurbit phloem exudate reveals a network of defence proteins. *Phytochemistry* **65**, 1795–1804.
9. Haebel, S. and Kehr, J. (2001). Matrix-assisted laser desorption/ionization time of flight mass spectrometry peptide mass fingerprints and post source decay: a tool for the identification and analysis of phloem proteins from *Cucurbita maxima* Duch. separated by two dimensional polyacrylamide gel electrophoresis. *Planta* **213,** 586–593.
10. Barnes, A., Bale, J., Constantinidou, C., Ashton, P., Jones, A., and Pritchard, J. (2004). Determining protein identity from sieve element sap in *Ricinus communis* L. by quadrupole time of flight (Q-TOF) mass spectrometry. *J. Exp. Bot.* **55,** 1473–1481.
11. Hayashi, H., Fukuda, A., Suzui, N., and Fujimaki, S. (2000). Proteins in the sieve element-companion cell complexes: their detection, localization and possible functions. *Aust. J. Plant Physiol.* **27,** 489–496.
12. Lucas, W. J. (1999). Plasmodesmata and the cell-to-cell transport of proteins and nucleoprotein complexes. *J. Exp. Bot.* **50,** 979–987.
13. Kehr, J., Haebel, S., Blechschmidt-Schneider, S., Willmitzer, L., Steup, M., and Fisahn, J. (1999). Analysis of phloem protein patterns from different organs of *Cucurbita maxima* Duch. by matrix-assisted laser desorption/ionisation time of flight mass spectroscopy combined with sodium dodecyl sulfate-polyacrylamide gel electrophoresis. *Planta* **207,** 612–619.
14. Yoo, B.-C., Lee, J.-Y., and Lucas, W. J. (2002). Analysis of the complexity of protein kinases within the phloem sieve tube system. Characterization of *Cucurbita maxima* calmodulin-like domain protein kinase 1. *J. Biol. Chem.* **277,** 15,325–15,332.
15. Iwai, H., Usui, M., Hoshino, H., et al. (2003). Analysis of Sugars in squash xylem sap. *Plant Cell Physiol.* **44,** 582–587.

16. Heizmann, U., Kreuzwieser, J., Schnitzler, J.-P., Brüggemann, N., and Rennenberg, H. (2001). Assimilate transport in the xylem sap of pedunculate oak (*Quercus robur*) saplings. *Plant Biol.* **3**, 132–138.
17. Lopez-Millan, A. F., Morales, F., Abad'a, A., and Abad'a, J. (2000). Effects of iron deficiency on the composition of the leaf apoplastic fluid and xylem sap in sugar beet. implications for iron and carbon transport. *Plant Physiol.* **124,** 873–884.
18. Alosi, M. C., Melroy, D. L., and Park, R. B. (1988). The regulation of gelation of phloem exudate from *Cucurbita* fruit by dilution, glutathione, and glutathione reductase. *Plant Physiol.* **86,** 1089–1094.
19. King, R. and Zeevaart, J. (1974). Enhancement of phloem exudation from cut petioles by chelating agents. *Plant Physiol.* **53,** 96–103.
20. Hoffmann-Benning, S., Gage, D. A., McIntosh, L., Kende, H., and Zeevaart, J. A. D. (2002). Comparison of peptides in the phloem sap of flowering and non-flowering Perilla and Lupine plants using microbore HPLC followed by matrix-assisted laser desorption/ionisation time-of-flight mass spectrometry. *Planta* **216,** 140–147.
21. Marentes, E. and Grusak, M. A. (1998). Mass determination of low-molecular-weight proteins in phloem sap using matrix-assisted laser desorption/ionization time of flight mass spectrometry. *J. Exp. Bot.* **49,** 903–911.
22. Kennedy, J. and Mittler, T. (1953). A method of obtaining phloem sap via the mouth parts of aphids. *Nature* **171,** 528.
23. Ishiwatari, Y., Honda, C., Kawashima, I., et al. (1995). Thioredoxin h is one of the major proteins in rice phloem sap. *Planta* **195,** 456–463.
24. Clark, A. M., Jacobsen, K. R., Bostwick, D. E., Dannenhoffer, J. M., Skaggs, M. I., and Thompson, G. A. (1997). Molecular characterization of a phloem-specific gene encoding the filament protein, Phloem Protein 1 (PP1), from *Cucurbita maxima. Plant J.* **12,** 49–61.
25. Laemmli, U. K. (1970). Cleavage of structural proteins during the assembly of the head of bacteriophage T4. *Nature* **227,** 680–685.
26. Read, S. M. and Northcote, D. H. (1983). Chemical and immunological similarities between the phloem proteins of three genera of the Cucurbitaceae. *Planta* **158,** 119–127.
27. Read, S. M. and Northcote, D. H. (1983). Subunit structure and interactions of the phloem proteins of Cucurbita maxima (pumpkin). *Eur. J. Biochem.* **134,** 561–569.
28. Christeller, J. T., Farley, P. C., Ramsay, R. J., Sullivan, P. A., and Laing, W. A. (1998). Purification, characterization and cloning of an aspartic proteinase inhibitor from squash phloem exudate. *Eur. J. Biochem.* **254,** 160–167.

5

Protein Extraction from Woody Plants

Christophe Plomion and Céline Lalanne

Summary

In this chapter we present a protocol for total protein extraction optimized for wood-forming tissue (differentiating secondary xylem). The protocol is then used for a series of other organs (root, leaf, pollen, bud, flower, cambium, and phloem) in broadleaf (oak and poplar) and conifer (pine) species. Proteins are first extracted from tissue powdered in liquid nitrogen using the TCA-acetone method and then solubilized in an optimized buffer. The resulting 2D gels can be viewed at http://cbi.labri.fr/outils/protic/index.php.

Key Words: Total protein extraction; two-dimensional polyacrylamide gel electrophoresis; wood; tree.

1. Introduction

In perennial plants, the successive addition of secondary xylem tissue differentiated from the vascular cambium gives rise to a unique tissue called wood *(1)*. Wood is composed of nonconducting and conducting elements implicated in the long-distance transport of water and nutrients in trees (see the ecotree web site at http://www.botany.uwc.ac.za/ecotree/ for more details). In gymnosperms, wood is comprised of two main cell types: tracheids and ray parenchyma. In angiosperms it also contains vessels and fibers. This simplicity hides the fact that it is also a highly variable material. The activity of the vascular cambium and the differentiation of newly divided cells are genetically controlled and also affected by environmental and ontogenic effects that ultimately influence wood and end-use properties.

Given the role wood plays both in the biosphere in terms of sinks for excess atmospheric CO_2 and in our daily life as a raw material for thousand of products and an endlessly renewable source of energy, it is surprising that our understanding of how wood develops is far from complete. Wood formation is a complex phenomenon driven by the coordinate expression of numerous genes

From: *Methods in Molecular Biology, vol. 335: Plant Proteomics: Methods and Protocols*
Edited by: H. Thiellement, M. Zivy, C. Damerval, and V. Méchin © Humana Press Inc., Totowa, NJ

especially involved in the biosynthesis and the assembly of polysaccharides, lignins, and cell wall proteins *(2)*. Up to now, the study of molecular mechanisms involved in the development of wood has mainly taken a transcriptomic approach, combining expressed sequence tag (EST) sequencing and transcript profiling *(3–6)*. Only one large-scale project has attempted to identify proteins from the differentiating secondary xylem of maritime pine *(7)*.

In this chapter we describe a protocol optimized to extract proteins from wood-forming tissue in three forest tree species: pine, poplar, and oak. The protocol was successfully extended to other organs, namely, root, leaf, pollen, bud, flower, cambium, and phloem. The resulting 2D gels can be viewed at http://cbi.labri.fr/outils/protic/index.php.

2. Materials

2.1. Tissue Sampling and Storage

2.1.1. Wood Forming Tissue

1. The rough outer bark is first removed using a bark shaver with a 30-cm blade.
2. The phloem and cambium are levered off the stem together by making incisions (two vertical cuts between two horizontal cuts) using a sharp knife into the exposed tissue.
3. Differentiating xylem (wood-forming tissue) is collected from the exposed log surface using a vegetable peeler or a knife, immediately frozen in liquid nitrogen and stored at –80°C until used for protein extraction.
4. Samples are always taken from adult trees during the active growing season (**Fig. 1**). Different types of wood (early and late wood, juvenile and mature wood, opposite and compression wood; *1*), differing in their chemical, anatomical, and physical properties, can be sampled within a single individual.

2.1.2. Other Tissues

For mapping the proteomes of different tissues/organs of the tree, as illustrated for poplar at http://cbi.labri.fr/outils/protic/PublicPopulus.php, leaves, roots, vegetative and sexual buds, pollen, phloem, and cambium are taken during the active growing phase of seedlings or adult trees.

2.2. Total Protein Precipitation (See Note 1)

1. Polycarbonate 10 mL oak ridge centrifuge tube (Nalgene, Rochester, NY).
2. Precipitation buffer (make fresh as required and store at –20°C): prepare a solution with 10% Trichloroacetic acid (TCA) a very effective protein precipitant, and 0.07% 2-mercaptoethanol in acetone.
3. Rinsing buffer (make fresh as required and store at –20°C): 0.07% 2-mercaptoethanol in acetone.

Fig. 1. Scraping of differentiating xylem taken with a knife on a pine tree.

2.3. Protein Solubilization (See Note 2)

1. Solubilization buffer: 7 *M* urea, 2 *M* thiourea, 0.4% Triton X-100 (*see* **Note 3**), 4% 3-[3-cholamidopropyl dimethylammonio]-1-propane sulfonate (CHAPS), 10 m*M* dithiothreitol (DTT), 1% immobilized ph gradient (IPG) buffer (Amersham Biosciences, Uppsala, Sweden) (*see* **Note 4**). Store at –20°C.

3. Methods

3.1. Total Protein Extraction (See Note 5)

1. Weigh an empty 10-mL centrifuge tube without the sealing cap. We used polycarbonate Oak Ridge centrifuge tubes for their physical strength.
2. Cell disruption: 500 mg fresh tissue (stored –80°C) is finely powdered in liquid nitrogen using a mortar and pestle.
3. Resuspend the powder in 8 mL of ice-cold precipitation buffer.
4. Transfer the resulting mix to the weighed centrifuge tube.
5. Rinse the mortar with 2 mL of ice-cold precipitation buffer.
6. Transfer the resulting mix to the same centrifuge tube.
7. Mix the tube contents by gently inverting the tube for 15 s, and then let the proteins precipitate for 1 h at –20°C.

8. Centrifuge the tube at 16,000*g* for 15 min (*see* **Note 6**).
9. Remove the supernatant by gently pouring down the sides of the tube.
10. Rinse the pellet (remove residual TCA) by adding 10 mL ice-cold rinsing buffer in the tube and let the tube sit for 1 h at –20°C.
11. Centrifuge the tube at 16,000*g* for 10 min (*see* **Note 6**).
12. Dry the pellet under vacuum for about 2 h.
13. Powder the pellet using a glass stick.
14. Weigh the centrifuge tube containing the dry pellet.

3.2. Protein Solubilization

1. Resuspend the powder in the solubilization buffer. We use between 10 and 30 µL of buffer per mg of powder. This buffer contains a mixture of chaotropes (urea, thiourea), detergents (Triton-X 100, CHAPS), reductants (DTT), and carrier ampholytes (IPG buffer).
2. Centrifuge the tube at 400*g* for 4 min at room temperature to precipitate any undisolved material (e.g., cellular fragments).
3. Pour the supernatant into a clean tube.
4. Centrifuge at 400*g* for 4 min at room temperature, and pour the supernatant into a clean tube.
5. Store the supernatant at –80°C.
6. Protein concentration is quantified over six replicated assays, using the protocol described by Ramagli et al. *(8)*. The mean concentration is then calculated and used to load 300 µg of proteins on an IPG-strip.

4. Notes

1. **Caution:** Because this buffer uses toxic compounds and generates unpleasant smell in the lab, it should be prepared under the hood.
2. It takes about 30 min to dissolve urea and thiourea. It is absolutely prohibited to heat the buffer to accelerate its solubilization. MilliQ water (18.2 MΩ-cm resistivity) is used to prepare the buffer.
3. A 20% working Triton X-100 solution is first prepared and stored at 4°C.
4. IPG buffer is to be used according to the chosen IPG strip pH gradient of the isoelectric focusing step.
5. Although the protocol is described for one replicated sample, three extractions are usually completed in parallel for each sample and pooled at the very end of the protocol to provide enough proteins to perform five replicated 2D gels.
6. Allow the temperature of the centrifuge to drop to 0°C before starting the centrifuge.

References

1. Lachaud, S., Catesson, A. M., and Bonnemain, J. L. (1999) Structure and functions of the vascular cambium *C. R. Acad. Sci. Paris* **322,** 633–650.
2. Plomion, C., Le Provost, G., and Stokes, A. (2001) Wood formation in trees. *Plant Physiol.* **127,** 1513–1523.

3. Allona, I., Quinn, M., Shoop, I., et al. (1998) Analysis of xylem formation in pine by cDNA sequencing. *Proc. Natl. Acad. Sci. USA* **95,** 9693–9698.
4. Sterky, F., Regan, S., Karlsson, J., et al. (1998) Gene discovery in wood forming tissues of poplar: Analysis of 5,692 expressed sequence tags. *Proc. Natl. Acad. Sci. USA* **95,** 13,330–13,335.
5. Hertzberg, M., Aspeborg, H., Schrader, J., et al. (2001) A Transcriptional roadmap to wood formation. *Proc. Nat. Acad. USA* **98,** 14,732–14,737.
6. Whetten, R., Sun, Y. H., Zhang, Y., and Sederoff, R. (2001) Functional genomics and cell wall biosynhesis in loblolly pine. *Plant Mol. Biol.* 47**,** 275-291.
7. Gion, J-M., Lalanne, C., Le Provost, G., et al. (2005) The proteome of maritime pine wood forming tissue. *Proteomics* **5,** 3731–3751.
8. Ramagli, L. S. and Rodriguez, L. V. (1985) Quantitation of microgram amounts of protein in two-dimensional polyacrylamide gel electrophoresis sample buffer. *Electrophoresis* **6,** 559–563.

6

Isolation of Chloroplast Proteins from *Arabidopsis thaliana* for Proteome Analysis

Klaas J. van Wijk, Jean-Benoit Peltier, and Lisa Giacomelli

Summary

This chapter describes a simple protocol for large-scale chloroplast purification for proteome analysis. The protocol has not been tested for protein import activity but is optimized for chloroplast proteome yield and purity and minimal protein degradation.

Key Words: Chloroplast; protein; *Arabidopsis thaliana*; purification.

1. Introduction

Chloroplasts are essential organelles in vegetative tissues in plants. Although best known for their role in photosynthesis, chloroplasts synthesize many essential compounds, such as plant hormones, fatty acids and lipids, amino acids, vitamins, purine and pyrimidine nucleotides, tetrapyrroles, and isoprenoids. Chloroplasts are required for nitrogen and sulfur assimilation *(1)* and also contain numerous protein chaperones and targeting components, cofactor chaperones, assembly factors, peptidases, and proteases involved in protein (complex) biogenesis and homeostasis. To facilitate chloroplast gene expression, chloroplasts contain proteins associated with plastid DNA and the plastid transcriptional and translation machinery, including many mRNA binding proteins involved in mRNA processing, stability, and translation *(2,3)*. It is predicted that all plastid types combined contain over 2000 to 4000 different proteins, with approx 550 α-helical proteins integral in thylakoid and inner envelope membranes and an unkown number peripherally associated with these membranes *(4)*. It is more than likely that the chloroplast stroma contains well over 1000 proteins. The thylakoid and envelope proteomes and associated pro-

From: *Methods in Molecular Biology, vol. 335: Plant Proteomics: Methods and Protocols*
Edited by: H. Thiellement, M. Zivy, C. Damerval, and V. Méchin © Humana Press Inc., Totowa, NJ

teins of chloroplasts from *Arabidopsis thaliana* have been analyzed in detail in a number of studies ***(5–13)***.

The method described here is based on a protocol developed for *A. thaliana* by Dr. David Christopher ***(14)*** and published protocols for isolation of chloroplasts from pea and spinach ***(15)***. The method was further developed in our lab for large-scale chloroplast purification for proteome analysis and was published in Peltier et al. ***(16)***. Specific protocols have been developed by other laboratories (e.g. by D. J. Schnell, K. Keegstra, and P. Jarvis) for purification of intact *Arabidopsis* chloroplasts for in vitro protein import essays ***(17)***. The protocol in this chapter has not been tested for protein import activity but is optimized for chloroplast proteome yield and purity and minimal protein degradation.

2. Materials

1. Medium A (grinding buffer; *see* **Note 1**): 50 m*M* HEPES-KOH (pH 8.0), 330 m*M* sorbitol, 2 m*M* EDTA-Na_2 (pH 8.0), 5 m*M* ascorbic acid, 5 m*M* cysteine; 0.05% bovine serum albumin (BSA) (can be omitted to minimize carryover to final intact chloroplast sample; *see* **Note 1**)
2. Medium B (wash medium): 50 m*M* HEPES-KOH (pH 8.0), 330 m*M* sorbitol, 2 m*M* EDTA-Na_2 (pH 8.0)
3. Medium C (lysis medium): 10 m*M* HEPES-KOH (pH 8.0), 5 m*M* $MgCl_2$, protease inhibitor cocktail (*see* **Table 1**).
4. 100% PF-Percoll: 0.8 g PEG 8000, 0.27 g Ficoll 400,000. Add Percoll to 27.5 mL end volume.
5. Percoll step gradients: PF Percoll (40% or 85%), 0.5 m*M* EDTA, 50 m*M* HEPES-KOH (pH 8.0), 330 m*M* sorbitol.
6. Concentration spin device with 3-kDa cutoff filter, e.g., from Millipore-Amicon 3 kDa.

3. Methods

For maximal yield of intact chloroplasts, grow the plants under short day length to stimulate vegetative growth and harvest the leaves during the first half of the light period. Prepare all media prior to harvesting of leaves and cool them down to 4°C in the cold room, preferentially on ice. Avoid exposing the chloroplasts to light (work in dim green light) as much as possible, and keep buffers, rotors, and centrifuge tubes ice-cold. Work as fast as possible to avoid breakage of the chloroplasts. Chloroplasts can be purified in 30 to 40 min.

1. Grow *A. thaliana* plants under a 10-h light/14-h dark cycle at 23°C/17°C. For maximum leaf material, grow the plants for 5 to 6 wk and harvest prior to bolting.
2. Prepare all media, collect centrifuge tubes, and cool on ice. Cool down centrifuge and rotors.
3. Collect rosettes and cut leaves from rosettes (*see* **Note 2**). Make sure there is no soil attached. Wash in distilled water if needed.

Table 1
Protease Inhibitor Cocktail

Inhibitor	Stock (mg/mL)	Medium	End-concentration (µg/µL)
Antipain	20	Water	50 (74 µ*M*)
Bestatin	1	0.15 *M* NaCl	40 (130 µ*M*)
Chymostatin	20	Dimethylsulfoxide	20
E64[a]	20	50% Ethanol	10
Leupeptin	1	Water	5
Phosphoramidon	20	Water	10
Pefabloc sc	100	Water	50
Aprotinin	10	Water	2

[a]Difficult to suspend.

4. Grind leaves 3X 10 s at half-speed in a Warren blender with sharp blades in ice-cold grinding medium (medium A). Use approx 10 g of leaves (about equivalent to five full grown rosettes) per 100 mL grinding medium.
5. Filter the homogenate through a double layer of nylon cloth (22 µm; *see* **Note 3**).
6. Collect crude chloroplasts by centrifugation for about 3 min at approx 1300*g* in a fixed angle rotor.
7. Resuspend the crude chloroplast pellet in medium B by "swirling" the suspension. Keep the volume as small as possible. Load the resuspended chloroplast on a Percoll step cushion (40–85%; *see* **Note 4**), and spin for 10 min at 3750*g* in a swing-out rotor.
8. Collect the intact chloroplasts from the 40/85% Percoll interface using a pipet, and add medium B to dilute the Percoll. Keep the volume low, with chlorophyll concentrations between 1 and 3 mg/mL (*see* **Note 5**). Spin the chloroplasts for 3 min at 1200*g*. Remove the supernatant. The pellet consists of intact chloroplasts.
9. For separation of thylakoid proteins and stromal proteins, add lysis buffer to about 1 to 3 mg chlorophyll/mL, and resuspend gently. Leave the suspension for 5 to 10 min on ice to swell the chloroplasts. These will burst and release stromal proteome. If needed, a Dounce homogenizer can be helpful to rupture the chloroplast. Be careful not to make too many strokes with the piston since this will partially rupture the thylakoid and release lumenal proteins into the stromal proteome.
10. Separate supernatant (containing mostly stroma) and pellet (containing the thylakoids) by a 20-min centrifugation at 10,000*g*. To remove envelopes and possible residual thylakoids from the stroma, carry out an additional spin of 25 min at 300,000*g* (*see* **Note 6**).
11. Concentrate stromal proteome in a concentration spin device with 3-kDa cutoff filter. The lysis buffer has a low salt and buffer concentration and is fully compatible with most protein separation steps used in proteome analysis [e.g., isoelec-

tric focusing, sodium dodecyl sulfate polyacrylamide gel electrophoresis (SDS-PAGE), blue native (BN)-PAGE or clear native (CN)-PAGE and in-solution digests].

12. For analysis of the thylakoid proteome, resuspend the pellet in the same low salt and weakly buffered medium, and, if needed, determine the chlorophyll concentration (*see* **Note 5**) and/or protein concentration. Subsequent thylakoid proteome analysis will require removal of pigments (chlorophylls and carotenoids) and other lipophilic molecules (e.g., plastoquinone) and removal of lipids. This can be done by extraction with organic solvents (e.g., 70% acetone) or by incubation with ionic (e.g., SDS) or nonionic detergents (OGG, DM). Recent examples of chloroplast membrane proteome analysis can be found in **refs.** ***7,8,10,11,18***, and ***19***.

4. Notes

1. How to make 1 L of medium A: 60.1 g sorbitol; 50 mL 1 *M* stock of HEPES-KOH (pH 8.0); 4 mL 0.5 *M* EDTA (autoclaved stock); fill up with double-distilled water to 1 L. Add fresh 0.88 g ascorbic acid; add 0.6 g cysteine; add 0.5 g BSA (can be omitted to minimize carryover to final intact chloroplast sample)
2. Use 40 to 50 full-grown rosettes per liter of grinding medium. For maximum yield, try to find an optimum grinding time and ratio between grinding medium/grams of leaves.
3. It is best to prewet the cloth with grinding medium.
4. To make a Percoll step gradient, load 40% on top of PF Percoll 85% or load 85% with a syringe below 40% Percoll.
5. Chlorophyll determinations were done according to **ref.** ***20***. Resuspend 2 to 5 µL (v) of green chloroplast or thylakoid solution in 1 mL 80% ice-cold acetone. Vortex and keep the suspension on ice for 10 min in the dark. Spin for 5 min at 15,000*g* at 4°C. Collect the (green) supernatant and measure light absorption at 663.2 and 646.8 nm in a spectrometer. The chlorophyll ($a + b$) concentration (µg/mL) = $(1 + v)/v \times (7.15 \times A_{663.2} + 18.71 \times A_{646.8})$.
6. An alternative centrifugation regime is to collect thylakoids and envelopes together in one spin for 25 min at 300,000*g*. However, these *g*-values can lead to fairly hard thylakoid pellets and possible some extra release of lumenal and peripheral thylakoid proteins into the supernatant, the stroma.

Acknowledgements

We thank Dr David Christopher for providing the starting protocol.

References

1. Buchanan, B., Gruissem, W., and Jones, R. L., eds. (2000) *Biochemistry and Molecular Biology of Plants*, American Society of Plant Physiologists, Rockville, MD.
2. Monde, R. A., Schuster, G., and Stern, D. B. (2000) Processing and degradation of chloroplast mRNA. *Biochimie* **82,** 573–582.
3. Barkan, A. and Goldschmidt-Clermont, M. (2000) Participation of nuclear genes in chloroplast gene expression. *Biochimie* **82,** 559–572.

4. Sun, Q., Emanuelsson, O., and van Wijk, K. J. (2004) Analysis of curated and predicted plastid subproteomes of *Arabidopsis*. Subcellular compartmentalization leads to distinctive proteome properties. *Plant Physiol.* **135,** 723–734.
5. Kleffmann, T., Russenberger, D., Von Zychlinski, A., et al. (2004) The *Arabidopsis thaliana* Chloroplast Proteome Reveals Pathway Abundance and Novel Protein Functions. *Curr. Biol.* **14,** 354–362.
6. Froehlich, J. E., Wilkerson, C. G., Ray, W. K., et al. (2003) Proteomic study of the *Arabidopsis thaliana* chloroplastic envelope membrane utilizing alternatives to traditional two-dimensional electrophoresis. *J. Proteome Res.* **2,** 413–425.
7. Ferro, M., Salvi, D., Brugiere, S., et al. (2003) Proteomics of the chloroplast envelope membranes from *Arabidopsis thaliana. Mol. Cell. Proteomics* **28,** 28.
8. Ferro, M., Salvi, D., Riviere-Rolland, H., et al. (2002) Integral membrane proteins of the chloroplast envelope: Identification and subcellular localization of new transporters. *Proc. Natl. Acad. Sci. USA* **99,** 11,487–11,492.
9. Ferro, M., Seigneurin-Berny, D., Rolland, N., et al. (2000) Organic solvent extraction as a versatile procedure to identify hydrophobic chloroplast membrane proteins. *Electrophoresis* **21,** 3517–3526.
10. Peltier, J. B., Ytterberg, A. J., Sun, Q., and van Wijk, K. J. (2004) New functions of the thylakoid membrane proteome of *Arabidopsis thaliana* revealed by a simple, fast, and versatile fractionation strategy. *J. Biol. Chem.* **279,** 49,367–49,383.
11. Friso, G., Giacomelli, L., Ytterberg, A. J., et al. (2004) In-depth analysis of the thylakoid membrane proteome of *Arabidopsis thaliana* chloroplasts: new proteins, new functions, and a plastid proteome database. *Plant Cell* **16,** 478–499.
12. Peltier, J. B., Emanuelsson, O., Kalume, D. E., et al. (2002) Central functions of the lumenal and peripheral thylakoid proteome of *Arabidopsis* determined by experimentation and genome-wide prediction. *Plant Cell* **14,** 211–236.
13. Schubert, M., Petersson, U. A., Haas, B. J., Funk, C., Schroder, W. P., and Kieselbach, T. (2002) Proteome map of the chloroplast lumen of *Arabidopsis thaliana. J. Biol. Chem.* **277,** 8354–8365.
14. Hoffer, P. H. and Christopher, D. A. (1997) Structure and blue-light-responsive transcription of a chloroplast psbD promoter from *Arabidopsis thaliana. Plant Physiol.* **115,** 213–222.
15. Cline, K. (1986) Import of proteins into chloroplasts. Membrane integration of a thylakoid precursor protein reconstituted in chloroplast lysates. *J. Biol. Chem.* **261,** 14,804–14,810.
16. Peltier, J. B., Ripoll, D. R., Friso, G., et al. (2004) Clp protease complexes from photosynthetic and non-photosynthetic plastids and mitochondria of plants, their predicted three-dimensional structures, and functional implications. *J. Biol. Chem.* **279,** 4768–4781.
17. Aronsson, H. and Jarvis, P. (2002) A simple method for isolating import-competent *Arabidopsis* chloroplasts. *FEBS Lett.* **529,** 215–220.
18. Seigneurin-Berny, D., Rolland, N., Garin, J., and Joyard, J. (1999) Technical advance: differential extraction of hydrophobic proteins from chloroplast envelope

membranes: a subcellular-specific proteomic approach to identify rare intrinsic membrane proteins. *Plant J.* **19,** 217–228.

19. Gomez, S. M., Nishio, J. N., Faull, K. F., and Whitelegge, J. P. (2002) The chloroplast grana proteome defined by intact mass measurements from liquid chromatography mass spectrometry. *Mol. Cell. Proteomics* **1,** 46–59.
20. Lichtenthaler, H. K. (1987) Chlorophylls and carotenoids: pigments of photosynthetic biomembranes. *Methods Enzymol.* **148,** 350–382.

7

Isolation and Subfractionation of Plant Mitochondria for Proteomic Analysis

Holger Eubel, Joshua L. Heazlewood, and A. Harvey Millar

Summary

Mitochondria carry out a variety of biochemical processes in plant cells. Their primary role is the oxidation of organic acids via the tricarboxylic acid cycle and the synthesis of ATP coupled to the transfer of electrons from reduced NAD^+ to O_2 via the electron transport chain. However, they also perform many important secondary functions such as synthesis of nucleotides, amino acids, lipids, and vitamins, they contain their own genome and undertake transcription and translation by some unique mechanisms, they actively import proteins and metabolites from the cytosol by a complex set of carriers and membrane channels, they influence programmed cell death of plants, and they respond to cellular signals such as oxidative stress. To understand the full range of mitochondrial functions in plants, the mechanisms that govern their biogenesis, and the way in which mitochondrial activity is perceived by the nucleus requires precise information about the protein components of these organelles. Isolation of mitochondria to identify their proteomes and the changes in these proteomes during development and environmental stress treatments is already under way. In this chapter we provide methods for isolating mitochondria from different plant tissue types, advice on assessing purity and storage of mitochondrial samples, and approaches to fractionate mitochondria to separate their membranes and soluble compartments from each other for proteome analysis.

Key Words: Cell fractionation; Percoll density gradients; inner membrane; outer membrane; matrix; intermembrane space.

1. Introduction

Plant mitochondria provide energy in the form of ATP, provide a high flux of metabolic precursors in the form of tricarboxylic acids to the rest of the cell for N-assimilation and biosynthesis of amino acids, and have well-studied roles in plant development, productivity, fertility, susceptibility to disease, and programmed cell death. Plant mitochondria house many hundreds of proteins

From: *Methods in Molecular Biology, vol. 335: Plant Proteomics: Methods and Protocols*
Edited by: H. Thiellement, M. Zivy, C. Damerval, and V. Méchin © Humana Press Inc., Totowa, NJ

identified by proteomics, and a total of 1000 to 1500 proteins are expected to be found in this cellular compartment. Techniques for the isolation of plant mitochondria without changing their morphological structure observed in vivo and maintaining their functional characteristics has required methods that avoid hyper- or hypoosmotic rupture of membranes and protection from harmful products released from other ruptured cellular compartments. Several extensive methodology reviews ***(1–3)*** and more specific methodology papers ***(4–6)*** are already available on plant mitochondrial purification. Study of the composition, metabolism, transport processes and biogenesis of plant mitochondria requires an array of procedures for the subfractionation of mitochondria to provide information on localization and on association of proteins or enzymatic activities. Here we present a basic appraisal of the classic methods for mitochondrial isolation and fractionation procedures to separate mitochondria into their four subcompartments.

2. Materials

1. Plant tissue: Mitochondria can be isolated from virtually any plant tissue. However, depending on the requirements for purity and yield, particular tissues have their advantages:
   a. Nonphotosynthetic fleshy root and tuber tissues allow for large-scale, but low-percentage-yield, plant mitochondrial isolation (approx 300 mg mitochondrial protein from 5–10 kg fresh weight [FW], 30–60 µg/g FW). Typically potato, sweet potato, turnip, and sugar beet are used for these purposes.
   b. Etiolated seedling tissues such as hypocotyls, cotyledons, roots, or coleoptiles are used, as they are free of dense chloroplast membranes, have lower phenolic content than green tissues, and provide high yields of functional mitochondria (approx 20 mg mitochondrial protein from 100 g FW, 200 µg/g FW).
   c. Interest in the function of mitochondria during photosynthesis and in plant species without abundant storage organs has led to methods for the isolation of chlorophyll -free mitochondria preparations from green tissues such as leaves and cotyledons (5–10 mg mitochondrial protein from 50–100 g FW, 100 µg/g FW).

   Once the plant material has been chosen, the tissue must be free from obvious fungal or bacterial contamination and should be washed in cold H_2O or in some cases surface-sterilized (for 5 min in a 1:20 dilution of a 14% w/v sodium hypochlorite stock solution). These steps ensure as little contamination as possible in further steps of the protocol, as bacteria will tend to copurify with mitochondria. All tissues should then be cooled to 2 to 4°C before proceeding.
2. Homogenization buffer solution: tissues should be processed as soon as possible following harvesting and cooling in order to maintain cell turgor and thus ensure maximal mitochondrial yields following homogenization. A basic homogenization medium consists of 0.3 to 0.4 *M* of osmoticum (sucrose or mannitol), 2 to

5 m*M* of divalent cation chelator (EDTA or EGTA), 25 to 50 m*M* of a basic pH buffer system (MOPS, TES or Na-pyrophosphate), and 5 to 20 m*M* of reductant (cysteine, ascorbate, dithiothreitol [DTT] or β-mercaptoethanol), which is added just prior to homogenization. The osmoticum maintains the mitochondrial structure and prevents physical swelling and rupture of membranes, the buffer prevents acidification from the contents of ruptured vacuoles, the EDTA inhibits the function of phospholipases and various proteases, and the reductant prevents damage from oxidants present in the tissue or produced on homogenization. Although this basic medium is sufficient for tuber tissues, a variety of additions have been suggested to improve yields and protect mitochondria from damage during isolation. Etiolated seedling tissues and green tissues often require addition of 0.2 to 1% (w/v) bovine serum albumin (BSA) to remove free fatty acids, and 1 to 2% (w/v) polyvinylpyrolidone to remove phenolics that damage organelles in the initial homogenate. As a result, we have routinely used a homogenization medium consisting of 0.4 *M* mannitol, 5 m*M* EGTA, 50 m*M* sodium pyrophosphate-KOH (pH 7.8), 10 m*M* cysteine, 0.5% BSA, and 1% PVP-40 with wide success in a variety of plant tissues. This media can be freshly prepared, stored overnight at 4°C or frozen at –20°C, and stored for many weeks. Filtrates from homogenization can be filtered through a wide variety of muslin fabrics, Miracloth (Calbiochem, La Jolla, CA), or disposable clinical sheeting available from medical suppliers.

3. Wash buffer solution: a standard wash buffer solution is then used for resuspension of organelle pellets, as the base media for Percoll gradients and for washing purified organelle pellets. We have successfully used 0.3 *M* mannitol, 10 m*M* TES-KOH, pH 7.2, and 0.1% w/v BSA in a variety of plants.
4. Percoll gradient solutions: A gradient of Percoll (*see* **Note 1**) is prepared in 0.3 *M* sucrose or 0.3 *M* mannitol supplemented with 10 m*M* TES, pH 7.2, 0.1% (w/v) BSA on the day of use (*see* **Notes 2** and **3**). For a continuous gradient a gradient maker is required (**Fig. 1A**), for a discontinuous step gradient simple inverted syringe bodies will suffice (**Fig. 1B**).
5. Cytochrome oxidase assays:
   a. For the oxygen consumption-based assay the following is required: O_2 electrode, 0.5 *M* Tris-HCl, pH 7.5, 5 m*M* Cyt c (oxidized), 0.5 *M* sodium ascorbate, 10% (w/v) Triton X-100, 0.1 *M* KCN
   b. For the spectrophotometric assay the following is required: Visible wavelength spectrophotometer, PD10 desalting column, 0.5 *M* Tris-HCl, pH 7.5, 5 m*M* Cyt c (reduced; *see* **Note 4**), 10% (w/v) Triton X-100, 0.1 *M* KCN.
6. Subfractionation of mitochondrial compartments: further fractionation of mitochondria will require: wash medium without BSA (0.3 *M* mannitol, 10 m*M* TES-KOH, pH 7.2), 2 *M* KCl stock solution, low osmotic buffer (50 m*M* sucrose, 2 m*M* EDTA, 10 m*M* MOPS pH 6.5), 2 *M* sucrose, sodium carbonate soultion (100 m*M* Na_2CO_3), and a Dounce homogenizer.

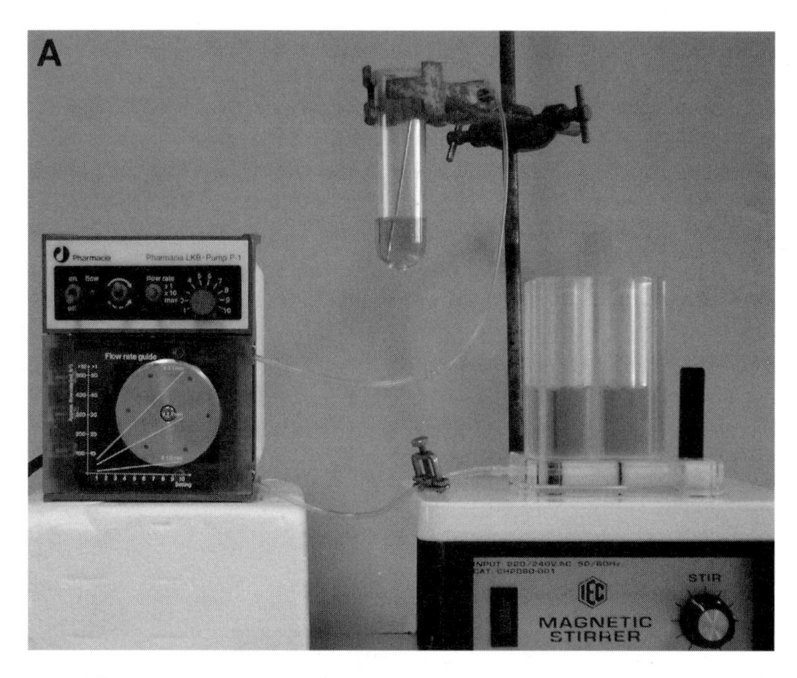

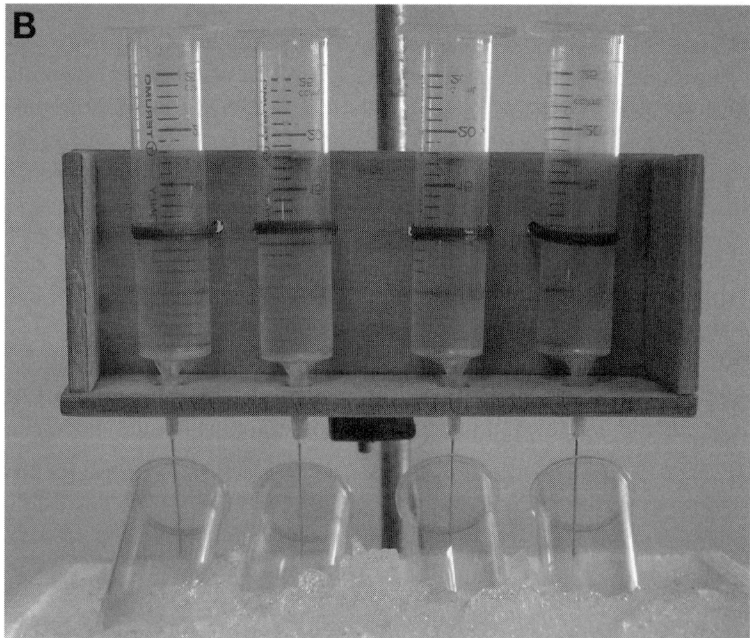

Fig. 1. Making density gradients. **(A)** A commercial apparatus for continuous gradient formation. **(B)** Home-made apparatus for discontinuous gradients (*see* explanation in **Note 3**).

3. Methods

Following selection of the plant material and the homogenization medium (above), the most critical steps in yield of plant mitochondria are often found to be the method of homogenization, the pH of the medium, the ratio of homogenization medium to plant tissue, and a combination of the temperature of solutions and the time taken (*see* **Note 5**).

3.1. Homogenization

Depending on the tissue, homogenization can be accomplished by one or a combination of several methods:

1. Grinding in a precooled mortar and pestle.
2. Homogenization in a beaker with either a Moulinex mixer for 20 to 60 s or a Polytron blender at 50% full speed for 5× 2 s.
3. Homogenization in a Waring blender for 15 s at high speed, 2× 15 s at low speed.
4. Juicing tuber material directly into 5X homogenization medium using a commercial vegetable/fruit juice extractor.
5. Grating of tissue by hand using a vegetable grater submerged under homogenization medium (usually required for mitochondrial extraction from soft fruits).

3.2. pH and Tissue Ratio

1. The pH value of the homogenization medium should be adjusted to approx 7.8 before homogenization to allow for acidification following cell breakage back to pH 7.0 to 7.6.
2. For some plant tissues, however, the amount of buffer may need to be increased to ensure that the pH of the final homogenate remains at or above pH 7.0, or the pH may need to be adjusted following homogenization with a stock of 5 *M* KOH or NaOH.
3. Homogenizing medium should be added in a ratio of 2 mL for each gram FW of non-green tissue or 4 mL per gram FW of green tissue. Decreases in these ratios can result in dramatic losses in yield.
4. Smaller ratios of medium to tissue (as low as 0.25 mL per gram FW) do work in the case of tuber tissues extracted by method 4 in **Subheading 3.1.** probably because of the low level of phenolics and other toxins in tuber material.

3.3. Filtering and Differential Centrifugation to Obtain a Crude Organelle Pellet

1. The final homogenate should be filtered through four layers of muslin at 4°C. Direct filtrate via a funnel into a 4°C cooled beaker or conical flask; the remainder of the fluid in the cloth can be collected by wringing the cloth into the funnel. This is best performed in a 4 to 10°C cold room.
2. Transfer filtered homogenate into 50-, 250-, or 500-mL centrifuge tubes, depending on the volume of the preparation, and centrifuged in a precooled rotor for 5 min at approx 1500*g* in a fixed angle rotor in a preparative centrifuge at 4°C.

3. Decant supernatant gently into another set of centrifuge tubes, taking care not to transfer the pellet material, which contains starch, nuclei, and cell debris. Centrifuge supernatant for 15 min at approx 18,000*g*; discard the resulting high speed supernatant. The tan, yellow, or green pellet in each tube contains an unwashed crude organelle pellet.
4. Resuspend the pellet in 50 to 10 mL of a standard wash medium (e.g., 0.3 *M* mannitol, 10 m*M* TES-KOH, pH 7.2, and 0.1% w/v BSA) with the aid of a clean soft-bristle paint brush. If sucrose is used in the preparation, then sucrose can be used to replace mannitol in the standard wash medium (again at 0.3 *M*).
5. Transfer resuspended organelles to 50-mL centrifuge tubes, adjust volume to 40 mL with more wash medium, and centrifuge samples at 1000*g* for 5 min.
6. Transfer supernatants into another set of tubes, and sediment the organelles by centrifugation at approx 18,000*g* for 15 min. The high-speed supernatant is once again discarded and the washed crude organelles pellets can be uniformly resuspended in a small volume of wash medium as in **step 5** (*see* **Note 6**).

3.4. Density Gradient Purification of Mitochondria

The mitochondrial preparation just described is often adequate for a variety of respiratory measurements, but, depending on the tissue, it is frequently contaminated by thylakoid or amyloplast membranes, peroxisomes, glyoxysomes, endoplasmic reticulum, and occasionally bacteria. Further purification can be carried out using Percoll (GE Health) density gradients.

1. Layer-washed mitochondria, from up to 80 g of etiolated plant tissue or up to 40 g green plant tissue, over 35 mL of the chosen gradient solution in a 50-mL centrifuge tube (*see* **Note 2**).
2. Centrifuge at approx 35,000*g* for 45 min in an angle rotor of a preparative centrifuge without braking on the deceleration.
3. After centrifugation, mitochondria form a buff-colored band below green chloroplast fragments or yellow-orange plastid membranes (**Fig. 2**). Aspirate the mitochondria with a Pasteur pipet, avoiding collection of the yellow or green plastid fractions. Dilute suspension with at least 4 vol of standard wash medium, and centrifuge at approx 18,000*g* for 15 min in 50-mL tubes.
4. The resultant loose pellets should be resuspended in wash medium and centrifuged again at approx 18,000*g* for 15 min. The mitochondrial pellet resuspended in wash medium at a concentration of 5 to 20 mg mitochondrial protein/mL. This can be determined using a Bradford or Lowry assay.

3.5. Purity Determinations

Marker enzymes for contaminants commonly found in plant mitochondrial samples can be used to access the purity of preparations. Peroxisomes can be identified by catalase, hydroxy-pyruvate reductase, or glycolate oxidase activities. Chloroplasts can be identified by chlorophyll content and etioplasts by carotenoid content and/or alkaline pyrophosphatase activity, and glyoxysomes

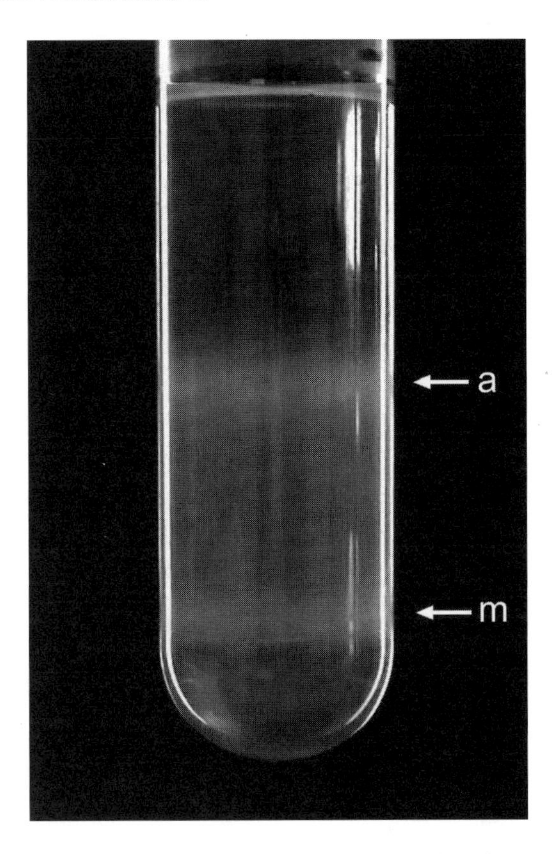

Fig. 2. Percoll gradient purification of plant mitochondria. Three-step Percoll gradient for the purification of potato mitochondria made from 6 mL of 45% Percoll, 18 mL of 26% Percoll, and 14 mL of 14% Percoll solution, from bottom to top. This gradient was ultracentrifuged at 70,000*g* in a fixed-angle rotor for 40 min. Amyloplast envelopes are concentrated in the 14 to 26% interphase (a), mitochondria in the 26 to 45% interphase (m). Residual starch formed a pellet on the bottom of the centrifuge tube. Mitochondrial fraction can be removed by a Pasteur pipet after carefully aspirating the top half of the gradient, including the 14 to 26% interphase.

by isocitrate lyase activity. Endoplasmic reticulum can be identified by antimycin-A-insensitive Cyt c reductase activity and plasma membranes by K+/ATPase activity. Cytosolic contamination is rare if density centrifugation is performed properly and care is taken in removal of mitochondrial fractions, but such contamination can easily be measured as alcohol dehydrogenase or lactate dehydrogenase activities. References to and methods for the assay of these enzymes are summarized in **ref.** ***7*** with several additions in **ref.** ***1***. Other considerations about mitochondrial purity are noted (*see* **Note 7**).

3.6. Integrity Determinations

A variety of assays can be used to determine the integrity of the membranes of plant mitochondrial samples, thus providing information on the structural damage (to these organelles) caused by the isolation procedure. Outer membrane integrity is often assessed by the degree of impermeability of the outer membrane to added Cyt c. The ratio of Cyt c oxidase activity before and after addition of a nonionic detergent (0.05% w/v Triton X-100) gives an estimate of the proportion of ruptured mitochondria in a sample. Two classic assays of its activity can easily be performed.

3.6.1. Rate of O_2 Consumption

Using an O_2 electrode, the rate of O_2 consumption in the presence of a reduced Cyt c regenerating system (to maintain a constant substrate concentration) can be measured.

1. Add ascorbate (20 m*M*) first to get rate of O_2 consumption in the presence of this reductant, then add Cyt c (25 μ*M*), then Triton X-100 (0.05%), and lastly KCN (1 m*M*) to inhibit the rate fully.
2. Cytochrome c oxidase activity is then defined as the rate in the presence of Cyt c and Triton minus the rate after the addition of KCN (this residue is also often the same rate as the ascorbate alone rate calculated before Cyt c addition).
3. The ratio "minus Triton" over "plus Triton" gives the proportion of broken mitochondria.

3.6.2. Oxidation of Reduced Cyt c

Using a spectrophotometer, the oxidation of reduced Cyt c can be measured at 550 nm.

1. Add mitochondrial protein to 0.1 *M* Tris-HCl solution containing 25 μ*M* Cyt c (reduced). Follow Cyt c oxidation at 550 nm, then add Triton X-100 (0.05%), follow the rate, and lastly add KCN (1 m*M*) to inhibit the rate fully.
2. Cytochrome c oxidase activity is then defined as the rate in the presence of Cyt c (reduced; *see* **Note 4**) and Triton minus the rate after the addition of KCN.
3. Again, the ratio "minus Triton" over "plus Triton" gives the proportion of broken mitochondria.

3.7. Storage

Once isolated by density gradient purification, mitochondria from most plant tissues can be kept on ice for 5 to 6 h without significant losses in membrane integrity and respiratory function. Longer term storage of functional mitochondria can be achieved by rapid-freezing small volumes of mitochondrial samples (>0.5 mL) in liquid N_2 following the addition of dimethylsulfoxide (DMSO) to 5% (v/v) or ethylene glycol to 7.5% (v/v). Frozen samples can be then kept at –80°C. In our hands, mitochondrial samples prepared in this way from a variety of plant species can be stored for months without significant losses in respiratory rate, coupling to ATP production, or outer membrane integrity following thawing of these samples on ice.

3.8. Subfractionation of Mitochondrial Compartments

Using a combination of osmotic shock and differential centrifugation, plant mitochondrial samples can be fractionated in four components comprising the two aqueous compartments, the matrix (MA) and the intermembrane space (IMS) and the two membrane compartments, the inner mitochondrial membrane (IM) and the outer mitochondrial membrane (OM).

1. Initially carry out a salt wash to remove nonspecific associated proteins from the mitochondrial outer membrane. Resuspend 5 to 10 mg mitochondrial pellet in 4 mL of wash medium without BSA and add 2 *M* KCl to give a final concentration of 200 m*M* KCl. Divide in to 2 × 2-mL microfuge tubes, and centrifuge at 20,000*g* for 15 min at 4°C.
2. Carefully remove supernatant (sacrifice some at the interface if necessary) and discard. Resuspend pellets in a low osmotic buffer to a total volume of 6 mL and add to a 10-mL conical flask, and place in an ice slurry on a stirring block for 15 min (*see* **Note 8**). This step swells the mitochondria, and the outer membrane bursts because of the expanding inner membrane. Rupture of the OM can be supported by applying some mechanical shearing force to the sample (e.g., by a glass Dounce homogenizer).
3. Return to standard osmotic conditions by adding 0.9 mL of 2 *M* sucrose (approx 1/8 dilution of sucrose). Dispense into 2-mL microfuge tubes, and centrifuge at 20,000*g* for 15 min at 4°C.
4. Carefully take supernatant, which contains the removed outer membrane vesicles (OM) and soluble inner membrane space proteins (IMS), and freeze at –80°C.
5. Remove residual supernatant and resuspend each pellet (mitoplasts) in 2 mL of 250 m*M* KCl. Centrifuge at 20,000*g* for 15 min at 4°C.
6. Again remove residual supernatant and resuspend each pellet (mitoplasts) in 0.5 to 1 mL double-distilled (dd)H_2O, pool the samples in a Dounce homogenizer, and homogenize for approx 5 min to help burst mitoplasts.
7. Aliquot homogenate into two 2-mL Eppendorf tubes, make up each to 2 mL with ddH_2O, and centrifuge at 20,000*g* for 15 min at 4°C.
8. Carefully remove supernatant, which contains the matrix (MA), and freeze at –80°C.
9. The pellets are comprised of inner membranes (IM) and unbroken mitoplasts. Resuspend pellets in 4 mL ddH_2O, and freeze-thaw again in liquid N_2 two or three times. Centrifuge samples at 20,000*g* for 15 min at 4°C and keep pellets. Some of these pellets (IM) can be carbonate-washed (in 100 m*M* Na_2CO_3) to enrich for integral membrane proteins if required.

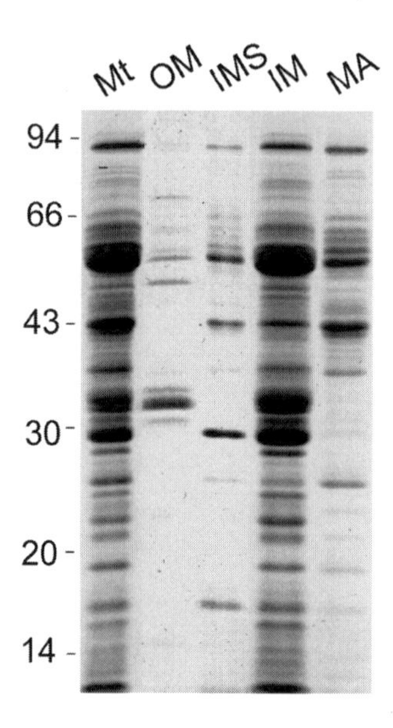

Fig. 3. SDS-PAGE separation of the protein profile of the four subfractions of potato mitochondria visualized by Coomassie staining. Mt , mitochondria; OM, outer membrane; IMS, intermembrane space; IM, inner membrane; MA, matrix space. Proteins solubilized in Laemmli buffer prior to electrophoresis.

10. Ultracentrifugation is required to separate OM and IMS and to remove membrane fragments from the MA sample. Thaw at 4°C the sample of MA and OM/IMS that has been stored at –80°C. Ultracentrifuge the OM and IMS mixture at 100,000*g* for 1 h at 4°C. The supernatant is the IMS fraction, and the pellet is the OM (retain and freeze). Ultracentrifuge the MA samples at 100,000*g* for 1 h at 4°C. Retain supernatant at the MA sample and freeze.
11. Sample purity can be assayed by measuring enzyme marker activities: cytochrome oxidase (COX) for IM, fumarase for MA, adenylate kinase for IMS, and antimycin A-insensitive NADH:cytochrome c oxidoreductase for OM. Antibodies to proteins can also be used in specific fractions. Separation of samples by sodium dodecyl sulfate-polyacrylamide gel electrophoresis (SDS-PAGE) show distinct protein patterns in each compartment (**Fig. 3**).

4. Notes

1. The colloidal silica sol Percoll allows the formation of isoosmotic gradients and through isopycnic centrifugation facilitates a range of methods for the density purification of mitochondria. The most common method is the sigmoidal, self-generating gradient obtained by centrifugation of a Percoll solution in a fixed-angle rotor. The density gradient is formed during centrifugation at >10,000*g* owing to the sedimentation of the poly-dispersed colloid (average particle size 29

nm diameter, average density ρ = 2.2 g/mL). The concentration of Percoll in the starting solution and the time of centrifugation can be varied to optimize a particular separation

2. Modification of the Percoll gradient is often required depending on the tissue being used. **Table 1** shows references to a range of papers using different density gradient methods for plant mitochondrial isolation from a range of plant tissue types. Figure 3 shows how to make such gradients.
3. Step gradients of Percoll are often used, as these aid the concentration of mitochondria fractions on a gradient at an interface between Percoll concentrations. Step gradients can easily be formed by setting up a series of inverted 20-mL syringes (fitted with 19-gauge needles) strapped to a flat block of wood, clamped to a retort stand over a rack at an angle of 45° containing the centrifuge tubes (**Fig. 3B**). The needles are lowered to touch the bevels against the inside, lower edge of the tubes. The step gradient solutions are then added (from bottom to top) to the empty inverted syringe bodies and each allowed to drain through in turn before the addition of the next step solution. Step gradients should, however, only be used once the density of mitochondria from a particular tissue and the density of contaminating components have been determined using continuous gradients. A range of density marker beads are available from GE Health for standardizing running conditions and establishing new protocols.
4. Cytochrome c (reduced) can be made by dilution of the oxidized form in 0.1 *M* Tris-HCl, pH 7.5, and addition of equimolar amounts of ascorbate, this reaction can be monitored spectrally at 400 to 600 nm. Alternatively, rather than monitoring the reaction by spectrometry (oxidized Cyt c is a red-brown colour when concentrated, and reduction leads to a general decrease in the coloration and a shift to red-pink), the progress of the reduction can be followed by eye. Excess ascorbate must be removed following reduction; this can be done using any standard desalting column (e.g., G25 or PD10), and reduced Cyt c can be stored at –20°C for many weeks.
5. Preparation of mitochondria should be undertaken as quickly as possible and without samples warming above 4°C or storage for extended periods between centrifugation runs. The time between homogenization and preparation of the washed crude pellet is the most critical for ensuring integrity and high yield.
6. Some researchers tend to omit **steps 5** and **6** and place the first high-speed organelle pellet obtained in **step 4** directly onto the density gradient. This choice largely depends on the success of the first low speed pellet in removing cell debris and the need to concentrate organelles from **step 4** further in order to place them on the density gradients.
7. Isolation procedures for mitochondria so far rely largely on the density of the organelles only, and this ultimately limits the purity of organelle fraction that can be obtained. Currently 2 to 5% of protein is routinely of nonmitochondrial origin even in the best mitochondrial preparations of the leading research groups. Additional steps for purity using different properties will be the next step to move purity to levels of 0.2 to 1.0% of nonmitochondrial protein in an organelle prepa-

Table 1
Density Gradient Purification Strategies for Isolation of Plant Mitochondria by Different Researchers

Species	Tissue	Purification method[a]	Reference
Non-green tissue			
Arabidopsis	Cell suspension	1, 3	***8***
	cultures	1, 2, 3	***9***
Barley	Seedlings	1, 3	***10***
Bean	Seedlings	1, 3	***11*** originally: ***12*** for potato
Cauliflower	Buds	1, 2	***6*** originally: ***4*** for potato
Maize	Embryos	1, 5, or 6	***13***
	Seedlings	1, 3	***14***
Pea	Seeds	1, 2, 3	***15***
Potato	Tubers	1, 3	***12***
		1, 2	***4***
		1, 2	***16***
	Stems	1, 3	***17*** originally: ***11*** for potato tubers
Rice	Seedlings	1, 2, 4	***18***
Rye	Seedlings	1, 3	***14***
Soybean	Roots	1, 4	***19*** originally: ***6*** for pea
	Cell suspension cultures	1	***19*** originally: ***20*** for tobacco
Tobacco	Cell suspension	1	***20***
	cultures	1, 4	***21*** originally: ***6*** for pea
Wheat	Seedlings	1, 3	***14***
		1, 3	***10***
Green tissue			
Arabidopsis	Leaves	1, 3	***22*** and ***7***
Pea	Leaves	1, 4	***6***
		1, 2	***23***
Soybean	Cotelydons	1, 4	***19*** originally: ***6*** for pea
Spinach	Leaves	1, 4	***24***
		1, 2, or 3	***25***
Tobacco	Leaves	1, 3	***26***

[a] Purification methods: 1, differential centrifugation; 2, continuous Percoll gradients; 3, discontinuous Percoll gradients; 4, combined continuous Percoll and PVP gradients; 5, discontinuous sucrose gradients; 6, continuous sucrose gradients.

ration. These approaches might use properties like surface charge of organelles or affinity capture techniques, but none have been routinely used to date by researchers.

8. The sucrose concentration required in the low osmotic buffer for mitoplast formation depends on the source of the mitochondria to be subfractionated. This osmotic strength can be altered to 10 to 80 m*M* to improve yields and purity significantly of the final fractions by maximizing OM rupture and minimizing MA protein release.

References

1. Neuburger, M. (1985) in *Higher-Plant Cell Respiration, vol. 18* (Douce, R. and Day, D. A., eds.), Springer-Verlag, Berlin, pp. 7–24.
2. Douce, R. (1985) *Mitochondria in Higher Plants: Structure, Function and Biogenesis,* American Society of Plant Physiologists.
3. Millar, A. H., Liddell, A., and Leaver, C. J. (2001) Isolation and subfractionation of mitochondria from plants. *Methods Cell Biol.* **65,** 53–74.
4. Neuburger, M., Journet, E. P., Bligny, R., Carde, J. P., and Douce, R. (1982) Purification of plant-mitochondria by isopycnic centrifugation in density gradients of Percoll. *Arch. Biochem. Biophys.* **217,** 312–323.
5. Leaver, C. J., Hack, E., and Forde, B. G. (1983) Protein synthesis by isolated plant mitochondria. *Methods Enzymol.* **97,** 476–484.
6. Day, D. A., Neuburger, M., and Douce, R. (1985) Biochemical characterization of chlorophyll-free mitochondria from pea leaves. *Austr. J. Plant Physiol.* **12,** 219–228.
7. Quail, P. (1979) Plant cell fractionation. *Ann. Rev. Plant Physiol.* **30,** 425–484.
8. Werhahn, W., Niemeyer, A., Jansch, L., Kruft, V., Schmitz, U. K., and Braun, H. P. (2001) Purification and characterization of the preprotein translocase of the outer mitochondrial membrane from *Arabidopsis*. Identification of multiple forms of TOM20. *Plant Physiol.* **125,** 943–954.
9. Millar, A. H., Sweetlove, L. J., Giege, P., and Leaver, C. J. (2001) Analysis of the *Arabidopsis* mitochondrial proteome. *Plant Physiol.* **127,** 1711–1727.
10. Focke, M., Gieringer, E., Schwan, S., Jansch, L., Binder, S., and Braun, H. P. (2003) Fatty acid biosynthesis in mitochondria of grasses: malonyl-coenzyme A is generated by a mitochondrial-localized acetyl-coenzyme A carboxylase. *Plant Physiol.* **133,** 875–884.
11. Eubel, H., Jansch, L., and Braun, H. P. (2003) New insights into the respiratory chain of plant mitochondria. Supercomplexes and a unique composition of complex II. *Plant Physiol.* **133,** 274–286.
12. Braun, H. P., Emmermann, M., Kruft, V., and Schmitz, U. K. (1992) Cytochrome c1 from potato: a protein with a presequence for targeting to the mitochondrial intermembrane space. *Mol. Genet. Genom.* **231,** 217–225.
13. Logan, D. C., Millar, A. H., Sweetlove, L. J., Hill, S. A., and Leaver, C. J. (2001) Mitochondrial biogenesis during germination in maize embryos. *Plant Physiol.* **125,** 662–672.
14. Borovskii, G. B., Stupnikova, I. V., Antipina, A. I., Vladimirova, S. V., and Voinikov, V. K. (2002) Accumulation of dehydrin-like proteins in the mitochondria of cereals in response to cold, freezing, drought and ABA treatment. *BMC Plant Biol.* **2,** 5.

15. Benamar, A., Tallon, C., and Macherel, D. (2003) Membrane integrity and oxidative properties of mitochondria isolated from imbibing pea seeds after priming or accelerated ageing. *Seed Sci. Res.* **13,** 35–45.
16. Diolez, P. and Moreau, F. (1983) Effect of bovine serum albumin on membrane potential in plant mitochondria. *Physiol. Plant.* **59,** 177–182.
17. Eubel, H., Heinemeyer, J., and Braun, H. P. (2004) Identification and characterization of respirasomes in potato mitochondria. *Plant Physiol.* **134,** 1450–1459.
18. Heazlewood, J. L., Howell, K. A., Whelan, J., and Millar, A. H. (2003) Towards an analysis of the rice mitochondrial proteome. *Plant Physiol.* **132,** 230–242.
19. Tanudji, M., Djajanegara, I. N., Daley, D. O., et al. (1999) The multiple alternative oxidase proteins of soybean. *Austr. J. Plant Physiol.* **26,** 337–344.
20. Vanlerberghe, G. C. and Mcintosh, L. (1992) Lower growth temperature increases alternative pathway capacity and alternative oxidase protein in tobacco. *Plant Physiol.* **100,** 115–119.
21. Norman, C., Howell, K. A., Millar, A. H., Whelan, J. M., and Day, D. A. (2004) Salicylic acid is an uncoupler and inhibitor of mitochondrial electron transport. *Plant Physiol.* **134,** 492–501.
22. Kruft, V., Eubel, H., Jansch, L., Werhahn, W., and Braun, H. P. (2001) Proteomic approach to identify novel mitochondrial proteins in *Arabidopsis. Plant Physiol.* **127,** 1694–1710.
23. Douce, R., Bourguignon, J., Brouquisse, R., and Neuburger, M. (1987) Isolation of plant mitochondria—general principles and criteria of integrity. *Methods Enzymol.* **148,** 403–415.
24. Pastore, D., Stoppelli, M. C., Di Fonzo, N., and Passarella, S. (1999) The existence of the K+ channel in plant mitochondria. *J. Biol. Chem.* **274,** 26,683–26,690.
25. Glaser, E., Knorpp, C., Hugosson, M., and vonStedingk, E. (1995) Macromolecular movement into mitochondria. *Methods Cell Biol.* **50,** 269–281.
26. Michalecka, A. M., Agius, S. C., Moller, I. M., and Rasmusson, A. G. (2004) Identification of a mitochondrial external NADPH dehydrogenase by overexpression in transgenic *Nicotiana sylvestris. Plant J.* **37,** 415–425.

8

Extraction of Nuclear Proteins from Root Meristematic Cells

Fernando González-Camacho and Francisco Javier Medina

Summary

Fractionation and extraction of nuclear proteins are techniques intended to facilitate dedicated plant proteomic studies. These techniques rely on subcellular fractionation, which makes it possible to define and characterize the proteome of a subcellular organelle, in this case the cell nucleus. Nuclear protein fractionation is proposed as a method to be carried out according to the solubility of proteins in buffers of increasing ionic strength. This physical criterion, accompanied in some steps by the use of additional reagents, such as detergents or enzymes, produces fractions that have been demonstrated to have functional significance. The proposed procedure yields five fractions, the first of them containing proteins associated with the nuclear envelope and remnants of the cytoskeleton. The second fraction, which is soluble at low ionic strength, contains ribonucleoproteins active in nuclear RNA metabolism. After increasing ionic strength and digesting with DNase, the result is the chromatin fraction. Finally, the fourth and fifth fractions correspond to the nuclear matrix and are obtained, respectively, by solubilization in high salt concentration and in the form of the residual pellet, which is only soluble in 7 *M* urea under sonication. This procedure offers a wide range of applicability, even in the cases in which the genome of the particular species investigated is not sequenced. In general, the functional criteria driving the extraction method described here will make this method capable of generating valuable and useful information.

Key Words: Isolated nuclei; protein fractionation; soluble ribonucleoproteins; nuclear matrix.

1. Introduction

The identification and characterization of large amounts of proteins in plants by proteomic methods currently faces the frequent problem of insufficient available data on plant genome sequences. Only two plant species, namely, *Arabidopsis* and rice, are fully sequenced, which means that only in very few

From: *Methods in Molecular Biology, vol. 335: Plant Proteomics: Methods and Protocols*
Edited by: H. Thiellement, M. Zivy, C. Damerval, and V. Méchin © Humana Press Inc., Totowa, NJ

cases has the mass spectrometry (MS) approach been used exclusively for protein identification, always with *Arabidopsis* as the biological model ***(1,2)***. In other cases, MS sequencing was accompanied by amino-terminal sequencing by the Edman reaction ***(3,4)***. Another strategy consisted of the identification of proteins after obtaining tag sequences by spectrometry, which could be then compared with the catalogue of known sequenced genes ***(1,3,5,6)***. In general, plant proteomics was largely facilitated by the existence of sequenced genomes for the biological models that were the object of the studies ***(7)***.

The method we describe here was designed to avoid, to some extent, the necessity of possessing the genome sequence of the considered plant species. The objective is to concentrate on a limited number of proteins that are enriched in different fractions known to have functional significance. Therefore, we will deal with the proteome of a subcellular organelle, in this case the cell nucleus, and we will divide it into functional subproteomes.

The first step is, obviously, to obtain the biological material for the study. Since we were interested in studying nuclear proteins related to, or involved in, cell proliferation events, we chose the root meristematic cell population. This material was homogenized and the organelles of interest (in our case nuclei) were purified. Nuclei isolation from meristematic cells is facilitated by the maximum relative cellular volume occupied by the nucleus in this cell type.

From these purified nuclei, a set of subproteomes was obtained after a stepwise procedure of protein fractionation based on protein solubility in buffers of increasing ionic strength ***(8)***. The procedure yields an initial washing fraction containing proteins of the nuclear envelope and remnants of the cytoskeleton. In the following fraction, which is soluble in a low ionic strength buffer containing EDTA, and is called S2, we obtain the most soluble nuclear proteins, which are ribonucleoproteins ***(9)***. In a third step, we extract the proteins of the chromatin, after incubation in DNase. Finally, the last two fractions contain proteins of the nuclear matrix, first by solubilization in a high salt buffer and then by recovering the insoluble pellet. These insoluble proteins are associated with the network of filaments that contain RNA ***(10)***.

After sequential protein extraction, each fraction is suspended in an appropriate buffer in order to separate proteins either in monodimensional sodium dodecyl sulfate-polyacrylamide gel electrophoresis (SDS-PAGE) or in bidimensional gels. In fact, the bidimensional gels constitute expression patterns that can be used for various purposes by comparing patterns obtained from cells in different physiological states, or extracted from certain mutants or genetic constructions, or subjected to drug treatments, or affected by diverse stimuli. In particular, we have used this approach to characterize nucleolar proteins involved in cellular proliferation events ***(11)***, as well as to determine expression changes induced by microgravity on board the International Space Station *(12)*.

2. Materials

2.1. Plant Material and Extraction Medium

1. Germination: for germination of onion (*Allium cepa* L.) bulbs, cylindrical glass receptacles containing filtered water (milli-RO4 water; Millipore, Bedford MA) are used.
2. Extraction medium: excised plant material is deposited in extraction medium, a blend of reagents intended to provide stability to subcellular fractions. The composition of this medium, which is a modification of the medium described by Greimers and Deltour ***(13)***, is as follows: 2% arabic gum (Sigma), 1.25% Ficoll (Sigma), 2.5% dextran sulfate (Fluka), 25 m*M* Tris-HCl, pH 7.4, 0.5 m*M* EDTA, 2.5 m*M* $MgCl_2$, 4 m*M* n-octanol, 8 m*M* β-mercaptoethanol (Sigma), 6.8 m*M* diethyl pyrocarbonate (DEPC; Sigma), 30% glycerol and a cocktail of protease inhibitors containing 1 μg/mL aprotinin, 1 μg/mL pepstatin, 1 μg/mL leupeptin and 0.1 m*M* phenylmethylsulfonyl fluoride (PMSF), all from Sigma (*see* **Notes 1** and **2**). **Caution:** Because two of the components of the medium, DEPC and β-mercaptoethanol, are toxic products, it is advisable to work in a fume hood and to wear chemical-resistant gloves.

2.2. Purification of Nuclei

1. Sample homogenization: performed in a high-speed Ultraturrax blender (IKA, Labortechnik, Staufen, Germany) and filtered through three layers of nylon cloth, with pores of 100, 50, and 30 μm, respectively.
2. Storage of purified nuclei: once nuclei are purified, they are stored in nuclei stock buffer (NSB), which is 10 m*M* Tris-HCl, pH 7.4, 10 m*M* HEPES, 10 m*M* KCl, 2 m*M* $MgCl_2$, 4 m*M* n-octanol, 0.1 m*M* $CaCl_2$, 0.24 *M* sucrose, 0.5 m*M* spermidine (Sigma), 0.15 m*M* spermine (Sigma), 0.02% sodium azide (Sigma), plus the cocktail of protease inhibitors mentioned from **Subheading 2.1.**, **item 2**.
3. This buffer can be stored at 4ºC for several months, although care must be taken to avoid fungal contamination. The cocktail of protease inhibitors must be added at the moment of use, not when preparing the solution.

2.3. Nuclear Protein Fractionation and Extraction

Throughout all these steps, the cocktail of protease inhibitors must supplement all solutions.

1. Fraction of membranes and remnants of cytoskeleton (**Fig. 1**, lane M): the first fraction is extracted by adding to NSB a solution of 1% (v/v) Nonidet NP-40 (Roche) and 0.5% (v/v) sodium deoxycholate (Sigma).
2. Soluble fraction (S2) (**Fig. 1**, lane S2): the second fraction is extracted with the low ionic strength buffer, which is 10 m*M* Tris-HCl, pH 8.0, and 1 m*M* EDTA.

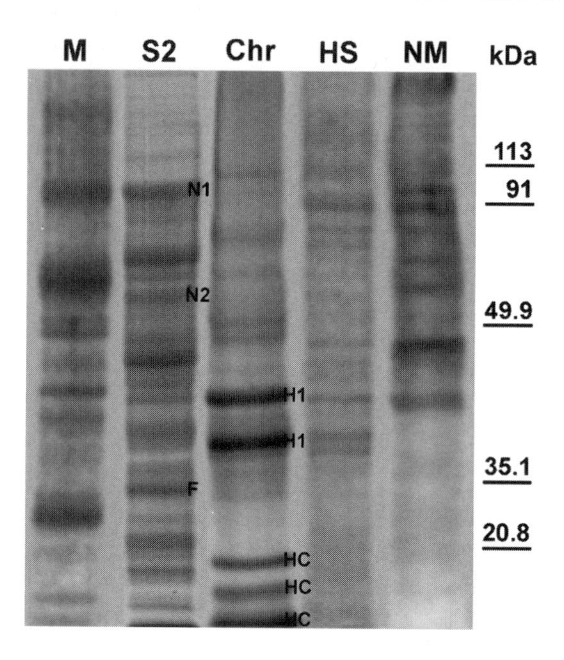

Fig. 1. Nuclear protein fractions obtained from onion root meristematic cells, separated by SDS-PAGE, and stained with Coomassie Brilliant Blue. A total of 10 μg of proteins was loaded in each well. Lane M, initial fraction containing membranes and remnants of the cytoskeleton. Lane S2, soluble ribonucleoprotein fraction, obtained by extraction in a low ionic strength EDTA buffer. Lane Chr, chromatin fraction, released after digestion with DNase I. Lane HS, first nuclear matrix fraction, soluble in a high ionic strength (high salt) buffer. Lane NM, insoluble part of the nuclear matrix fractions. kDa, molecular weight markers, in kilodaltons. Some major bands are identified, corresponding to characterized proteins, such as the nucleolin-like proteins NopA100 (N1; *see* **ref. *11***) and NopA64 (N2; *see* **ref. *16***), fibrillarin (F; *see* **ref. *14***), histone H1 (H1), and histones of the core nucleosome (HC).

3. Chromatin fraction (**Fig. 1**, lane Chr): for extraction of the third fraction, the buffer used is composed of NSB to which 100 μg/mL RNase-free DNase I (Sigma) and 0.5% Triton X-100 are added. After DNA digestion, 0.25 *M* ammonium sulfate is incorporated.
4. High salt fraction (**Fig. 1**, lane HS): extracted with NSB containing 2 *M* NaCl.
5. Insoluble fraction or nuclear matrix fraction (**Fig. 1**, lane NM); can only be solubilized in urea buffer, which is 20 m*M* MES, pH 6.6, 1 m*M* EGTA, 0.1 m*M* $MgCl_2$, 1% β-mercaptoethanol, 0.02% sodium azide, 0.5% Triton X-100, 8 *M* urea, 2 *M* thiourea. This buffer is stable during several months at 4°C, but, since urea may form crystals in cold, it is advisable to transfer it to room temperature some time before its use, or, alternatively, to warm it up slightly, not exceeding 37°C, at the moment of use.

2.4. Protein Preparation for SDS-PAGE and for 2D-Electrophoresis

1. Precipitation step: proteins from each fraction are precipitated with 7% trichloroacetic acid (TCA). The pellet is washed in ethanol/ether 3:1 (v/v).
2. Preparation for SDS-PAGE: if proteins are to be separated by SDS-PAGE, dry pellets of nuclear protein fractions are resuspended in Laemmli buffer, which is 10% glycerol, 0.2 *M* Tris-HCl, pH 6.8, 0.002% bromophenol blue (Bio-Rad), 4% SDS, 5% β-mercaptoethanol.
3. Preparation for 2D-electrophoresis: dry pellets of nuclear protein fractions are resuspended in rehydration buffer, which is 40 m*M* Tris-HCl, 2 *M* thiourea, 7 *M* urea, 4% Triton X-100, 100 m*M* dithiothreitol (DTT), 2% carrier ampholytes (Bio-Rad) (pH 7–4 carrier ampholytes were mixed with pH 3–10 ampholytes at a 2:1 ratio), and 0.001% bromophenol blue (Bio-Rad). The buffer can be stored at room temperature, without DTT and the ampholytes, which should be added at the moment of protein loading onto the gel.
4. Quantitative assessment of the protein concentration. For this purpose, the Bradford Protein Assay reagent (Bio-Rad) is used. The optical density of the reaction is measured in a spectrophotometer, at a wavelength of 595 nm. The machine is calibrated to zero with a blank containing 750 µL of water, 250 µL of the Bradford reagent, and 1 µL of electrophoresis buffer (Laemmli or rehydration) without proteins and bromophenol blue. The value obtained is compared with a standard graphic, previously calculated, composed from known concentrations of BSA, ranging from 1 to 140 µg/mL.

3. Methods

3.1. Plant Material

1. Peel the onion *(Allium cepa L.)* bulbs, removing the brown dry shell, wash them with tap water, and place each one on top of a cylindrical glass tube containing approx 90 mL of filtered water, so that only the base remains submerged in the water (*see* **Note 3**). The water should be continuously aerated by bubbling air at 10 to 20 mL/min. Renew water every 24 h (*see* **Note 4**).
2. Two or 3 d after initiation of the culture, many roots have sprouted from most bulbs; a normal germination process produces more than 40 roots per bulb, of at least 2 cm in length, in 95% of bulbs, after 3 d of culture. They are ready for collection.
3. With the aid of a scalpel or a razor blade, remove the root cap and, with a pair of forceps, dissect the first 3 mm of the root tip; this portion is the root meristem (*see* **Notes 5 and 6**). Take approx 500 mg of meristems.
4. Deposit samples in a glass Petri dish, 4 cm in diameter, containing 6 mL of Extraction medium, to which the protease inhibitor PMSF has been added just before use.
5. Collect root meristems up to an amount of 1 g of sample.
6. The Petri dish must be placed on ice.
7. **Caution:** the work should be performed under a fume hood.

8. With the help of a vacuum pump, degas the collected sample for 10 min, to facilitate the infiltration of the medium within the sample (*see* **Note 7**).

3.2. Purification of Nuclei

1. After incubation of root meristems in extraction medium (may be extended from 15 min to overnight, at 4°C), homogenize them with the Ultraturrax blender and filter the homogenate through three nylon cloth layers with pores of 100, 50, and 30 μm, respectively.
2. Collect the remaining material retained in the nylon with a spatula, resuspend this material in extraction medium, and repeat the filtration procedure twice more. Each homogenization step, involving the operation of the blender, consists of the introduction of the rotating stick of the blender into the tube containing the sample up to its bottom and the activation of the blender motor, at 16,000 rpm, in three strokes of 5 s each, separated by 5-s intervals.
3. Collect and unify the filtrates and wash the resulting sample by centrifugation at 800*g*, for 10 min at 4°C.
4. Suspend the pellet again in 1 mL of extraction medium. Facilitate good suspension with the help of pipet, and then centrifuge twice more. At the end of this procedure, the pellet must contain clean isolated nuclei.
5. Transfer the pellet to an 1.5-mL Eppendorff tube containing 1 mL of nuclei stock buffer (NSB). Check the identity of the nuclear fraction and the abundance and purity of nuclei under the light microscope. Phase contrast may be used for examination, but a simple staining agent, such as methyl green/pyronin, or toluidine blue, or 4,6-diamidino-2-phenylindole (DAPI), can be of help in identifying nuclei (*see* **Notes 8 and 9**).

3.3. Protein Extraction by Nuclear Fractionation

For the design of this procedure, we used the procedure described by Penman and coworkers ***(10)*** and adapted it to our requirements ***(14,15)***. The most important modification was the introduction of a step of extraction of the most soluble proteins, because it was reported in the literature that the nuclear protein fraction soluble in a low ionic strength buffer was functionally significant because it was enriched in ribonucleoproteins active in the nuclear RNA metabolism ***(8,9,16,17)***.

In general, the procedure consists of a sequential fractionation of proteins using solubility as the fundamental criterion. The sample is transferred through successive buffers of increasing astringency or ionic strength. The successive steps are as follows:

1. Fraction of membranes and remnants of cytoskeleton:
   a. The first step is the incubation of purified nuclei, contained in NSB, in a buffer rich in detergents.

b. Add 1% (v/v) NP-40 and 0.5% (v/v) sodium deoxycholate to the nuclei sample and incubate for 10 min at 4°C in an orbital shaker; then, shake the tube with a vortex in two strokes, 20 s each.
c. Centrifuge the sample at 1000*g* for 10 min, at 4°C and collect the supernatant, which is the first fraction (membranes and remnants of the cytoskeleton; **Fig. 1**, lane M).

2. Soluble fraction (S2): Resuspend the pellet in 1 mL of the low ionic strength buffer with a vortex, incubate for 1 h, at 4°C in the orbital shaker, and then centrifuge at 1000*g*, for 10 min at 4°C. The supernatant is the second fraction, called the S2 extract, containing ribonucleoproteins (**Fig. 1**, lane S2).
3. Chromatin fraction:
   a. Suspend the pellet again in 400 μL of NSB supplemented with 100 μg/mL RNase-free DNase I and 0.5% Triton X-100 with a vortex, and incubate for 30 min at room temperature; this will digest DNA.
   b. Extract digested DNA by adding ammonium sulfate (400 μL) and incubating for 5 min at room temperature (*see* **Note 10**). As in previous steps, incubations are made under stirring in an orbital shaker.
   c. Centrifuge at 2000*g* for 10 min to obtain in the supernatant the fraction of proteins associated with chromatin (**Fig. 1**, lane Chr).
4. High salt fraction:
   a. Add 400 μL of NSB to the pellet, resuspend it with a vortex, and increase the ionic strength of the buffer gradually with NaCl to a final concentration of 2 *M* (*see* **Note 11**).
   b. Leave the sample in this buffer for 5 min in the orbital shaker, at room temperature, and centrifuge at 10,000*g* for 10 min.
   c. The supernatant is the fourth nuclear protein fraction, corresponding to a first nuclear matrix extract, called high salt fraction (**Fig. 1**, lane HS).
5. Insoluble fraction or nuclear matrix fraction: the remaining pellet contains the insoluble proteins of the nuclear matrix. However, in order to make this pellet soluble, incubate it in 1 mL of the urea buffer, after vigorous vortexing, for 45 min at room temperature in the orbital shaker, followed by additional vortexing in two strokes and sonication for 30 s (**Fig. 1**, lane NM).

3.4. Protein Preparation for 1D- or 2D-Electrophoresis

1. Protein precipitation.:
   a. For each protein fraction,obtained in the supernatant in each step of the fractionation procedure, precipitate proteins with 7% TCA for 2 h, at 0°C.
   b. Centrifuge at 10,000*g* for 30 min, and discard the supernatant.
   c. Wash the pellet, to eliminating the rest of the TCA, in precooled ethanol-ether at –20°C for a minimum of 2 h, although it can be extended to overnight.
2. Resuspension in the electrophoresis buffer:
   a. Collect proteins by centrifugation at 20,000*g* for 30 min and dry the pellet under vacuum.

b. Add to the dry pellet 60 µL of either Laemmli buffer (if SDS-PAGE is to be performed) or rehydration buffer in the case of 2D-electrophoresis.
c. To facilitate protein dissolution in rehydration buffer, it is advisable to sonicate three times, 2 s each, on ice (*see* **Notes 12** and **13**).

3. Assessment of protein concentration:
   a. Quantify protein concentration by the Bradford reaction ***(18)***. Take 1 µL of the sample contained in the buffer (without bromophenol blue) and add 749 µL of water and 250 µL of the Bradford Protein Assay reagent.
   b. Mix well with a vortex and leave for 15 min to develop color.
   c. Measure optical density in the spectrophotometer and compare the value obtained with the standard graphic (see **Note 14).**

An example of the results of the nuclear protein fractionation procedure and further electrophoretic separation of proteins in one dimension by SDS-PAGE is shown in **Fig. 1**.

4. Notes

1. Extraction medium can be prepared the day before use, but PMSF should not be added until the moment before you begin. The medium is stored in the refrigerator, conveniently closed and labeled.
2. To prepare PMSF, add 0.3484 g of PMSF to 20 mL of isopropanol. The solution must be kept in the refrigerator; it is quickly degraded at room temperature. Under storage, it tends to precipitate, but it is easily dissolved again by gentle shaking.
3. The dry brown shell of the onion bulbs must be totally removed, and the bottom crown from which roots sprout out must be carefully cleaned to remove dry root remnants. Washing of clean bulbs is carried out under tap water flow for a minimum of 30 min. Onions have a seasonal germination rhythm; the optimal seasons for root growth are autumn and winter. Commercial onions, purchased in a market or greengrocery, are usually suitable for research, but the possibility that the bulbs have received a physical or chemical treatment to inhibit root growth should be considered.
4. Continuous aeration can be obtained by means of a simple air pump (for example, the type used in aquaria) connected to a system of rubber tubes, each one directed to a water-filled bulb receptacle, in which it is immersed by means of a glass tip, conveniently bent.
5. Root cap removal and meristem dissection can be obtained simultaneously by using a system of two razor blades assembled with Parafilm, so they leave a space of 3 to 5 mm between them.
6. When data from meristematic (proliferating) cells are to be compared with a nonmeristematic (nonproliferating) cell population, differentiated root cells can be obtained at the same time of the meristem dissection, by discarding a root fragment of 3 mm immediately next to the meristem and collecting the rest of the root.
7. The degassing procedure of extraction medium containing root samples should be at a pressure such that bubbles begin to appear from the medium. As for the rest of

the nuclei purification procedure, this step must be carried out in a cold environment (i.e., on ice).

8. When different samples are to be compared (e.g., from different treatments or conditions), the number of nuclei is the factor of data normalization, allowing comparisons. Nuclei are counted under the light microscope (toluidine blue staining is helpful for this purpose) using a graticule.
9. If flow cytometry is to be used as a method of analysis, the extraction medium must be especially crystal clear, to eliminate possible precipitates that could interfere with the analysis. For this purpose, an additional centrifugation at 1200*g* for 30 min could be helpful.
10. A stock 0.5 *M* solution of ammonium sulfate should be prepared in advance. This solution is very stable, even at room temperature, for years. From this solution, small amounts are added to the sample (four times) to reach a final concentration of 0.25 *M*. Each addition is followed by vigorous shaking of the tube.
11. As in the previous case, the final 2 *M* concentration of NaCl is obtained from an initial 4 *M* stock solution, which is also very stable at room temperature. The addition, in this case, is practically dropwise, since it is done eight times. A total of 400 µL of NaCl is used.
12. Protein concentration is evaluated on an aliquot not containing bromophenol blue. In any case, a yellowish tonality of the sample, once the buffer is added, is indicative of incomplete elimination of TCA. To overcome this, 2 to 3 µL of Tris-HCl saturated in distilled water are added.
13. Alternatives to TCA for precipitating proteins exist, such as ammonium sulfate and alcohols. Furthermore, precipitation can be suppressed by mixing the sample directly with the buffer in a proportion of 1:1. The usefulness of precipitation is the possibility of concentrating proteins and of evaluating their concentration.
14. In the standard graphic used to evaluate protein concentration, the range selected (1–140 µgr/mL of BSA) is the interval in which a linear relationship exists between optical density and protein concentration. If the value measured is above the limits of this graphic, dilute the protein before the measurement, and then multiply the result by the dilution factor.

Acknowlegments

Work performed in the authors' laboratory was supported by the Spanish Plan Nacional de Investigación Científica, Desarrollo e Innovación Tecnológica, Grants. ESP2001-4522-PE and ESP2003-09475-C02-02.

References

1. Gallardo, K., Job, C., Groot, S. P. C., et al. (2002) Proteomics of *Arabidopsis* seed germination. A comparative study of wild-type and gibberellin-deficient seeds. *Plant Physiol.* **129,** 823–837.
2. Gallardo, K., Job, C., Groot, S. P. C., et al. (2001) Proteomic analysis of *Arabidopsis* seed germination and priming. *Plant Physiol.* **126,** 835–848.

3. Prime, T. A., Sherrier, D. J., Mahon, P., Packman, L. C., and Dupree, P. (2000) A proteomic analysis of organelles from *Arabidopsis thaliana*. *Electrophoresis* **21,** 3488–3499.
4. Shen, S., Jing, Y., and Kuang, T. (2003) Proteomics approach to identify wound-response related proteins from rice leaf sheath. *Proteomics* **3,** 527–535.
5. Mathesius, U., Keijzes, G., Natera, S. H. A., Weinman, J. J., Djordevic, M. A., and Rolfe, B. G. (2001) Establishment of a root proteome reference map for the model legume *Medicago truncatula* using the expressed sequence tag database for peptide mass fingerprinting. *Proteomics* **1,** 1424–1440.
6. Chivasa, S., Ndimba, B.K., Simon, W.J., et al. (2002) Proteomic analysis of the *Arabidopsis thaliana* cell wall. *Electrophoresis* **23,** 1754–1765.
7. Shevchenko, A., Jensen, O.N., Podtelejnikov, A.V., et al. (1996) Linking genome and proteome by mass spectrometry: large-scale identification of yeast proteins from two dimensional gels. *Proc. Natl. Acad. Sci. USA* **93,** 14,440–14,445.
8. Busch, H., Ballal, N. R., Rao, M. R. S., Choi, Y. C., and Rothblum, L. I. (1978) Factors affecting nucleolar rDNA readouts, in *The Cell Nucleus*, vol. 5, *Chromatin*, part B. (Busch, H., ed.), Academic Press, New York, pp. 416–468.
9. Bourbon, H. M., Bugler, B., Caizergues-Ferrer, M., and Amalric, F. (1983) Role of phosphorylation on the maturation pathways of 100 KD nucleolar protein. *FEBS Lett.* **155,** 218–222.
10. He, D., Nickerson, J. A., and Penman, S. (1990) Core filaments of the nuclear matrix. *J. Cell Biol.* **110,** 569–580.
11. González-Camacho, F. and Medina, F. J. (2004) Identification of specific plant nucleolar phosphoproteins in a functional proteomic analysis. *Proteomics* **4,** 407–417.
12. Matía, I., González-Camacho, F., Marco, R., Kiss, J.Z., Gasset, G., and Medina, F.J. (2005) Nucleolar structure and proliferation activity of *Arabidopsis* root cells from seedlings germinated on the International Space Station. *Adv. Space Res.* **36,** 1244–1253.
13. Greimers, R. and Deltour, R. (1981) Organization of transcribed and nontranscribed chromatin in isolated nuclei of *Zea mays* root cells. *Eur. J. Cell Biol.* **23,** 303–311.
14. De Cárcer, G., Cerdido, A., and Medina, F. J. (1997) NopA64, a novel nucleolar phosphoprotein from proliferating onion cells, sharing immunological determinants with mammalian nucleolin. *Planta* **201**, 487–495.
15. De Cárcer, G., Lallena, M. J., and Correas, I. (1995) Protein 4.1 is a component of the nuclear matrix of mammalian cells. *Biochem. J.* **312,** 871–877.
16. Schnapp, A., Pfliderer, C., Rosenbauer, H., and Grummt, I. (1990) A growth-dependent transcription initiation factor (TIF-IA) interacting with RNA polymerase I regulates mouse ribosomal RNA synthesis. *EMBO J.* **9,** 2857–2863.
17. Dunham, V. L. and Bryant, J. A. (1983) Nuclei, in *Isolation of Membranes and Organelles from Plant Cells* (Hall, J. L. and Moore, L., eds.), Academic Press, London, pp. 237–275.
18. Bradford, M. M. (1976) A rapid and sensitive method for the quantitation of microgram quantities of protein utilizing the principle of protein-dyebinding. *Anal. Biochem.* **7,** 248–254.

9

Extraction of Nuclear Proteins

Setsuko Komatsu

Summary

The integrity of a subcellular proteome such as the nucleus, is largely dependent on purification of the isolated compartment away from other cellular contaminants. The separation of high-purity nuclei from plants is a difficult task. However, successful purification has been achieved through a series of fractionation processes. Initially, centrifugation in a 2.0 *M* sucrose density gradient *(1)* or a percoll density gradient *(2)* was used to isolate nuclei from cultured rice suspension cells. A modified version of the sucrose gradient method described in Morre and Anderson *(1)* has proved to be more rapid and efficient for the isolation of nuclei from cultured rice suspension cells. The nuclei are uniform spheres with an average diameter of approx 20 μm. The nuclear proteins were prepared from the purified nuclei using lysis buffer *(3)* or SDS sample buffer *(4)*. The purity of the isolated nuclear fraction was evaluated by Western blot analysis using antihistone H1 antibody, a specific antibody for nuclear proteins. Histone H1 was found in the nuclear fraction, but not in the supernatant fraction, suggesting that the preparation is enriched in nuclear proteins.

Key Words: Rice; subcellular fractions; nuclear proteins; sucrose density gradient; histone H1.

1. Introduction

The eukaryotic nucleus is surrounded by two membranes containing phospholipids. The inner nuclear membrane defines the nucleus itself. In many cells, the outer nuclear membrane is continuous with the rough endoplasmic reticulum. The space between the inner and outer nuclear membranes is continuous with the lumen, or inner cavity, of the rough endoplasmic reticulum. The membranes appear to fuse at the nuclear pores, a particularly vivid fusion when the nucleus is viewed by the freeze-fracture technique. In a nucleus that is not dividing, the chromosomes are elongated, are only about 25 nm thick, and can-

From: *Methods in Molecular Biology, vol. 335: Plant Proteomics: Methods and Protocols*
Edited by: H. Thiellement, M. Zivy, C. Damerval, and V. Méchin © Humana Press Inc., Totowa, NJ

not be observed by the light microscope. However, a suborganelle of the nucleus is easily recognized under the light microscope. The nucleolar organizer, a region of one or more chromosomes in the nucleolus, contains many copies of the DNA that directs the synthesis of ribosomal RNA. Most of the cell's ribosomal RNA is synthesized in the nucleus; some ribosomal proteins are added to ribosomal RNAs within the nucleolus as they pass through a nuclear pore into the cytoplasm ***(5)***.

The various organelles of a cell differ in both size and density. Most fractionation procedures begin with rate-zonal centrifugation, or differential-velocity centrifugation, a technique in which an ultracentrifuge generates most powerful sedimenting forces. Initially, 2.0 *M* sucrose density gradient ***(1)*** and percoll density gradient ***(2)*** centrifugation techniques were adopted to isolate nuclei from cultured rice suspension cells. A modified version of the sucrose gradient method described in Morre and Anderson ***(1)*** has proved to be more rapid and efficient for the isolation of nuclei from cultured rice suspension cells. The nuclei were uniform spheres with an average diameter of approximately 20 μm ***(6)***. Pea nuclei were also observed as uniform spheres with an average diameter of about 20 to 30 μm, but *Arabidopsis* nuclei were smaller in diameter (5–10 μm) and nonspherical ***(2)***. In plants, proteomic profiling of nuclear proteins has been reported for rice ***(6)*** and *Arabidopsis* ***(7)***. In rice, a total of 549 nuclear proteins was resolved within pI values of 4 to 10 by 2D gel electrophoresis ***(6)***. This is in agreement with work on *Arabidosis* nuclear proteins reported by Bae et al. ***(7)***; they detected around 544 proteins in a *pI* range of 3 to 10.

In this chapter, protocols are given for preparation of nuclear proteins from plant tissues.

2. Materials

1. Cultured rice suspension cells used for the isolation of nuclei were cultured in N6 liquid medium ***(8)*** supplemented with 1 mL/L 2,4-dichlorophenoxyacetic acid (2,4-D) under continuous shaking in an incubator at 22°C.
2. Homogenization medium: 50 m*M* HEPES (pH 7.4), 10 m*M* KCl, 1 m*M* EDTA, 10 m*M* ascorbate, 0.1% bovine serum albumin, 20 m*M* dithiothreitol, 400 m*M* sucrose, and 1 m*M* phenylmethylsulfonyl fluoride (PMSF).
3. 2.0 *M* sucrose cushion: 37.5 m*M* Tris-maleate (pH 6.5), 5 m*M* $MgCl_2$, and 1% dextrin T500 (Amersham Biosciences, Piscataway, NJ).
4. Lysis buffer (for 2D-polyacrylamide gel electrophoresis [PAGE]): 9.5 *M* urea, 2% NP-40, 2% Ampholine (GE healthcare) (pH 3.5–10.0), 5% mercaptoethanol, and 0.05% PVP-40 ***(3)***.
5. Sodium dodecyl sulfate (SDS) sample buffer (for SDS-PAGE): 0.06 *M* Tris-HCl (pH 6.8), 2% SDS, 10% glycerol, and 5% mercaptoethanol ***(4)***.
6. Phosphate-buffered saline (PBS), pH 7.6: 80 m*M* Na_2HPO_4, 20 m*M* NaH_2PO_4, 100 m*M* NaCl.

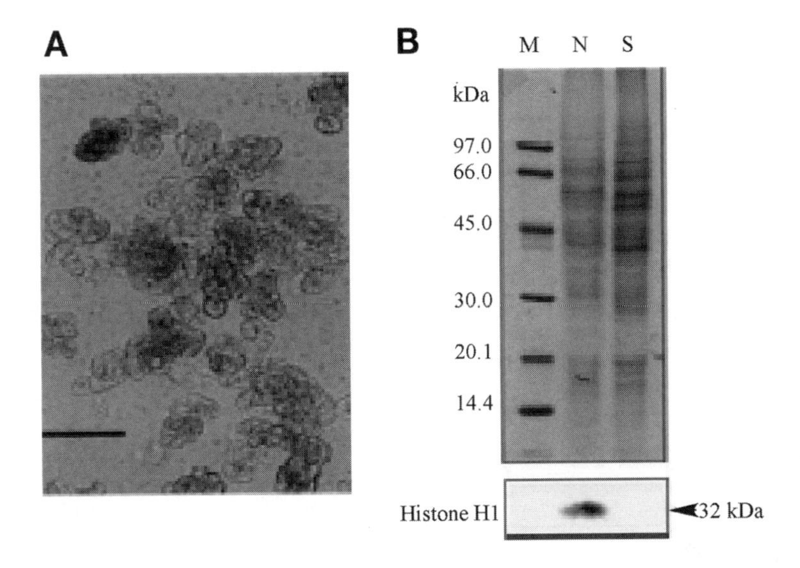

Fig. 1. Isolation of rice nuclear proteins. **(A)** Micrograph of isolated rice nuclei. Scale bar 50 μm. **(B)** Westerm blot analysis of fractionated proteins with antihistone H1 antibody. Proteins were separated by SDS-PAGE and stained by Coomassi brillant blue (CBB) (upper panel). After SDS-PAGE, proteins were blotted with a PVDF membrane, and reacted with antihistone H1 antibody (lower panel). M, low molecular marker; N, nuclear fraction from sucrose density gradient centrifugation; S, supernatant fraction from sucrose density gradient centrifugation *(6)*. (Modified from **ref. *6***.)

3. Methods

1. Nuclei are prepared from cultured rice suspension cells following the sucrose density gradient method described by Morre and Andersson *(1)* with some modifications. All the steps for isolation of nuclei are performed on ice or at 4°C.
2. Approximately 3 g of cultured rice suspension cells are pelleted at 3000*g* for 5 min and resuspended with 5 mL ice-cold PBS. The suspension is centrifuged again at 3000*g* for 5 min. The pellet is resuspended in 5 mL homogenization medium in a glass mortar and homogenized by a pestle.
3. The homogenates are passed two times through a double-layer Miracloth (Calbiochem, Darmstadt, Germany).
4. The homogenate is transferred to a 15 ml falcon tube and centrifuged at 1000*g* for 10 min.
5. The pellet is gently resuspended in the homogenization medium and layered on top of a 2.0 *M* sucrose cushion prepared in 37.5 m*M* Tris-maleate (pH 6.5), 5 m*M* $MgCl_2$, and 1% dextrin T500 and centrifuged at 50,000*g* for 30 min at 4°C.
6. The supernatant solution is removed carefully, and the pellet is gently resuspended again in 5 mL homogenization medium.
7. The suspension is again layered on top of a 2.0 *M* sucrose cushion and centrifuged at 50,000*g* for 30 min. The pellet containing the nuclei is resuspended in 100 μL homogenization medium and is observed under a light microscope (**Fig. 1A**).

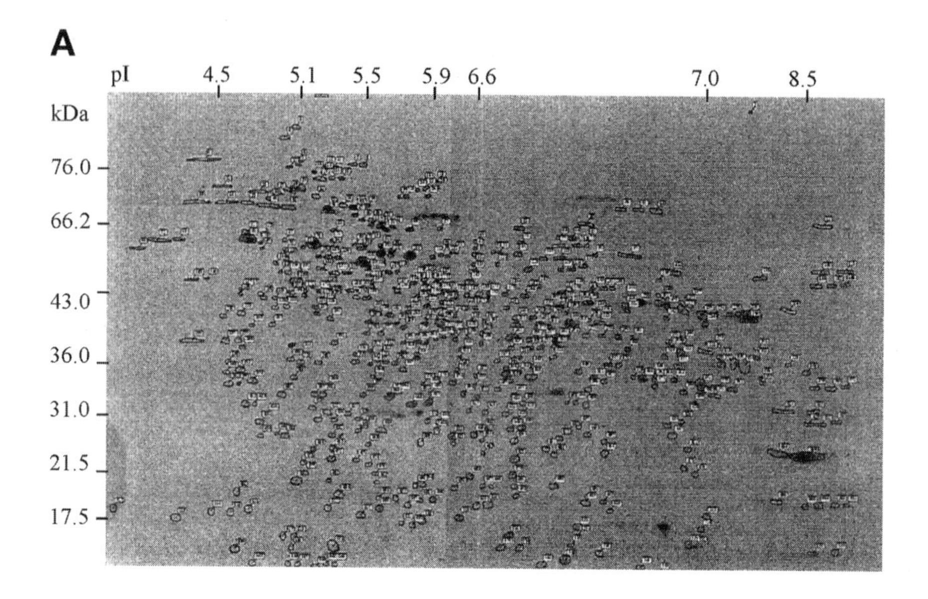

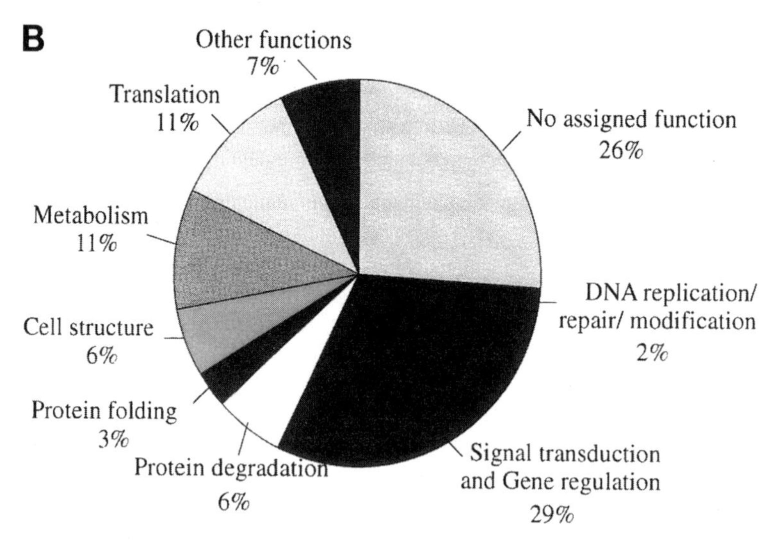

Fig. 2. 2D-PAGE map of rice nuclear proteins and assignment of the identified nuclear proteins to functional categories. **(A)** Proteins (200 μg) were extracted from the purified nuclei of cultured rice suspension cells and separated by 2D-PAGE with isoelectric focusing (IEF) and immobilized pH gradient (IPG) in the first dimension and SDS-PAGE in the second dimension. The 2D maps separately constructed from IEF and IPG tube gels were overlapped at around p*I* 6.0. After detection by Coomassie brillant blue (CBB) staining, protein spots were analyzed by Image-Master 2D Elite software. Numbers indicate the different nuclear protein spots ***(6)***.

8. The suspension containing the nuclei is employed for the extraction of nuclear proteins using lysis buffer *(3)* for 2D-PAGE or SDS sample buffer *(4)* for SDS-PAGE. Nuclei from 20 µL of the suspension ($4 \leftrightarrow 10^5$ nuclei) are solubilized in 500 µL of lysis buffer or SDS sample buffer using a glass homogenizer.
9. The purity of the nuclear proteins is tested using Western blot analysis with antihistone H1 antibody (Santa Cruz Biotechnology, Santa Cruz, CA) (**Fig. 1b**).
10. Proteins (200 µg) extracted by lysis buffer from purified nuclei are separated by 2D-PAGE (**Fig. 2** and *see* **Note 1**).

4. Notes

1. The protein concentration is determined using the Bradford assay *(9)*.

References

1. Morre, D. J. and Andersson, B. (1994) Isolation of all major organelles and membranous cell components from a single homogenate of green leaves. *Methods Enzymol.* **228,** 412–419.
2. Folta, K. M. and Kaufman, L. S. (2000) Preparation of transcriptionally active nuclei from etiolated *Arabidopsis thaliana. Plant Cell Rep.* **19,** 504–510.
3. O'Farrell, P. H. (1975) High resolution two-dimensional electrophoresis of proteins. *J. Biol. Chem.* **250,** 4007–4021.
4. Laemmli, U. K. (1970) Cleavage of structural proteins during the assembly of the head of bacteriophage T4. *Nature* **227,** 680–685.
5. Darnell, J., Lodish, H., and Baltimore, D. (1990) in *Molecular Cell Biology*, Scientific American Books, New York.
6. Khan, M. and Komatsu, S. (2004) Rice proteomics: recent developments and analysis of nuclear proteins. *Phytochemistry* **65,** 1671–1681.
7. Bae, M. S., Cho, E. J., Choi, E.-Y., and Park, O. K. (2003) Analysis of the Arabidopsis nuclear proteome and its response to cold stress. *Plant J.* **36,** 652–663.
8. Murashige, T. and Skoog, F. (1962) A revised medium for rapid growth and bioassay with tobacco tissue culture. *Physiol. Plant* **15,** 473–497.
9. Bradford, M. (1976) A rapid and sensitive method for the quantitation of microgram quantities of protein utilizing the principle of protein-dye binding. *Anal. Biochem.* **72,** 248–254.

Fig. 2. *(continued)* A detailed list of the nuclear proteins identified is provided at the Rice Proteome Database web site, http://gene64.dna.affrc.go.jp/RPD/main.html. **(B)** The percentages of proteins in each functional category from the total number (190) of identified proteins are shown *(6)*. To understand the function of the nuclear proteins, the proteins identified were sorted into different categories. Of the 190 proteins identified by Edman sequencing and mass spectrometric analysis, the most abundant category was involved in signal transduction and gene regulation (29%). (Modified from **ref. *6***.)

10

Isolation of Cell Wall Proteins from *Medicago sativa* Stems

Bonnie S. Watson and Lloyd W. Sumner

Summary

Plant cell walls are highly dynamic and chemically active components of plant cells. Cell walls consist primarily of polysaccharides, with proteins comprising approx 10% of the cell wall mass. These proteins are difficult to isolate with a high degree of purity from the complex carbohydrate matrix. This matrix traps proteins and is a source of contamination for subsequent 2-DE analysis. Mature plant tissues provide a further challenge owing to the formation of secondary walls that contain phenolic compounds. This chapter discusses protein extraction from cell walls and presents a specific method for the isolation of proteins from *Medicago sativa* stem cell walls. The method includes cell disruption by grinding, copious washes with both aqueous and organic solutions to remove cytosolic proteins and small molecule contaminants, and two different salt extractions that provide a highly enriched cell wall protein fraction from alfalfa stem cell walls. Following treatment with a commercial clean-up kit, the protein extracts yield high-quality and high-resolution 2-DE separations from which proteins can be readily identified by mass spectrometry.

Key Words: Cell wall; proteins; proteomics; extraction; two-dimensional gel electrophoresis; 2-DE; SDS-PAGE; *Medicago sativa*; $CaCl_2$; LiCl.

1. Introduction

As in all subcellular proteomics, a critical factor in isolating the cell wall proteome is the extraction of a protein fraction uncontaminated by proteins from other cellular locales. Generally, two approaches have been utilized to isolate cell wall proteins: nondisruptive and disruptive protein extraction. Both methods have advantages and drawbacks, especially concerning issues of contamination and to the development of the current method. A second important factor for cell wall proteomic efforts that include two-dimensional electrophoresis (2-DE) is the removal of polysaccharide and polyphenolic contaminants that

From: *Methods in Molecular Biology, vol. 335: Plant Proteomics: Methods and Protocols*
Edited by: H. Thiellement, M. Zivy, C. Damerval, and V. Méchin © Humana Press Inc., Totowa, NJ

cause problems during electrophoresis. In the past, cell cultures have provided a source of homogeneous and relatively "clean" tissue for isolating cell wall proteins. Differentiated plant tissue, although often more biologically relevant, is more complex and difficult to work with, and large-scale proteomic efforts are relatively new for these tissues. This introduction includes a short review of current cell wall protein extraction methods followed by the description of a specific method for the extraction of cell wall proteins from *Medicago sativa* stems.

Perhaps the oldest and gentlest method of isolating cell wall proteins involves the direct extraction from intact cultured cells by suspending or washing the cells in an appropriate solution. Smith et al. *(1)* used $CaCl_2$ gradients to elute extensin precursors from tomato suspension cultures poured into small columns, and Scott and O'Neil *(2)* extracted extracellular proteins from carrot suspension cells by suspending them in different concentrations of $CaCl_2$ and Triton X-100 as well as in 0.1 *M* LiCl. Carrot suspension cultures extracted with 0.1% Triton X-100 or 0.1 *M* $CaCl_2$ could then be returned to culture medium with little change in their growth patterns.

One of the first attempts to isolate and identify large numbers of cell wall–associated proteins involved extractions of live suspension cultures containing only primary walls from five plant species *(3)*. The cells were harvested by filtration, and proteins were extracted by stirring the cells in five different buffers. The buffers included 0.2 *M* $CaCl_2$, 50 m*M* cyclohexane diaminotetraacetic acid (CDTA) in 50 m*M* sodium acetate, pH 6.5, 2 m*M* dithiothreitol, 1 *M* NaCl, and 0.2 *M* borate, pH 7.5, and were used in both sequential and nonsequential extraction regimes. Blee et al. *(4)* employed a suspension culture of transformed tobacco cells that contained both primary and secondary walls as the source of cell wall proteins extracted directly from living cells. Proteins were extracted with 0.2 *M* $CaCl_2$ or 50 m*M* CDTA.

A modification of this method uses vacuum infiltration of protoplasts *(5)* or whole plant tissues *(6,7)* to elute apoplastic and cell wall proteins. This method involves an infusion step whereby a solution of salts or other solutions enters the intercellular space by the use of gentle reduced pressure. Roger et al. *(5)* used a solution of 1 *M* NaCl and 0.4 *M* $CaCl_2$ to infiltrate immobilized protoplasts derived from hypocotyls of flax plants, and harvested the apoplastic proteins by centrifugation and filtration. Apoplastic fluids from potato tubers were harvested by infiltrating the tissue with a buffer containing 0.6 *M* NaCl, then centrifuging to collect the extract *(6)* apoplastic fluid from leaves has been collected using this method with a solution of 20 m*M* ascorbic acid, 20 m*M* $CaCl_2$ *(7)*. Root apoplastic fluid was collected by simply cutting and centrifuging the tissue to harvest the exudate *(8)*.

A major disadvantage of the nondisruptive methods is that the yield of cell wall proteins is usually low. The current genomic-based assessments predict many hundreds of secreted proteins ***(9)***; however, only a fraction of this number is found in the nondisruptive studies of plant cell wall proteins. A second disadvantage is the possibility of rupturing the plasma membrane, which would lead to contamination of the cell wall fraction with cytosolic and membrane proteins. Because contamination is a formidable obstacle, methods of ensuring cell wall purity, such as assays for marker enzyme activities or Western blots for typical contaminant proteins, should be included.

Another method to elute cell wall proteins from live cells utilizes plasmolysis of the plasma membrane to shrink it away from the cell wall. Borderies et al. ***(10)*** used live *Arabidopsis* suspension culture cells to extract loosely bound cell wall proteins. The cultures were first plasmolyzed with 50% glycerol and then extracted sequentially with various buffers. Two different extraction regimes using NaCl, EDTA, and 0.2 *M* $CaCl_2$ or NaCl, EDTA, and 2 *M* LiCl were utilized to optimize the method and decrease the problem of membrane permeability caused by the extraction treatments. Several of these treatments resulted in contamination of the cell wall proteins with intracellular proteins, so transmission electron microscopy was employed to verify an intact plasma membrane.

Disruptive methods are the second major approach to cell wall protein extractions, and plasmolysis has also been used with this method. After treating live *Arabidopsis* suspension cells with a solution containing glycerol and mannitol, the cells were passed through an N_2 disruption bomb ***(11)***. The wall fraction was than recovered and examined by microscopy. Proteins were extracted from the cell wall preparations with 2% sodium dodecyl sulfate (SDS) and analyzed for contaminating intracellular proteins by Western blots. As with the nondisruptive methods, contamination is a significant problem, and a means of ensuring cell wall purity is necessary.

Homogenized *Arabidopsis* suspension cells were also used as a source for cell wall proteins in two papers from the same group ***(12,13)***. In both cases, cells were washed, filtered, and passed through a cell disrupter. The homogenate was layered on a glycerol solution, and after sedimentation by gravity the cell wall fraction was collected and washed. The purified cell walls were examined by electron microscopy and checked for contaminating intracellular or plasma membrane proteins by immunofluorescence and immunoblots ***(12)***. The results supported the authors' claim that the cell wall fractions were free of contaminating proteins. In both papers, proteins were extracted from the cell wall fraction first with 0.2 *M* $CaCl_2$ and then with a urea buffer.

Cell wall proteins have been extracted from live and homogenized lignin-producing suspension cultures of Norway spruce ***(14)***. Ionically bound cell wall proteins were eluted from washed and filtered cells by incubating them in a 1 *M* NaCl buffer. Separated cells were homogenized, and proteins that were liberated, but ionically bound to the cell walls, were extracted with the same 1 *M* NaCl buffer and then discarded. The cell walls were isolated, washed, homogenized, and suspended in extraction buffer. These walls were then lyophilized and treated with a mixture of cellulase/macerozyme to digest cell wall carbohydrates and release covalently bound proteins.

Traditionally, protein mixtures extracted from the cell walls of plant tissues are used as the source for isolating single proteins or classes of proteins. For example, 0.2 *M* $CaCl_2$ was used to extract a hydroxyproline-rich cell wall glycoprotein (HPRG) from the seed coat of soybean seeds ***(15)***. Similarly, a proline-rich cell wall protein was isolated from the cell walls of soybean seedlings ***(16)***. The protein was detected in the supernatant remaining after the cell wall proteins were precipitated by 10% trichloroacetic acid (TCA). Cell wall proteins involved in extension were purified by McQueen-Mason et al. ***(17)*** from salt-extractable cell wall proteins found in cucumber seedlings. McDougall ***(18)*** used a single tissue, xylem, to isolate differentially expressed oxidases from cell walls of the wood of compression and noncompression branches of spruce.

Large-scale cell wall proteomics has been used less frequently with differentiated plant tissues, primarily owing to limitations imposed by protein, polysaccharide, and polyphenolic contaminants. A method is described here that was developed as a large-scale approach for comparative cell wall proteomics of plant tissues and that addresses the major contaminant issues. It is based on experience (*see* **Notes**) and on information found at http://cellwall.genomics.purdue.edu. The method involves cell disruption, a series of stringent washes to remove contaminating proteins, and extraction of cell wall proteins with $CaCl_2$ and LiCl. The steps are monitored for purity by Bradford assays and SDS-polyacrylamide gel electrophoresis (PAGE) gels. After treatment with a clean-up kit, the proteins are ready to separate by 2-DE. The method has been used to obtain reproducible separations by 2-DE of a large and varied population of proteins from the isolated cell walls of lignified alfalfa (*Medicago sativa*) stems. The difficulties associated with removing polysaccharides and polyphenolics that interfere with focusing in the first dimension and cause streaking and background staining in the second dimension have been minimized, resulting in protein preparations that are reasonably free of contaminating materials and that contain several hundred proteins. The protein spots can easily be identified by matrix-assisted laser desorption ionization-time of flight mass spectrometry (MALDI-TOF MS) or by liquid chromatography (LC)/MS/MS ***(19)***.

One drawback of the method is the apparent lack of cell wall proteins often observed in other types of cell wall preparations. These include highly glycosylated proteins and proteins associated with the plasma membrane (PM). The glycosylated proteins are more soluble than other wall proteins and are therefore likely to be lost during the stringent washes necessary for removing cytosolic proteins ***(19)***. The PM-associated proteins are not as tightly bound to the cell wall as other proteins and are similarly lost in the stringent washes. It is probable that these proteins could be concentrated and separated by 2-DE, or a complementary, nondisruptive technique could be used to isolate them. As with other cell wall protein extraction methods, a requirement of this method is the need to show that the isolated cell walls are free of contaminating proteins. SDS-PAGE or immunoblots of marker enzymes should be done on the washes to ascertain protein extract purity.

The extraction salts chosen in this method have been commonly used for eluting cell wall proteins. Calcium chloride is thought to release charged molecules through ion exchange ***(1)***, has been predicted to be a more efficient ion exchanger than NaCl ***(1)***, has been shown to release more proteins from the cell wall than 0.1% Triton X-100 ***(2)*** in live cultures and has been used extensively to extract cell wall protein from plant cell walls ***(1–5,7,10,12,13,16, 18,19)***. Membrane integrity is known to be interrupted by 0.2 *M* $CaCl_2$ ***(10)***; thus we chose 0.2 *M* $CaCl_2$ for extracting cell wall proteins from disrupted tissue.

The method reported here also utilizes LiCl, a common cell wall protein extractant ***(10,20,21)***. Voight ***(22)*** used concentrations of LiCl ranging from 0.1 *M* to 8 *M* on *Chlamydomonas* cells and found that the optimal concentration to extract cell wall proteins from all cell-cycle stages was at least 2 *M*, with no additional proteins extracted at concentrations above 4 *M* LiCl. As the concentration of LiCl increased, greater amounts of carbohydrates were extracted along with the proteins. Our method uses 3 *M* LiCl for extractions to harvest the greatest amount of protein without coextracting an excess of carbohydrates.

Although the washes preceding the cell wall protein fractions isolated using this method appear to be clean by 1D SDS-PAGE and by Western blot, only approx 50% of the identified proteins have a signal peptide. This proportion of classically secreted proteins seems reasonable compared with other proteomic studies of cell wall proteins ***(10,12)*** and may reflect the inherent difficulties in purifying a subset of proteins from an "organelle" without a surrounding membrane. It may also suggest a nonclassical pathway(s) for protein secretion ***(23,24)***.

2. Materials

2.1. Tissue Disruption, Washes, and Gel or Western Assay for Purity

1. Liquid N_2.
2. Mortar and pestle, chilled to 4°C.

3. Buchner funnel, side arm flask, and vacuum tubing.
4. House vacuum or vacuum pump.
5. Nylon mesh membrane (nitex, 47 μm^2), cut to fit the Buchner funnel.
6. Spatula.
7. 1 *M* Na acetate, pH 5.5, stock.
8. 1 *M* NaCl, stock.
9. 1 *M* Ascorbic acid (store in dark bottle), stock.
10. Grinding buffer: 50 m*M* Na acetate, pH 5.5, 50 m*M* NaCl, 30 m*M* ascorbic acid (*see* **Note 1**) in Milli Q water, 4°C.
11. Wash buffer 1: 100 m*M* NaCl in Milli Q water, 4°C.
12. Milli Q water, 4°C.
13. Acetone stored at –20°C.
14. Wash buffer 2: 10 m*M* Na acetate, pH 5.5, in Milli Q water, 4°C.
15. 25% TCA.
16. Bio-Rad Criterion 10 to 20% acrylamide gel (cat. no. 345-0042) or other commercial SDS-PAGE gel.
17. Coomassie Blue R-250 stain, 1 g/L in 40% methanol, 5% acetic acid in Milli Q water, or other suitable stain.
18. Destain: 40% methanol, 5% acetic acid in Milli Q water or other suitable destain.
19. Millipore Immobilon-P (cat. no. 1PVH20200) or other membrane for Western blots.
20. Transfer solution for Western blotting: 25 m*M* Tris-HCl, 192 m*M* glycine, 20% methanol or other suitable transfer buffer.
21. Antibody to known cytosolic protein such as caffeoyl-CoA-*O*-methyl transferase (CCoAOMT).
22. Amersham ECL kit (cat. no. RPN2109) or other commercial detection kit for Western blots.

2.2. Protein Extraction from Cell Walls

1. 1 *M* $CaCl_2$, stock.
2. 5 *M* LiCl, stock.
3. 1 *M* Na-acetate, pH 5.5, stock.
4. Extraction buffer 1: 200 m*M* $CaCl_2$, 50 m*M* Na-acetate, pH 5.5, in Milli Q water, 4°C.
5. Extraction buffer 2: 3 *M* LiCl, 50 m*M* Na-acetate, pH 5.5, in Milli Q water , 4°C.

2.3. Protein Concentration and Desalting

1. Amicon Ultra-15 centrifugal filtration devices (Millipore cat. no. UFC900524) or other concentration devices.
2. Distilled or Milli Q water, 4°C.

2.4. Determination of Concentration, Protein Clean-up, and Rehydration of Sample

1. Bio-Rad protein determination (cat. no. 500-0006) or other protein concentration determination method.

2. Bovine serum albumin (BSA) for a standard.
3. Distilled or Milli Q water.
4. Bio-Rad ReadyPrep 2-D Cleanup Kit (cat. no. 163-2130).
5. Resuspension buffer: 8 *M* urea, 4% 3[3-cholaminopropyl diethylammonio]-1-propane sulfonate (CHAPS), 20 m*M* dithiothieitol (DTT), 0.2% ampholytes in Milli Q water; SDS sample buffer; or other appropriate buffer.

3. Methods

3.1. Tissue Disruption, Washes, and Gel to Assay for Purity

1. Harvest 7 to 8 g of mature (internodes four to six) alfalfa stems. The yield of cell wall protein for that amount of mature stem tissue should be more than 1 mg (*see* **Note 2**). Store at –80°C after harvest or process immediately.
2. Keep all solutions on ice.
3. Break stems into small pieces approximately 1 to 2 cm in length to make grinding easier.
4. Thaw frozen tissue (*see* **Note 3**).
5. Grind with mortar and pestle until the tissue is foamy (*see* **Note 4**). Make sure the tissue is ground well. First grind without buffer, then add 10 mL of grinding buffer, and grind again.
6. Pour ground tissue onto the nylon mesh filter in the Buchner funnel and vacuum filter. Use a spatula to spread the tissue evenly over the filter. Rinse the mortar and pestle with 15 mL more of grinding buffer and use it to wash/filter over cell debris (*see* **Note 5**). Rinse debris pellet three times more with 25 mL of buffer each time for a total of 100 mL per 7 to 8 g (approx 14 mL buffer/g) of starting tissue (*see* **Note 6**). Remove the washes by applying a vacuum. Save a portion of the wash to assay for purity (*see* **Note 7**). SDS-PAGE (**Fig. 1A**, lane 1) or Western blots with an antibody to a known cytosolic protein (**Fig. 1B**, lane 2) are appropriate methods for purity assays.
7. Wash the debris with 50 mL of wash buffer 1 (NaCl wash; *see* **Note 8**). Pour it over the cell debris on the nylon filter and remove by vacuum filtration. Turn off the vacuum while adding washes and make sure the washes are in contact with all the debris. Save a portion of the wash to assay for purity (**Fig. 1A**, lane 2; **Fig. 1B**, lane 3).
8. Repeat the washes with 2X 50 mL of distilled water (**Fig. 1A**, lane 3; **Fig. 1B**, lane 4).
9. Wash with 5X 50 mL of ice cold acetone (*see* **Note 9**; **Fig. 1A**, lane 4).
10. Repeat **step 9** (**Fig. 1A**, lane 5).
11. Wash a final time with 50 mL of wash buffer 2 (10 m*M* Na-acetate; **Fig. 1A**, lane 6; **Fig. 1B**, lane 5).

3.2. Extraction of Cell Wall Proteins

1. Place the cell debris in a 50-mL tube on ice and add 7.5 mL extraction buffer 1 ($CaCl_2$ buffer). Lay the tube on its side to increase the surface area of debris in contact with the extraction buffer. Set on a shaker and shake gently for 30 to 45 min.

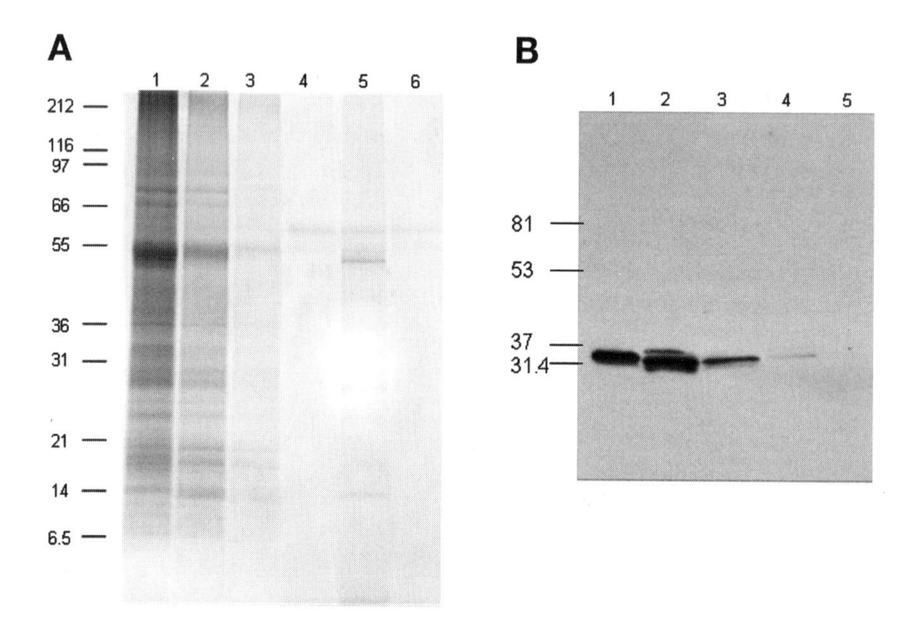

Fig. 1. **(A)** SDS-PAGE and **(B)** CCoAOMT western blot analysis of sequential wash steps in a cell wall preparation. **(A)** Lane 1, grinding buffer (soluble fraction), **Subheading 3.1.6.**; lane 2, wash buffer 1 (NaCl wash), **Subheading 3.1.7.**; lane 3, first water wash, **Subheading 3.1.8.**; lane 4, acetone wash, **Subheading 3.1.9.**; lane 5, second water wash, **Subheading 3.1.10.**; lane 6, wash buffer 2 (Na-acetate wash), **Subheading 3.1.11.** Lane 1 contains 25 μg of protein as determined by Bio-Rad protein concentration assay, lane 2 contains 0.2% of the wash, lane 3 contains 0.4% of the wash, and all other lanes contain 1% of their respective washes precipitated with TCA. Proteins were electrophoresed on a 10 to 20% acrylamide gel and stained with Coomassie. **(B)** Lane 1, positive control, recombinant CCoAOMT, a known cytosolic protein; lane 2, grinding buffer (soluble fraction), **Subheading 3.1.6.**; lane 3, wash buffer 1 (NaCl wash), **Subheading 3.1.7.**; lane 4, first water wash, **Subheading 3.1.8.**; lane 5, wash buffer 2 (Na-acetate wash), **Subheading 3.1.11.** Lane 1 contains 0.25 μg of recombinant CCoAOMT, and lane 2 contains 0.2% (15 μg) of the soluble fraction; other lanes contain 1% of their respective washes precipitated with TCA. The proteins were transferred to PVDF from a 10 to 20% acrylamide gel, probed with anti-CCoAOMT antibody, diluted 1:10,000, and anti-horseradishperoxidase (HRP), diluted 1:10,000. Detection was by chemiluminescence (Amersham ECL). **(A)**, adapted from Phytochemistry, 65, Watson, B. S., Lei, Z., Dixon, R. A., and Sumner, L. W. (2004) Proteomics of *Medicago sativa* cell walls. *Phytochemistry* **65,** 1709-1720, (with permission from Elsevier.)

2. Remove the protein extraction solution (*see* **Note 10**) and save on ice. Repeat the extraction with a second aliquot of extraction buffer 1 for 30 to 45 min (*see* **Note 11**).

3. Remove the second protein extract and combine with the first extract. Add 15 mL of extraction buffer 2 (LiCl buffer) and extract for at least 45 min on ice or overnight in a cold box (*see* **Notes 12** and **13**).
4. Centrifuge gently (approx 1000*g*), at 4°C to remove particulate matter before placing extracts in concentrators.

3.3. Protein Concentration and Desalting

1. Concentrate with a centrifugal concentrator according to the manufacturer's directions (*see* **Note 14**). We like the 15-mL Amicon Ultra Centrifugal Filter Device. The extracts can be kept separate or combined in this step. Concentrate to a volume of 100 to 150 μL.
2. The ReadyPrep 2-D Cleanup kit from Bio-Rad accomodates salt concentrations up to 1 *M*. Since the LiCl extracts are in 3 *M* LiCl, samples are desalted by adding 2 vol of distilled water to the concentrated extract and concentrating again. Remove the protein solution to a microfuge tube. To remove the sample from the concentrator, it may help to cut the end off a 200-μL pipet tip. Rinse the membrane of the concentrator with a small volume (approx 100 μL) of distilled water or buffer and combine with the concentrated proteins.

3.4. Protein Clean-up, Rehydration of Sample, and Determination of Concentration

1. Determine protein concentration using the Bradford method or equivalent.
2. Cell wall preps at this stage can be used for SDS-PAGE without any further clean-up (**Fig. 2**, lanes 1 and 2), but the protein bands will not be as sharp as those processed with a commercial clean-up kit and resuspended in SDS sample buffer (**Fig. 2**, lanes 3 and 4).
3. For 2-DE (**Fig. 3A**, $CaCl_2$ and LiCl extracts combined; **Fig. 3B**, $CaCl_2$ extract; **Fig. 3C**, LiCl extract; *see* **Subheading 3.3.1.**), use The ReadyPrep 2-D Cleanup kit from Bio-Rad (*see* **Note 15**) or another commercial clean-up kit (*see* **Note 16**) and then resolubilize in rehydration buffer. The Bio-Rad kit works for 1 to 500 μg of protein in a volume of 100 μL.

4. Notes

1. The acidic pH mimics the pH of the cell wall and reduces protease activity, the NaCl helps prevent cytoplasmic proteins from associating with the cell walls, and the ascorbic acid helps inactivate polyphenol oxidases.
2. Approximately 3 g of young stems (internodes one to three) yield a similar amount of protein
3. Stems that are ground after they are frozen and thawed will probably show a slightly different profile of proteins than those that are ground fresh or those that are ground while frozen. Decide which method suits you best and then stick with it. Consistency while grinding is important.
4. Grinding with a mortar and pestle rather than disrupting with a glass homogenizer has yielded more reproducible results in our hands.

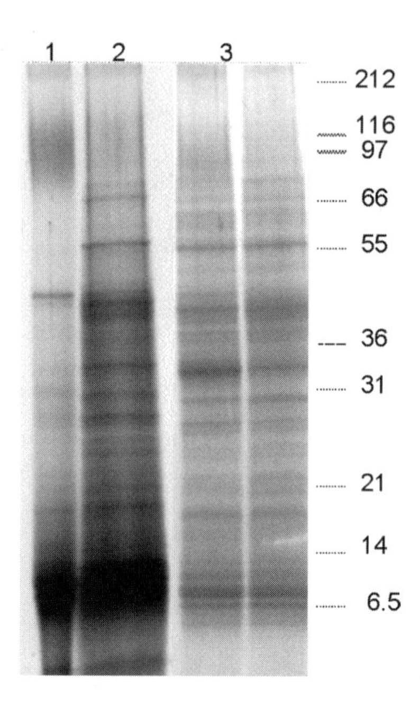

Fig. 2. SDS-PAGE of cell wall proteins. Lanes 1 and 2 are protein extracts before treatment (**Subheading 3.4.2.**), and lanes 3 and 4 are protein extracts after treatment with the ReadyPrep 2-D Cleanup kit (**Subheading 3.4.3.**). Lanes 1 and 3 contain $CaCl_2$-extracted proteins (extraction buffer 1), and lanes 2 and 4 contain LiCl-extracted proteins (extraction buffer 2). Twenty five micrograms of protein in each lane were electrophoresed on a 10 to 20% acrylamide gel and stained with Coomassie.

5. Filtration has several advantages over centrifugation. The nylon mesh 47-μm^2 nitex used in filtering allows starch grains to pass through but retains cell walls, and low-speed centrifugation does not pellet the debris well. Higher speed centrifugation would form a firmer pellet, but it would also pull proteins out of the supernatant.
6. Use sufficient volume and sufficient number of washes to ensure clean cell walls. Recommended ratios are approx 6.7–7.0 mL/g fresh weight for salt and sodium acetate, twice (14 mL/g) that for each H_2O wash, and five times that (35 mL/g) for acetone wash. Monitor the prep steps (by SDS-PAGE, Bradford, Western blot or chosen method) to ensure that the washes become progressively cleaner.
7. Twenty-five micrograms of the soluble proteins and 1% of each wash precipitated with TCA is enough to assess purity by SDS-PAGE stained with Coomassie Blue. If you are using Western blots to assess purity, either less protein can be used or an equivalent amount of protein will yield a higher assurance of purity.
8. This is a short, moderate salt concentration wash to remove membrane associated proteins or cytosolic proteins stuck to the cell wall. Higher salt concentration or a prolonged exposure will extract cell wall proteins.

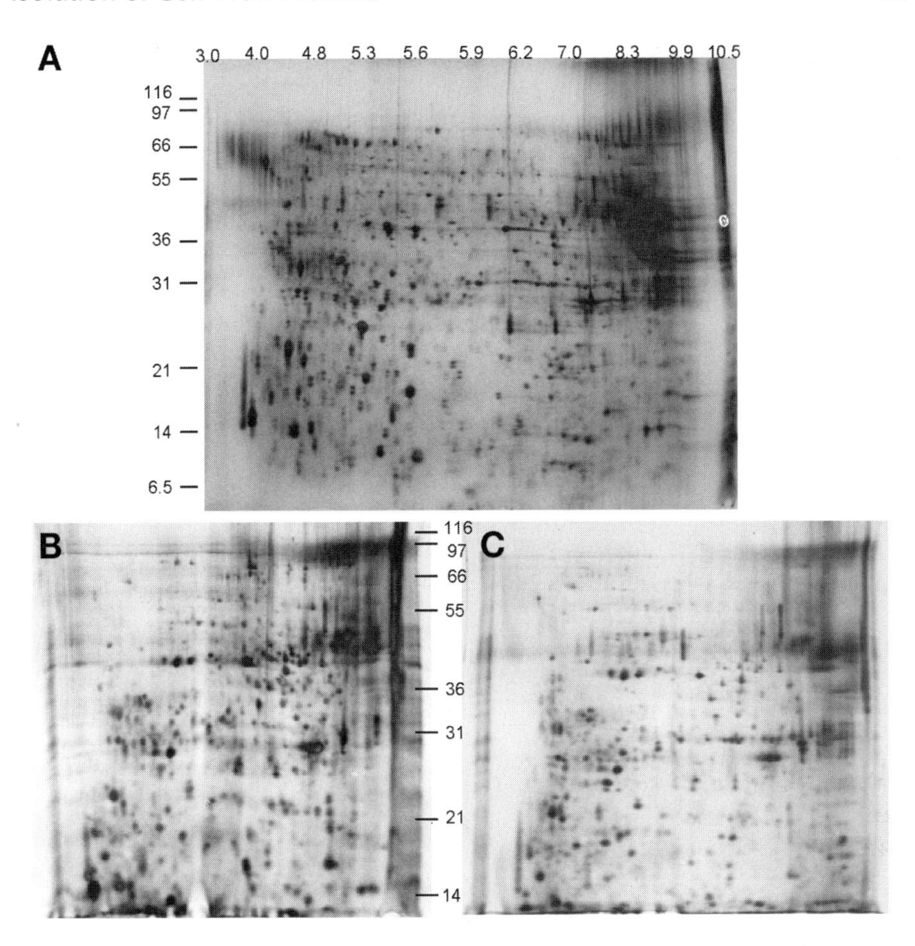

Fig. 3. 2-DE of **(A)** combined $CaCl_2$ and LiCl extracted (extraction buffers 1 and 2), **(B)** $CaCl_2$ (extraction buffer 1) extracted, and **(C)** LiCl (extraction buffer 2) extracted cell wall proteins. **(A)** Five hundred micrograms of protein were focused by IEF to 64,000 Vh on a p*I* 3 to 10 NL, 24-cm strip. Molecular weight markers are listed on the side of the gel, and p*I* markers are noted at the top. Two-hundred-fifty micrograms protein **(B)** and 150 μg protein **(C)** were focused by IEF to 30,000 Vh on p*I* 3 to 10 NL, 11-cm strips. The MW markers are the same as in **(A)**. Twelve percent acrylamide gels were used for the second dimension; all gels were silver stained.

9. Acetone disrupts membranes and allows the removal of membrane proteins, chlorophyll, and other contaminating substances.
10. Use a pipet to compress the cell debris against the side of the tube and squeeze out the extraction buffer. Filtration could also be used to separate the debris from the extracted proteins, but this could entail a loss of proteins. We recommend extraction with at least 2 mL of each extraction buffer per gram of starting tissue.

11. Extract twice with $CaCl_2$ in case the first extraction is diluted by residual wash.
12. Protein yield is higher if the cell walls are extracted overnight in the cold, but a reasonable protein yield can still be obtained with a shorter extraction (45 min).
13. Several other extraction procedures have been tried but have yielded lower quality data:
    a. The $CaCl_2$ extraction was followed by a urea extraction. Although both extracts contained proteins visible on SDS-PAGE and 2-DE gels, the 1D gels of the urea extracts were smeary, and the 2-DE gels from both samples were very streaky and ugly.
    b. TCA/acetone or phenol/EtOH was used to precipitate the samples. Samples precipitated with TCA/acetone were difficult to focus using isoelectric focusing (IEF) and yielded 2-DE gels streaked owing to cell wall polysaccharides or other contaminants. The phenol/ETOH-precipitated samples were easier to focus, and the gels were not as streaky, but there was a great loss of protein.
    c. Extracts were also treated with cellulose and pectinase, with and without protease inhibitors, to remove polysaccharides and other substances. This helped a little in focusing the $CaCl_2$ samples, but the gels were still ugly, and the proteins extracted with urea were almost completely lost after treatment with cellulose and/or pectinase.
    d. The cell wall residue was boiled in SDS and separated on SDS-PAGE gels; this yielded a few protein bands that were overwhelmed by streaks and smears.
    e. Additionally, a high-speed centrifugation of the protein solution to remove polysaccharides yielded little benefit.
14. Other methods of concentrating the proteins were attempted and include precipitations (*see* **Note 12**) and Amicon Centricons; however, the Amicon Ultra-15s have worked best for us.
15. Without using the 2-D Clean-up kit, the first dimension gels were hard (often impossible) to focus, and the second dimension gels were streaky, smeary, and ugly unless the protein loads (and therefore contaminant loads) were very low (<50 μg). Also, when the concentration of proteins loaded onto the gel was increased, the number of proteins visible on the gel did not increase proportionally, probably owing to the streaks and smears.
16. We have only tried the Bio-Rad kit and have had good experiences with it. Only once have we had problems, with a kit that had been on the shelf too long.

References

1. Smith, J. J., Muldoon, E. P., and Lamport, D. T. A. (1984) Isolation of extensin precursors by direct elution of intact tomato cell suspension cultures. *Phytochemistry* **23,** 1233–1239.
2. Scott, T. K., and O'Neill, R. A. (1984) Cell wall proteins extracted from suspension-cultured cells, in *Structure, Function and Biosynthesis of Plant Cell Walls: proceedings of the Seventh Annual Symposium in Botany, January 12–14, 1984, University of California, Riverside* (Dugger, W. M. and Bartnicki-Garcia, S., eds.), American Society of Plant Physiologists, Rockville, MD, pp. 269–283.

3. Robertson, D., Mitchell, G. P., Gilroy, J. S., et al. (1997) Differential extraction and protein sequencing reveals major differences in patterns of primary cell wall proteins from plants. *J. Biol. Chem.* **272,** 15,841–15,848.
4. Blee, K. A., Wheatley, E. R., Bonham, V. A., et al. (2001) Proteomic analysis reveals a novel set of cell wall proteins in a transformed tobacco cell culture that synthesises secondary walls as determined by biochemical and morphological parameters. *Planta* **212,** 404–415.
5. Roger, D., David, A., and David, H. (1996) Immobilization of flax protoplasts in agarose and alginate beads. *Plant Physiol.* **112,** 1191–1199.
6. Olivieri, F., Godoy, A. V., Escande, A., et al. (1998) Analysis of intercellular washing fluids of potato tubers and detection of increased proteolytic activity upon fungal infection. *Physiol. Plant.* **104,** 232–238.
7. Hiilovaara-Teijo, M., Hannukkala, A., Griffith, M., et al. (1999) Snow-mold-induced apoplastic proteins in winter rye leaves lack antifreeze activity. *Plant Physiol.* **121,** 665–673.
8. Yu, Q., Tang, C., Chen, Z., et al. (1999) Extraction of apoplastic sap from plant roots by centrifugation. *New Phytol.* **143,** 299–304.
9. Lee, S. J., Saravanan, R. S., Damasceno, C. M. B., et al. (2004) Digging deeper into the plant cell wall proteome. *Plant Physiol. Biochem.* **42,** 979–988.
10. Borderies, G., Jamet, E., Lafitte, C., et al. (2003) Proteomics of loosely bound cell wall proteins of *Arabidopsis thaliana* cell suspension cultures: a critical analysis. *Electrophoresis* **24,** 3421–3432.
11. Bayer, E., Thomas, C. L., and Maule, A. J. (2004) Plasmodesmata in *Arabidopsis thaliana* suspension cells. *Protoplasma,* **223,** 93–102.
12. Chivasa, S., Ndimba, B. K., Simon, W. J., et al. (2002) Proteomic analysis of the *Arabidopsis thaliana* cell wall. *Electrophoresis* **23,** 1754–1765.
13. Ndimba, B. K., Chivasa, S., Hamilton, J. M., et al. (2003) Proteomic analysis of changes in the extracellular matrix of *Arabidopsis* cell suspension cultures induced by fungal elicitors. *Proteomics* **3,** 1047–1059.
14. Kärkönen, A., Koutaniemi, S., Mustonen, M., et al. (2002) Lignification related enzymes in *Picea abies* suspension cultures. *Physiol. Plant.* **114,** 343–353.
15. Cassab, G. I., Nieto-Sotelo, J., Cooper, J. B., et al. (1985) A developmentally regulated hydroxyproline-rich glycoprotein from the cell walls of soybean seed coats. *Plant Physiol.* **77,** 532–535.
16. Kleis-San Francisco, S. M., and Tierney, M. L. (1990) Isolation and characterization of a proline-rich cell wall protein from soybean seedlings. *Plant Physiol.* **94,** 1897–1902.
17. McQueen-Mason, S., Durachko, D. M., and Cosgrove, D. J. (1992) Two endogenous proteins that induce cell wall extension in plants. *Plant Cell* **4,** 1425–1433.
18. McDougall, G. J. (2000) A comparison of proteins from the developing xylem of compression and non-compression wood of branches of Sitka spruce *(Picea sitchensis)* reveals a differentially expressed laccase. *J. Exp. Bot.* **51,** 1395–1401.
19. Watson, B. S., Lei, Z., Dixon, R. A., et al. (2004) Proteomics of *Medicago sativa* cell walls, *Phytochemistry* **65,** 1709–1720.

20. Melan, M. A. and Cosgrove, D. J. (1988) Evidence against the involvement of ionically bound cell wall proteins in pea epicotyl growth. *Plant Physiol.* **86,** 469–474.
21. Voigt, J. and Frank, R. (2003) 14-3-3 Proteins are constituents of the insoluble glycoprotein framework of the *Chlamydomonas* cell wall. *Plant Cell* **15,** 1399–1413.
22. Voight, J. (1985) Extraction by lithium chloride of hydroxyproline-rich glycoproteins from intact cells of *Chlamydomonas reinhardii. Planta* **164,** 379–389.
23. Bendtsen, J. D., Jensen, L. J., Blom, N., et al. (2004) Feature-based prediction of non-classical and leaderless protein secretion. *Protein Eng. Design* **17,** 349–356.
24. Pitarch, A, Sanchez, M., Nombela, C., et al. (2002) Sequential fractionation and two-dimensional gel analysis unravels the complexity of the dimorphic fungus *Candida albicans* cell wall proteome. *Mol. Cell. Proteomics* **1,** 967–982.

11

Plant Plasma Membrane Protein Extraction and Solubilization for Proteomic Analysis

Véronique Santoni

Summary

The plasma membrane (PM) exists as the interface between the cytosol and the environment in all living cells and is one of the most complex and differentiated membrane. The identification and characterization of membrane proteins (either extrinsic or intrinsic) is a crucial challenge since many of these proteins are involved in essential cellular functions such as cell signaling, osmoregulation, nutrition, and metabolism. Methods to isolate PM fractions vary according to organisms, tissues, and cell type. This chapter emphasizes isolation, from the model plant *Arabidopsis thaliana*, of PM fractions from a microsomal membrane fraction by two-phase partioning, a methodology that utilizes the different surface properties of membranes. PM proteins that do not span the lipid bilayer are generally well recovered after 2D gel electrophoresis. By contrast, the recovery of transmembrane proteins requires first the depletion of the PM fraction from soluble proteins, being either cytosolic contaminants or functionally associated proteins, and second, to the use of specific solubilization procedures. This chapter presents protocols to strip PM based on alkaline treatment of membranes and to solubilize hydrophobic proteins to increase their recovery on 2D gels. Aquaporins that are highly hydrophobic proteins are used to probe the relevance of the procedures.

Key Words: Aquaporin; *Arabidopsis thaliana*; hydrophobic protein; plasma membrane; protein solubilization; two-dimensional gel electrophoresis; two-phase partitioning.

1. Introduction

The plant plasma membrane (PM) regulates the exchange of information and substances between the cell and its environment. This interfacial position gives to PM proteins critical roles in cellular processes such as signal transduction, metabolite and ion transport, and endocytosis, as well as response to pathogens and cell wall assembly, which are specific plant functions.

From: *Methods in Molecular Biology, vol. 335: Plant Proteomics: Methods and Protocols*
Edited by: H. Thiellement, M. Zivy, C. Damerval, and V. Méchin © Humana Press Inc., Totowa, NJ

The differences in physicochemical properties of cellular membranes are the basis for their separation. Plasma membranes can be purified according to three major procedures:

1. Differences in size and density between PM and other membranes is one feature used to isolate PM. This protocol requires centrifugation of a microsomal fraction through a continuous or discontinuous density gradient *(1)*.
2. The free-flow electrophoresis procedure separate cellular membranes according to their charge *(2)*. The PM is more negatively charged than other membranes and thus can be efficiently separated from other membranes. However this approach requires a specific free flow electrophoresis instrument.
3. The isolation of PM fraction by two-phase partition relies on differences in surface properties between membrane vesicles of different origins *(3)*. The underlying principle of this technique is that numerous water-soluble, high-molecular-weight polymers do not mix above a certain concentration but will form separate phases, each composed of more than 85% water. This makes the polymer-containing phases suitable for biological material. Added membranes separate between the aqueous polymer phases according to differences in surface properties. Originally described by Larsson et al. *(3)*, this technique is now widely used in the scientific plant community since it gives rise to a high yield of PM proteins with few endomembrane contaminations. In addition, it is a simple procedure that does not require specific equipment.
   Membrane proteins can be classified into two groups: those that span the lipid bilayer are defined as intrinsic or integral proteins, and the others are associated with membranes. Associations are mediated by posttranslational modifications such as a glycolipid anchor, by the grafting of a fatty acid, or by protein–protein interaction. This suggests the occurrence of dynamics of protein association with the PM. In a PM-enriched fraction, a large proportion (60–80%) of proteins are soluble, either extrinsic proteins functionally associated with the PM or cytosolic contaminants *(4)*. Thus, direct solubilization of a PM fraction, even with the most efficient detergent, which is sodium dodecyl sulfate (SDS), may be inefficient in recovering intrinsic proteins, owing to their low representation in the PM fraction. Methods based on the use of detergents *(5)*, organic solvents *(5–7)*, or alkaline treatments *(5,8,9)* of membranes are used to categorize intrinsic proteins. Among them, alkaline extraction has gained widespread popularity as an easy and efficient method for selectively stripping extrinsic proteins off membranes without affecting the disposition of integral components *(10)*. A procedure based on alkaline-urea treatment of the PM is described in this chapter, allowing the recovery of very hydrophobic proteins such as aquaporins *(11–13)*.

Increasing data show that function and subcellular localization of membrane proteins are regulated by posttranslational modifications. 2D gel electrophoresis constitutes one of the more general methods able to separate modified forms of proteins posttranslationally. Thus glycosylation, carbamylation, and deamidation appear as trains of spots that differ in isoelectric point (p*I*) and/or apparent molecular mass ***(14,15)***; phosphorylation ***(16)*** and N^{α}-acetylation ***(17)*** shift the pI of proteins toward acidic pH and basic pH, respectively, without modification in the apparent molecular mass. More generally, any covalent modification of charged residues may modify the net charge of the protein, which in turns shifts its apparent pI. The limitations in 2D gel electrophoresis analysis of intrinsic proteins occur through their high hydrophobicity, which requires use of particular solubilization protocols ***(18)*** (*see* Chapter 12 this volume) and have generally low abundance. This chapter describes procedures for purifying PM from root, leaf, and suspension cells from the model plant *Arabidopsis thaliana* and for solubilizing PM proteins for proteomic purposes.

The plasma membrane intrinsic protein (PIP) aquaporins, which are 26 to 35-kDa hydrophobic proteins with six transmembrane α-helices ***(19–22)***, are used as pilot proteins to assess the relevance of extraction and solubilization procedures.

2. Materials

2.1. Biological Material

1. *Arabidopsis thaliana* L. (Heynh.) plants (ecotype Wassilewskija or Columbia) are cultivated in hydroponic conditions as described in Santoni et al. ***(13)***.
2. *Arabidopsis thaliana* L. (Heynh.), ecotype Columbia, suspension cells are cultured at 24°C under continuous light as described in Gerbeau et al. ***(23)***.

2.2. Equipment

1. Ultrapure water (double-distilled, deionized) is used for all reagent preparations.
2. Reagent grades should be the highest quality as appropriate for intended use (cell culture, ultrapure electrophoretic).
3. Waring blender: to disrupt plant material mechanically (leaf and root) and to pregrind suspension cells (*see* **Subheading 3.1.1.**).
4. Cell disrupter: to disrupt *Arabidopsis* suspension cells (*see* **Subheading 3.1.1.**; from Constant System, Warwick, UK).
5. Immobilized pH gradient (IPG; linear and nonlinear pH gradient from 3 to 10, 18 cm length; Amersham Pharmacia Biotech).
6. IPGphor apparatus: for isoelectrofocalization of proteins (Amersham Pharmacia Biotech).
7. Protean II: for sodium dodecyl sulfate polyacrylamide gel electrophoresis (SDS-PAGE) electrophoresis (Bio-Rad).

2.3. Reagents

2.3.1. Products and Stock Solutions

1. Dextran T-500 and polyethylene glycol (PEG) 3350 are from Amersham and Union Carbide, respectively. Stock solutions of Dextran T-500 and PEG 3350 are made on a weight basis and prepared as 20 and 40% stock solutions in water, respectively (*see* **Note 1**).

2. 0.2 *M* EDTA disodium salt dihydrate (EDTA-Na_2) stock solution: dissolve 37.2 g EDTA-Na_2 in 500 mL water (final volume). Adjust the pH with TRIZMA® base (Sigma) in powder until pH 8 is reached. Store at room temperature for several months.
3. 0.2 *M* EGTA stock solution: dissolve 38.1 g EGTA in 500 mL water (final volume). Adjust the pH with TRIZMA base powder pH 8 until is reached. Store at room temperature for several months.
4. Leupeptin (Sigma, cat. no. L 2884) is prepared as a 10 m*M* stock solution in water (5 mg in 1.05 mL water) and stored at –20°C for up to 6 mo.
5. 0.5 *M* NaF stock solution: dissolve 10.5 g NaF in 500 mL water. Store at room temperature for maximum 1 mo.
6. 0.2 *M* phosphate buffer pH 7.8 stock solution: dissolve 1.4 g KH_2PO_4 in 50 mL water. Dissolve 17.4 g K_2HPO_4 in 500 mL water. Mix the two solutions, which gives 550 mL of a 0.2 *M* phosphate buffer, pH 7.8. Store at –20°C for several months.

2.3.2. Buffers

1. Washing buffer for suspension cells (WBSC): 20 m*M* KCl, 5 m*M* EDTA. This buffer is prepared from a 10 times concentrated solution made of 200 m*M* KCl (14.9 g in 1 L) mixed together with 50 m*M* EDTA (18.6 g in 1 L). It is stored for several months at –20°C.
2. Homogenization medium: the minimum requirements of a homogenization medium are the presence of an osmoticum to minimize swelling and rupture of organelles, a buffer at pH 7 to 8 to minimize activities of hydrolytic enzymes and to counteract the low pH of the vacuoles, and means to control the level of free divalent metal cations. β-glycerophosphate, Na-orthovanadate, and phenantroline constitute a cocktail of phosphoprotein phosphatase inhibitors that were shown to help maintain specific cellular activities such as water transport activity at the PM level ***(23)***. NaF is a phospholipase inhibitor, which, together with EGTA and EDTA, reduces phospholipase activities ***(24)***. Leupeptin is a protease inhibitor. Ascorbic acid is an antioxidant, and dithiothreitol (DTT) is a sulfhydryl group protectant. Polyvinylpyrrolidone (PVP) is added to adsorb phenolic compounds. The composition is given in **Table 1**.
3. Microsome buffer: the micrososomal pellet is resuspended in a buffer compatible with the phase partitioning procedure. The composition is given in **Table 2**.
4. Phase partitioning: the two polymers used for PM isolation are dextran T-500 and polyethylene glycol 3350, 6.4% (w/w) each. The two-phase system also contains 5 m*M* potassium phosphate, pH 7.8 to buffer the system, sucrose as an osmoticum, and KCl, which modifies phase properties. The composition of the phase partitioning system, used to isolate PM from *Arabidopsis* suspension cell, leaf, and root from *Arabidopsis*, is given in **Table 3** (*see* **Notes 1** and **2**).
5. PM washing buffer: PMs are cleaned of any polymers after phase partitioning by dilution in a washing buffer; the composition is given in **Table 4**.

Table 1
Homogenization Buffer[a]

Component	Mixture	Final concentration (mM)
TRIZMA base	6.05 g	50
Sucrose	167 g	500
Glycerol	125.7 g	10% (w/v)
EDTA-Na_2	100 mL of 0.2 M stock solution	20
EGTA	100 mL of 0.2 M stock solution	20
NaF	100 mL of 0.5 M stock solution	50
β-glycerophosphate	1.08 g	5
Phenantroline	0.198 g	1
PVP[b]	6 g	0.6% (w/v)
Ascorbic acid	1.76 g	10
H_2O make to	1 L	

[a] The homogenization medium is adjusted at pH 8.0 with 2-morpholinoethane sulfonic acid (MES), and stored at –20°C for several months. Just before use, leupeptin 1 mM, DTT 5 mM, and Na-orthovanadate 1 mM are added to the homogenization medium.

[b] PVP: average molecular weight from 30,000 to 40,000 (ICN, reference 102786).

Table 2
Microsome Resuspension Buffer

Component	Mixture	Final concentration (mM)
Phosphate buffer, pH 7.8	12.5 mL of 0.2 M stock solution	5
Sucrose	56 g	330
DTT	154 mg	2
NaF	10 mL of 0.5 M stock solution	10
H_2O, make to	500 mL	

[a] The microsome resuspension buffer is stable for several months at –20°C.

Table 3
Composition of the Two-Phase System Used for PM Purification

Component	Mixture	Final concentration (mM)
Dextran T-500 20% (w/w)	11.82 g	6.4%
PEG-3350 40% (w/w)	5.76 g	6.4%
Phosphate buffer 0.2 M, pH 7.8	0.9 mL	5
KCl, 2 M [a]	0.09 mL	5
Sucrose 1.6 M [b]	6.74 mL	300
H_2O make to 27 g		

[a] KCl, 2 M: dissolve 74.6 g KCl in 500 mL water. Store at –20°C for several months.

[b] Sucrose 1.6 M: dissolve 273.8 g sucrose in 500 mL water. Store at –20°C for several months.

Table 4
Composition of the Plasma Membrane (PM) Washing Buffer[a]

Components	Mixture	Final concentration (m*M*)
TRIZMA base	1.21 g	10
Boric acid	0.61 g	10
Sucrose	102 g	300
KCl	0.67 g	9
EDTA-Na_2	25 mL of 0.2 *M* stock solution	5
EGTA	25 mL of 0.2 *M* stock solution	5
NaF	100 mL of 0.5 *M* stock solution	50
H_2O make to	1 L	

[a] The pH should be 8.3. If pH is too acidic it is adjusted with TRIZMA® base; if it is too basic it is adjusted with 2-morpholinoethane sulfonic acid (MES). The PM washing buffer is stored aliquoted at –20°C for several months. PMs are stored in a PM preserving buffer composed of PM washing buffer made up with 10 μ*M* leupeptin and 5 m*M* dithiothreitol (DTT; *see* **Subheading 3.1.2.**). This solution is prepared freshly.

Table 5
Composition of Buffer A[a]

Components	Mixture	Final concentration (m*M*)
TRIZMA base	1 mL of 1 *M* stock solution pH 9.5	5
EDTA-Na_2	5 mL of 0.2 *M* stock solution	5
EGTA	5 mL of 0.2 *M* stock solution	5
Urea	48 g	4
H_2O make to	200 mL	

[a] The pH is adjusted to 9.5 with NaOH 6 *N*. The buffer is stored for several months at –20°C.

6. Extraction of intrinsic PM proteins by a urea-NaOH treatment: compositions of buffers required for extraction of intrinsic proteins are given in **Tables 5** and **6**. Urea is used as a chaotropic reagent.
7. Solubilization buffers: proteins are resuspended in sample buffer (SB2X) ***(25)*** (**Table 7**) for subsequent SDS-PAGE electrophoresis. DTT is used to reduce proteins. SDS is a detergent used to solubilize, denature, and impart a strong negative charge to proteins. Glycerol is added to help sample loading. Proteins are resuspended in 2D solution for subsequent 2D gel electrophoresis. Thiourea is a chaotrope, and a urea-thiourea mixture exhibits a solubilizing power superior to that of urea alone. Extensive analyses of nonionic and zwitterionic detergents have revealed that Triton X-100, β-dodecylmaltoside, and ASB14 provide efficient recovery of hydrophobic proteins on 2D gels ***(26,27)***. Solubilisation buffers, once thawed, should not be refrozen.

Table 6
Composition of Buffer B[a]

Components	Mixture	Final concentration (m*M*)
TRIZMA base	1 mL of 1 *M* stock solution, pH 8.0	5
EDTA-Na_2	2 mL of 0.2 *M* stock solution	2
EGTA	2 mL of 0.2 *M* stock solution	2
H_2O make to	200 mL	

[a] The buffer is stored for several months at –20°C.

Table 7
Composition of Sample Buffer 2X (SB 2X)[a]

Components	Mixture	Final concentration
TRIZMA base	1.52 g	0.125 m*M*
Glycerol	20 g	20% (w/v)
DTT	3 g	0.2 *M*
SDS	20 mL of 20% stock solution	4% (w/v)
Bromophenol blue	1 mg	1% (w/v)
H_2O make to	100 mL	

[a] The sample buffer is stored for several months at –20°C. DTT, dithiothreitol; SDS, sodium dodecyl sulfate.

3. Methods

3.1 Plasma Membrane Isolation

3.1.1. Isolation of a Microsomal Fraction

The isolation of a PM fraction from mechanically disrupted leaf or root tissues or suspension cells first requires isolation of a PM-containing microsomal membrane fraction. PM is then separated from the microsomal fraction by two-phase partitioning. All the procedures are carried out at 4°C.

Arabidopsis Leaf and Root

1. Leaves and roots are quickly harvested, put on moist paper on ice and then briefly rinsed with ice-cold distilled water. The fresh weight of the material is measured. The homogenization medium to tissue ratio is 2:1 (mL medium/g fresh weight). The entire amount of homogenization medium is added to tissues in a Waring blender and homogenized for 10 s at low speed and then for 4 × 10 s at high speed.
2. The resulting homogenate is filtered through a nylon cloth (100-µm diameter) to discard any cell wall debris.

3. The filtered homogenate is then centrifuged at 26,000*g* max. for 25 min, which allows pelleting of chloroplasts and mitochondria.
4. The resulting pellet is discarded, and the supernatant is filtered through two successive filters, 63- and 34-µm diameter each.
5. A microsomal pellet is obtained from the supernatant by centrifugation at 84,000*g* max. for 25 min. The pellet is resuspended to a total volume of 9 mL microsomal buffer/20 to 40 g fresh weight (**Table 2**).

Arabidopsis Suspension Cells

1. Suspension cells are washed with cold washing buffer (1:1, WBSC/suspension cells) (*see* **Subheading 2.3.2.**) through a porosity 2 glass fiber.
2. The fresh weight of the material is measured.
3. Suspension cells are incubated for 10 min in homogenization medium (2.5:1, mL medium/g fresh weight).
4. Cells are then prehomogenized in a Waring blender for 10 s at low speed and then fully disrupted in a cell disrupter that allows cells to be ground by a high-pressure cavitation process. The principle is to impose high pressure (0.54 kbar for *Arabidopsis* suspension cells) to force cells through a 180-µm-diameter channel. During transfer from the high-pressure compartment to the low-pressure compartment cells acquire high speed. Projection of cells associated with the cavitation effect induces cell bursting.
5. The cell homogenate is then centrifuged at 10,000*g* max for 10 min.
6. The supernatant is filtered through a 100-µm-diameter nylon cloth.
7. A microsomal pellet is obtained from the supernatant by centrifugation at 50,000*g* max for 35 min. The pellet is resuspended in a total volume of 9 mL microsomal buffer/20 to 40 g initial fresh weight (**Table 2**).

3.1.2. PM Isolation by Two-Phase Partitioning

1. The same protocol is used for PM isolation from *Arabidopsis* suspension cells, leaves, and roots. Two phase partition is temperature dependent, so it is best to work consistently in a cold room.
2. A 27 g phase system is mixed with 9 g of a microsomal suspension, which is equivalent to a maximum of 40 g fresh weight. Overloading the phase partitioning system will prevent a correct phase separation.
3. The tube is shaken vigorously 15 to 20 times. It is essential that the contents of the tube mix and that a cushion of dextran does not slide along the wall of the tube without getting mixed into the system.
4. The system is allowed to separate out into two phases by centrifugation at 2,000*g* for 10 min. PM-derived vesicles partition preferentially to the PEG-upper phase (up1), whereas membrane vesicles originating from other membranes partition to the lower phase (low1, **Fig. 1**).
5. To purify the PM from the upper phase, this phase is repartitioned against a fresh lower phase (low2, **Fig. 1**).

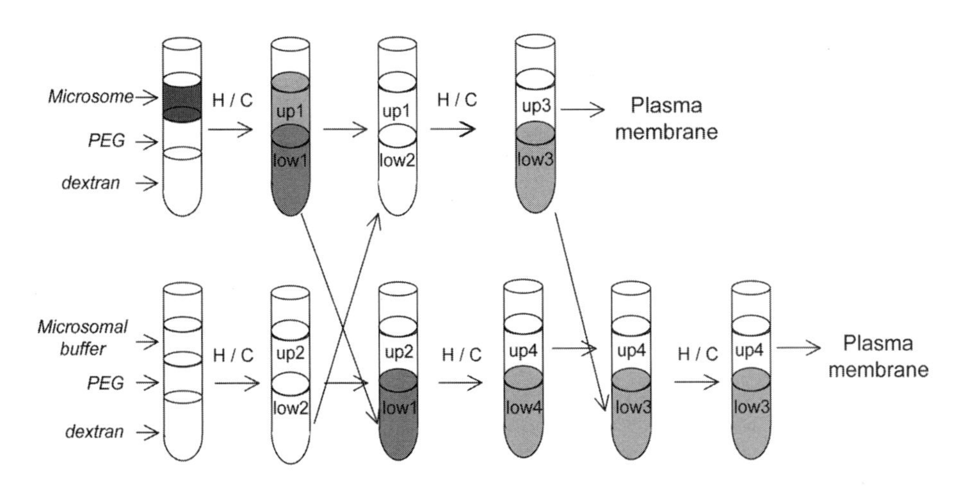

Fig. 1. Schematic representation of PM purification by two-phase partitioning. The microsomes are mixed with the two-phase partitioning system (**Table 3**) composed of PEG 3350 and Dextran T-500 (*see* **Subheading 3.1.2.**). After homogenization and centrifugation (H/C), the upper and lower phases become enriched in PM and endomembranes, respectively. Each phase (up1 and low1) is enriched once again using complementary fresh phases (up2 and low2). The upper phase up3 is collected. The upper phase up4 is washed before being collected.

6. Upper and lower fresh phases are prepared by mixing 9 g of microsomal buffer with a 27-g two-phase system.
7. After homogenization and centrifugation at 2,000*g* for 10 min, two fresh phases are obtained, up2 and low2. Phases up1 and low1 are then mixed with fresh complementary phases low2 and up2, respectively.
8. After another homogenization/centrifugation step, phases up3, low3, up4, and low4 are obtained (**Fig. 1**).
9. The upper phase up3 is collected.
10. The upper phase up4 is mixed with the lower phase low3.
11. After homogenization/centrifugation, the washed upper phase up4 is collected.
12. The two upper phases up3 and up4 are diluted in 60 mL of PM washing buffer (**Table 4**) before pelleting for 35 min at 85,000*g* max. in the case of suspension cells and at 176,000*g* max. for 30 min in the case of leaf and root.
13. Pelleted PM proteins are resuspended in PM washing buffer complemented with 10 μ*M* leupeptin and 5 m*M* DTT, frozen in liquid nitrogen, and stored for a long term at –80°C.
14. The protein yield is between 1.5 and 2.5 mg protein/100 g fresh weight (*see* **Note 3**).

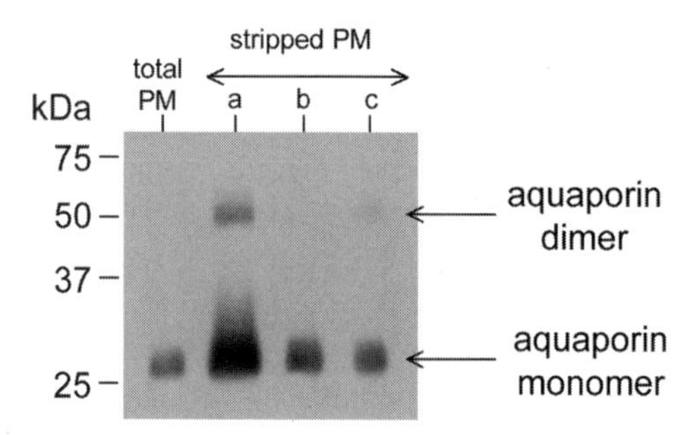

Fig. 2. Detection of aquaporin PIPs in the plasma membrane of *Arabidopsis* roots. Total plasma membrane (PM) and PM stripped with a urea-NaOH treatment (lane a) (*see* **Subheading 3.2.** and **Note 6**), carbonate washing (lane b) ***(8)***, and 0.2% (w/v) Triton X-100 (lane c) ***(5)*** were separated by SDS-PAGE (11% acrylamide gel) and probed with antibodies raised against PIP2;1 ***(13)***. Ten micrograms of protein were loaded per lane. The immunodetected signal at 28 kDa corresponds to the monomer form of aquaporin. The signal at 50 kDa corresponds to the dimer of aquaporin. The urea-NaOH treatment strongly enhances the immunodetected signal.

Despite a carefully adapted two-phase system, complete purification is almost impossible; thus it is necessary to estimate the relative contamination of the PM fraction by other membranes and soluble proteins (*see* **Notes 4** and **5**).

3.2. Enrichment in Hydrophobic Proteins

Among alkaline treatment, urea-NaOH treatment of PM gives rise to a better recovery of hydrophobic proteins aquaporins (*see* **Note 6** and **Fig. 2**).

1. PM proteins (0.5 mg) are incubated in 15 mL of buffer A (**Table 5**) for 5 min on ice before being centrifuged for 10 min at 100,000*g* max.
2. The supernatant is discarded, and the subsequent pellet is resuspended in 20 m*M* NaOH and centrifuged at 100,000*g* max for 10 min.
3. Pelleted proteins are then washed with buffer B (**Table 6**) and then centrifuged at 100,000*g* max for 10 min.
4. The final pellet is resuspended in PM preserving buffer (**Table 4**) for protein assay (*see* **Note 3**).
5. For SDS-PAGE electrophoresis, the PM protein fraction is diluted twice in sample buffer SB2X (**Table 7**) (**Fig. 3** and *see* **Note 7**).

3.3. Protein Separation by 2-D Gel Electrophoresis

3.3.1. Protein Solubilization

1. PM proteins (120 μg; either total PM or urea-NaOH-treated PM) are pelleted at 100,000*g* max. for 15 min
2. The pellet is resuspended in 350 μL of 2D solution (**Table 8**) for 1 h under constant shaking.

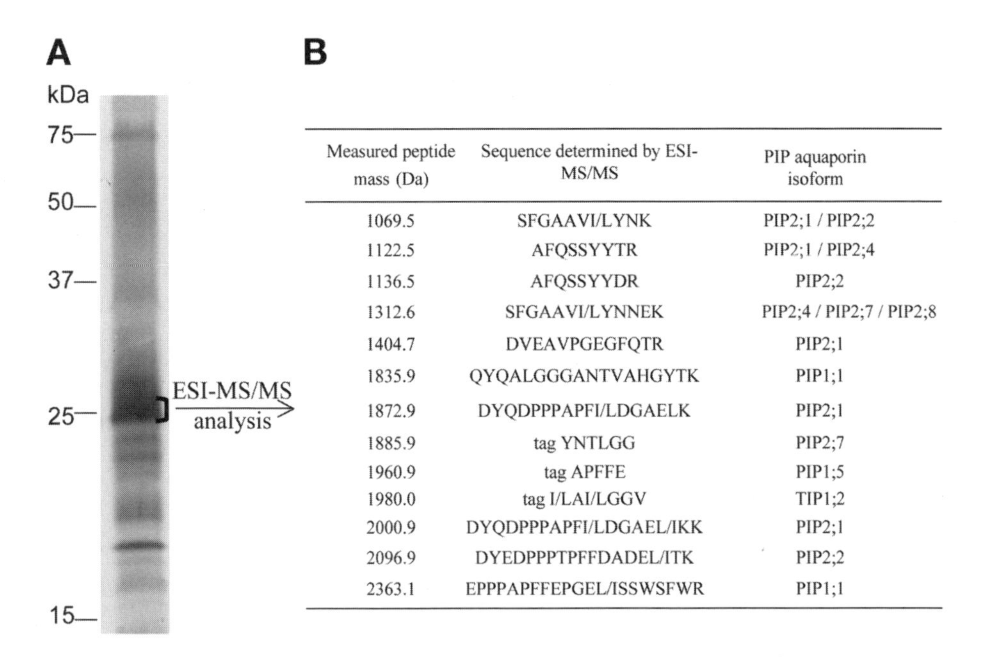

Measured peptide mass (Da)	Sequence determined by ESI-MS/MS	PIP aquaporin isoform
1069.5	SFGAAVI/LYNK	PIP2;1 / PIP2;2
1122.5	AFQSSYYTR	PIP2;1 / PIP2;4
1136.5	AFQSSYYDR	PIP2;2
1312.6	SFGAAVI/LYNNEK	PIP2;4 / PIP2;7 / PIP2;8
1404.7	DVEAVPGEGFQTR	PIP2;1
1835.9	QYQALGGGANTVAHGYTK	PIP1;1
1872.9	DYQDPPPAPFI/LDGAELK	PIP2;1
1885.9	tag YNTLGG	PIP2;7
1960.9	tag APFFE	PIP1;5
1980.0	tag I/LAI/LGGV	TIP1;2
2000.9	DYQDPPPAPFI/LDGAEL/IKK	PIP2;1
2096.9	DYEDPPPTPFFDADEL/ITK	PIP2;2
2363.1	EPPPAPFFEPGEL/ISSWSFWR	PIP1;1

Fig. 3. Mass spectrometry (MS) analysis of trypsin-cleaved 28-kDa band immunodetected by anti-PIP aquaporin antibody in the root PM *(13)*. (**A**) SDS-PAGE of PM stripped with urea and NaOH. Two hundred micrograms were loaded on the gel, electrophoresed, and revealed by Coomassie G-250 staining. The bracket indicates the excised band. (**B**) Electrospray ionization (ESI)-MS/MS analysis of SDS-PAGE band immunodetected by anti-plasma membrane intrinsic membrane (PIP) antibody *(13)*. Different aquaporin PIP isoforms are present in the 28-kDa band (*see* **Note 7**).

Table 8
Composition of 2D Solution

Components	Mixture	Final concentration
Detergent[a]	1 g	2% (w/v)
Urea	21 g	7 *M*
Thiourea	7.6 g	2 *M*
Triton X-100	0.25 g	0.5% (w/v)
Pharmalytes (3-10)[b]	0.6 mL	1.2% (w/v)
DTT	0.16 g	20 m*M*
H_2O make to	50 mL	

[a] Detergent is either β-dodecylmaltoside (Sigma) or ASB14 (Calbiochem).
[b] Pharmalytes are from Amersham (cat. no. 17-0456-01).

3. After centrifugation at 10,000*g* for 10 min, the supernatant is carefully collected and readjusted to 350 µL with 2D solution (**Table 8**).

3.3.2. 2-D gel Electrophoresis

1. Isoelectric focusing (IEF) is performed with commercially immobilized pH gradients (linear or nonlinear pH gradient from 3 to 10, 18 cm length; Amersham Pharmacia Biotech), using a IPGphor apparatus (Amersham Pharmacia Biotech).
2. The gel is rehydrated in the presence of 350 µL of sample protein for 4 h without voltage and then for 7 h under constant voltage (50 V).
3. The gel is then subjected to focusing with the following running conditions: from 0 to 300 V in 1 min, 300 V for 3 h, from 300 V to 3500 V in 1 h, 3500 V until 80 kV.h are reached.
4. After the IEF run, the first-dimension gel is successively incubated at room temperature in solutions containing 2% DTT and 2.5% iodoacetamide according to Chevallet et al. *(26)*.
5. The second dimension (SDS-PAGE) is carried out on homogeneous 11% T gels (Protean II, Bio-Rad).
6. The first-dimension gel is sealed at the top of the running gels with low-melting-point agarose *(28)*.
7. Electrophoresis is conducted at 20 mA for 1 h and then at 40 mA for 4 to 5 h.
8. For analytical purposes or subsequent MS analysis, gels are stained with silver or colloidal Coomassie G-250, respectively (**Figs. 4** and **5** and *see* **Note 8**).

4. Notes

1. Preparation of dextran T-500: because dextran T-500 is hygroscopic, the determination of the exact concentration should be determined. First, 220 g of dextran T-500 are dissolved in 780 g of water under gentle shaking. Complete dissolution requires about 2 h. Then 5 g of this solution are precisely diluted in 25 mL water, and the optical rotation is measured at 589 nm. As the specific rotation is +199°/mL/g/dm, the final concentration (%, w/w) is given by the following equation: optical rotation × 25 × 100/199 × 5.

   To ensure uniformity of the dextran T-500 used in a series of experiments, a large batch should be made up at one time; this becomes especially necessary if a polarimeter is not used to correct for the water content of the dextran. Here, the stock solution is made assuming a dextran T-500 water content of 5%. The molecular weight distribution differs between different lots of dextran T-500. This means that with every new lot of dextran T-500, membrane separation in the phase system must be evaluated and necessary adjustments of the composition of the phase system made.
2. It is practical to prepare stock solutions of the phase system components and store them in aliquots at –20°C. A two-phase system is constituted on a weight basis. Its size depends on how much membrane material is to be loaded on the system. As an example, **Table 3** shows how to build a 27 g sample system that will be subsequently mixed with 9 g of microsomes, thus leading to a final 36 g phase system. After preparation, the two-phase systems can be stored at –20°C.

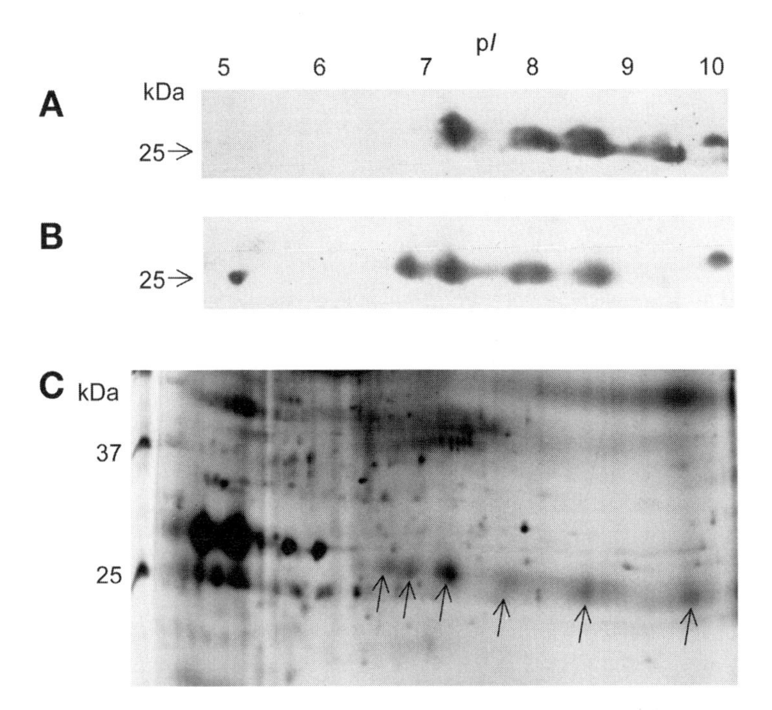

Fig. 4. Protein separation by 2D gel electrophoresis reveals multiple forms of aquaporins expressed in the PM of *Arabidopsis* root. Protein solubilization was carried out as described in **Subheading 3.3.1.** 2D gel electrophoresis was carried out as described in **Subheading 3.3.2.** using 3 to 10 linear IPG gels *(13)*. **(A)** The antibody raised against the PIP1;1 aquaporin isoform can recognize four PIP1 homologs but surprisingly immunodetected at least seven spots. **(B)** A similar result was obtained with the anti-PIP2;2 antibody, which can recognize three PIP2 homologs and yet revealed at least six distinct spots. These results suggest that modified forms of PIP1 and PIP2 homologs, in addition to unmodified ones, are present in the root PM sample. **(C)** Several of the immunodetected spots could be assigned to proteins revealed by silver staining gel and are shown by an arrow. MALDI-TOF-MS analysis of these spots has confirmed the presence of PIP aquaporins in these spots *(13)*.

3. Protein concentrations are estimated using a modified Bradford procedure *(29)* and bovine serum albumin as the protein standard.
4. Adaptation of the two-phase system: the two-phase partitioning protocol described in this chapter is adapted for *Arabidopsis* leaf and root and *Arabidopsis* suspension cells. Starting with any other plant material will require modification of phase partitioning. By varying the polymer concentration and the salt composition of the phase system, optimal conditions for separation can be obtained. Using green material enables one to follow the partitioning of intracellular membranes by eye and to evaluate the most suitable polymer and salt concentrations quickly.

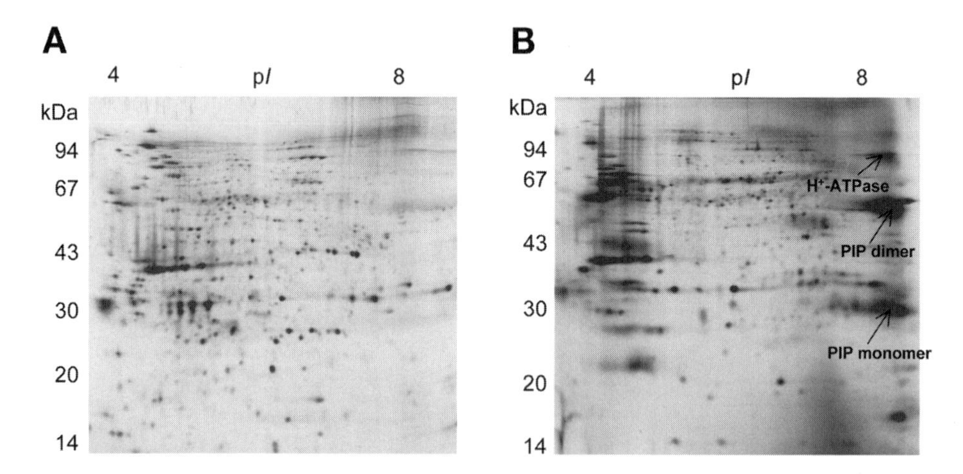

Fig. 5. 2D gel electrophoresis of total PM proteins (**A**) and urea-NaOH treated PM proteins (**B**) of *Arabidopsis* leaf (*see* **Note 8**). Protein solubilization was carried out as described in **Subheading 3.3.1.** 2D gel electrophoresis was carried out as described in **Subheading 3.3.2.** using 3 to 10 nonlinear IPG gels. Gels were stained with colloidal Coomassie G-250. Arrows indicate strong enrichment in aquaporins (PIP monomer and PIP dimer) and ATPase (H+-ATPase) after urea-NaOH treatment of membrane. Proteins were identified based on immunological detection and MS analysis.

As a starting point, a full-scale experiment should be done according to the protocol described in this chapter. If the upper phases up3 and up4 (**Fig. 1**) are green, a series of small-phase systems (10% of the phase system described in **Table 3**) are prepared without changing the KCl concentration. The lowest polymer concentration giving chlorophyll partitioning in the lower phase should be chosen. On the other hand, if the upper phases up3 and up4 (**Fig. 1**) are whitish, with a very low protein yield, a corresponding experiment is done with polymer concentrations below those used. Appropriate enzymatic markers are then measured according to Widell and Larsson 1990 ***(30)*** to assess the purity of the PM fraction (*see* **Note 5**).

5. Quality criteria for PM purity: estimating the purity of a PM fraction can be hard work since such a proteome is not static; in addition to its basic plasticity ***(31)***, the PM proteome may vary in the same tissue according to development and changing environment, and even through protein translocation to another compartment. However, enzymatic markers specific for membranes are currently used to evaluate roughly the presence of other membranes. Protocols for such an evaluation are not the purpose of this chapter; procedures and critical evaluation of enzymatic markers are detailed in Widell and Larsson ***(30)***. The sensitivity of the Mg-ATPase activity to vanadate, oligomycin, and KNO_3 is commonly used to create markers for PM, mitochondria, and tonoplast, respectively, and the presence of Golgi membranes is assessed by IDPase activity ***(32)***. PMs purified with

the protocol described in this chapter exhibit activities sensitive to vanadate, oligomycin, and KNO_3 amounting to 83%, less than 1%, and 13% of the total ATPase activity, respectively *(4,13,33)*. IDPase activity represented 4% of the total ATPase activity *(4,13,33)*. These features are similar to those obtained for various other PM-enriched fractions prepared by phase partitioning *(3,30)*.

6. A comparative analysis of protocols to extract PM hydrophobic proteins has revealed qualitative and quantitative differences in proteomes according to the kind of extraction, suggesting that different sets of proteins are specific for extraction procedures *(5)*. In particular, alkaline treatment allowed a better recovery of PM hydrophobic proteins than treating PM with detergents (Triton X-114, Triton X-100 alone, and Triton X-100 coupled to organic solvents) *(5)*. Comparison of alkaline treatments revealed that carbonate treatment of PM *(8)* gives a higher protein yield (30%) than urea-NaOH-treated PM (20%), but the latter treatment gives rise to a higher enrichment in hydrophobic proteins as revealed by immunodetection of aquaporins (**Fig. 2**).
7. Because of the high solubilization efficiency of SDS, SDS-PAGE electrophoresis remains one of the best methods to separate hydrophobic proteins. Mass spectrometry analysis of SDS-PAGE bands then allows identification of proteins even if several proteins are mixed in the same band. Thus, analysis by ESI-MS/MS of the 28-kDa band of a urea-NaOH-treated PM reveals the presence of at least five PIP aquaporin isoforms (**Fig. 3**). This result reveals the power of mass spectrometry, which allows one to distinguish between highly homologous proteins. Thus, SDS-PAGE electrophoresis combined with mass spectrometry analysis constitutes a efficient strategy to build an inventory of hydrophobic proteins. However, SDS-PAGE protein separation does not allow one to determine the occurrence of posttranslational modifications that affect protein p*I* without altering the molecular weight. 2D gel electrophoresis can help to complement this gap *(13)*, providing good recovery of hydrophobic proteins on the 2D gel (**Fig. 4**).
8. A comparative analysis of solubilization buffers revealed differences in detergent efficiency according to the extraction procedure. In particular, detergents with an aromatic core (e.g. C8ϕ) were shown to delipidate poorly but to break protein aggregates efficiently *(5,27)*. By contrast, linear detergents (e.g., ASB14) delipidate efficiently but do not break aggregates as well *(5,27)*. In a optimized protocol, hydrophobic pelleted proteins are solubilized in the presence of urea, thiourea, and detergent, which is either β-dodecylmatoside or ASB14 (**Table 8** and **Fig. 5**). A critical evaluation of solubilization procedures for hydrophobic proteins is given in Chapter 12.

References

1. Hodges, T. K. and Mills, D. (1986) Isolation of the plasma membrane. *Methods Enzymol.* **118,** 41–54.
2. Bardy, N., Carrasco, A., Galaud, J. P., Pont-Lezica, R., and Canut, H. (1998) Free-flow electrophoresis for fractionation of *Arabidopsis thaliana* membranes. *Electrophoresis* **19,** 1145–1153.

3. Larsson, C., Widell, S., and Kjellbom, P. (1987) Preparation of high-purity plasma membranes. *Methods Enzymol.* **148,** 558–568.
4. Santoni, V., Rouquié, D., Doumas, P., et al. (1998) Use of a proteome strategy for tagging proteins present at the plasma membrane. *Plant J.* **16,** 633–641.
5. Santoni, V., Kieffer, S., Masson, F., Desclaux, D., and Rabilloud, T. (2000) Membrane proteomics: use of additive main effects with multiplicative interaction model to classify plasma membrane proteins according to their solubility and electrophoretic properties. *Electrophoresis* **21,** 3329–3344.
6. Ferro, M., Salvi, D., Rivière-Rolland, H., et al. (2002) Integral membrane proteins of the chloroplast envelope: Identification and subcellular localization of new transporters. *Proc. Natl. Acad. Sci. USA* **99,** 11,487–11,492.
7. Marmagne, A., Rouet, M. A., Ferro, M., et al. (2004) Identification of new intrinsic proteins in *Arabidopsis* plasma membrane proteome. *Mol. Cell. Proteomics* **3,** 675–691.
8. Santoni, V., Rabilloud T., Doumas P., et al. (1999) Towards the recovery of hydrophobic proteins on two-dimensional electrophoresis gels. *Electrophoresis* **20,** 705–711.
9. Millar, A. H. and Heazlewood, J. L. (2003) Genomic and proteomic analysis of mitochondrial carrier proteins in *Arabidopsis*. *Plant Physiol.* **131,** 443–453.
10. Fujiki, Y., Hubbard, A. L., Fowler, S., and Lazarow, P. B. (1982) Isolation of intracellular membranes by means of sodium carbonate treatment: application to endoplasmic reticulum. *J. Cell Biol.* **93,** 97–102.
11. Hasler, L., Walz, T., Tittmann, P., Gross, H., Kistler, J., and Engel, A. (1998) Purified lens major intrinsic protein (MIP) forms highly ordered tetragonal two-dimensional arrays by reconstitution. *J. Mol. Biol.* **279,** 855–864.
12. Fotiadis, D., Jenös, P., Mini, T., et al. (2001) Structural characterization of two aquaporins isolated from native spinach leaf plasma membranes. *J. Biol. Chem.* **276,** 1707–1714.
13. Santoni, V., Vinh, J., Pflieger, D., Sommerer, N., and Maurel, C. (2003) A proteomic study reveals novel insights into the diversity of aquaporin forms expressed in the plasma membrane of plant roots. *Biochem. J.* **373,** 289–296.
14. Sarioglu, H., Lottspeich, F., Walk, T., Jung, G., and Eckerskorn, C. (2000) Deamidation as a widespread phenomenon in two-dimensional polyacrylamide gel electrophoresis of human blood plasma proteins. *Electrophoresis* **21,** 2209–2218.
15. Packer, N. H., Pawlak, A., Kett, W. C., Gooley, A. A., Redmond, J. W., and Williams, K. L. (1997) Proteome analysis of glycoforms: a review of strategies for the microcharacterisation of glycoproteins separated by two-dimensional polyacrylamide gel electrophoresis. *Electrophoresis* **18,** 452–460.
16. Towbin, H., Özbey, Ö., and Zingel, O. (2001) An immunoblotting method for high-resolution isoelectric focusing of protein isoforms on immobilized pH gradients. *Electrophoresis* **22,** 1887–1893.
17. Kimura, Y., Takaoka, M., Tanaka, S., et al. (2000) N^{α}-acetylation and proteolytic activity of the yeast 20S proteasome. *J. Biol. Chem.* **275,** 4635–4639.

18. Santoni, V., Molloy, M., and Rabilloud, T. (2000) Membrane proteins and proteomics: un amour impossible? *Electrophoresis* **21,** 1054–1070.
19. Fu, D., Libson, A., Miercke, L. J. W., et al. (2000) Structure of a glycerol-conducting channel and the basis for its selectivity. *Science* **290,** 481–486.
20. Sui, H., Han, B. G., Lee, J. K., Walian, P., and Jap, B. K. (2001) Structural basis of water-specific transport through the AQP1 water channel. *Nature* **414,** 872–878.
21. Johanson, U., Karlsson, M., Johansson, I., et al. (2001) The complete set of genes encoding major intrinsic proteins in *Arabidopsis* provides a framework for a new nomenclature for major intrinsic proteins in plants. *Plant Physiol.* **126,** 1–12.
22. Quigley, F., Rosenberg, J. M., Shachar-Hill, Y., and Bohnert, H. J. (2002) From genome to function: the *Arabidopsis* aquaporins. *Genome Biol.* **3,** 1–17.
23. Gerbeau, P., Amodeo, G., Henzler, T., Santoni, V., Ripoche, P., and Maurel, C. (2002) The water permeability of *Arabidopsis* plasma membrane is regulated by divalent cations and pH. *Plant J.* **30,** 71–81.
24. Whitman, C. E. and Travis, R. L. (1985) Phospholipid composition of a plasma membrane-enriched fraction from developing soybean roots. *Plant Physiol.* **79,** 494–498.
25. Laemmli, U. K. (1970) Cleavage of structural proteins during the assembly of the head of bacteriophage T4. *Nature* **222,** 680–865.
26. Chevallet, M., Santoni, V., Poinas, A., et al. (1998) New zwitterionic detergents improve the analysis of membrane proteins by two-dimensional electrophoresis. *Electrophoresis* **19,** 1901–1909.
27. Luche, S., Santoni, V., and Rabilloud, T. (2003) Evaluation of non-ionic and zwitterionic detergents as membrane protein solubilizers in two-dimensional electrophoresis. *Proteomics* **3,** 249–253.
28. Adessi, C., Miege, C., Albrieux, C., and Rabilloud, T. (1997) Two-dimensional electrophoresis of membrane proteins: a current challenge for immobilized pH gradients. *Electrophoresis* **18,** 127–135.
29. Stoscheck, C. M. (1990) Quantitation of proteins. *Methods Enzymol.* **182,** 50–68.
30. Widell, S. and Larsson, C. (1990) A critical evaluation of markers used in plasma membrane purification, in *The Plant Plasma Membrane–Structure, Function and Molecular Biology* (Larsson, C. and Moller, I. M., eds.), Springer-Verlag, Berlin, pp. 16–43.
31. Masson, F. and Rossignol, M. (1995) Basic plasticity of protein expression in tobacco plasma membrane. *Plant J.* **8,** 77–85.
32. Santoni, V., Vansuyt, G., and Rossignol, M. (1990) Differential auxin sensitivity of proton translocation by plasma membrane H^+-ATPase from tobacco leaves. *Plant Sci.* **68,** 33–38.
33. Santoni, V., P, D., Rouquié D, Mansion M, Rabilloud T, and Rossignol, M. (1999) Large scale characterization of plant plasma membrane proteins. *Biochimie* **81,** 655–661.

12

Detergents and Chaotropes for Protein Solubilization before Two-Dimensional Electrophoresis

Thierry Rabilloud, Sylvie Luche, Véronique Santoni, and Mireille Chevallet

Summary

Because of the outstanding separating capabilities of two-dimensional electrophoresis for complete proteins, it would be advantageous to be able to apply it to all types of proteins. Unfortunately, severe solubility problems hamper the analysis of many classes of proteins, but especially membrane proteins. These problems arise mainly in the extraction and isoelectric focusing steps, and solutions are sought to improve protein solubility under the conditions prevailing during isoelectric focusing. These solutions deal mainly with chaotropes and new detergents, which are both able to enhance protein solubility. The input of these compounds in proteomics analysis of membrane proteins is discussed, as well as future directions.

Key Words: Hydrophobic protein; membrane proteins; protein solubilization; isoelectric focusing; chaotropes; detergents; zwitterionic detergents.

1. Introduction

The solubilization of proteins for 2D electrophoresis-based proteomics is a difficult task. For example, proteins must reach their p*I*, which is also the minimum for solubility, and still stay soluble at the p*I*. Moreover, as protein mobility decreases when proteins get close to their p*I*, isoelectric focusing (IEF) is performed under strong electric fields (200 V/cm is not uncommon compared with the 10–20 V/cm used for sodium dodecyl sulfate-polyacrylamide gel electrophoresis [SDS-PAGE]). This means, in turn, that salts and ionic compounds in general are almost precluded in IEF. Moreover, any solubilizing agent used prior to IEF must not change the original p*I* of the proteins. Consequently, this precludes the use of strong ionic detergents such as SDS. However, low amounts (up to 0.03% w/v) of ionic detergents can be used, provided that con-

From: *Methods in Molecular Biology, vol. 335: Plant Proteomics: Methods and Protocols*
Edited by: H. Thiellement, M. Zivy, C. Damerval, and V. Méchin © Humana Press Inc., Totowa, NJ

ditions favoring the exchange of SDS for other, nonionic detergents are used in IEF *(1–3)*. This ensures removal of bound SDS from the proteins but also means that the benefits of SDS are lost for the IEF dimension. However, the use of SDS has been often recommended as a way to ensure a complete initial solubilization before IEF.

Apart from these problems arising from the nature of proteins, other problems are frequently encountered in many biological samples arising from the nonproteinaceous compounds that can be present in the sample. A canonical example is that of nucleic acids, which completely blur the 2D electrophoresis pattern when present at too high a concentration *(4)*. Nucleic acids act as mobile ion exchangers at the low ionic strength required by IEF, thereby creating severe artifacts. Other classes of compounds (lipids, salts, and so on) can be encountered in many samples and create their artifacts. This is especially true for plant-derived samples, owing to the ability of plant tissues to synthetize a host of compounds with varied structures. For example, phenolics and chlorophyll can be very abundant in some plant tissues and can completely ruin protein separation in proteomics techniques.

There are thus different problems depending on the starting material. When one starts from whole tissues, the main problem is generally the interference arising from nonproteinaceous compounds. These aspects are described in other chapters in this book. This chapter will therefore focus on the other aspect of the problem, i.e., the intrinsic solubilization of proteins, and will mainly deal with the chaotropes and detergents used for initial solubilization and for IEF.

As mentioned earlier, the constraints present in IEF limit the choice to chemicals showing no net electric charge in solution over the pH range used for IEF, i.e., to nonionic or zwitterionic compounds. This narrows the choice for chaotropes to the amide and urea families, as guanidines and amidines are charged below pH 12. Among the possible chaotropes, urea has been used for quite a long time. More recently, the addition of thiourea to urea as an additional chaotrope has shown interesting features for protein solubilization *(5)* but also to limit protease action *(6)*. The role of chaotropes in the solubilization process is to break the noncovalent interactions between the various molecules present in the sample (e.g., hydrogen bonds, dipole-dipole interactions, and hydrophobic interactions) and to unfold the proteins. Although ionic bonds are not directly affected by nonionic chaotropes such as urea and thiourea, the influence of these chaotropes on the dielectric constant of water also alters the strength of the ionic bonds.

On the detergent side, it is quite clear that the uncharged detergents are clearly much less efficient than the ionic ones. Ionic detergents make a charged "coat" on the protein molecules, so that the protein-detergent complexes repel each other via ionic interactions, thereby preventing protein aggregation.

Unfortunately, ionic detergents cannot be used for IEF-based 2D separation. However, they can be used in methods using differential zone electrophoresis, and a short example will be given in this chapter.

Among the wide choice of commercially available uncharged detergents, two subfamilies can be distinguished. Nonionic detergents have no charges on the molecule, whereas zwitterionic detergents have an equal number of negative and positive charges on their molecules. Depending on the p*K* of the ionizable groups on the molecules, some detergents can be ionic in a certain pH range (in which at least one group is titrated) and zwitterionic in another pH range, whereas other detergents can be zwitterionic on the complete pH range. As an example of the two cases, classical betaines (bearing a quaternary ammonium and a carboxylic group) are positively charged at low pH, when the carboxylic group is not fully deprotonated. When the carboxylic group is fully deprotonated, i.e., more than 2 pH units above the p*K*, they behave as a zwitterionic detergent. In contrast, sulfobetaines, bearing a quaternary ammonium and a sulfonic group, are zwitterionic over the 0 to 14 pH range, as both groups are ionized in this range. As a matter of fact, only detergents completely zwitterionic over the pH range of interest can be used for IEF.

The two "historical" detergents used for 2D electrophoresis are Triton X-100 (or NP-40), a nonionic detergent, and the zwitterionic detergent 3[3-cholaminopropyl diethylammonio]-1-propane sulfate (CHAPS). Both have been used extensively in combination with urea and have not proved very efficient for the solubilization of sparingly soluble proteins, e.g., membrane proteins ***(7)***. However, recent work has shown that either specially designed zwitterionic detergents ***(8–10)***, or carefully selected nonionic detergents ***(11)*** can solubilize membrane proteins. It is interesting to note that Triton X-100 is not very efficient when used with urea alone and much more efficient in urea-thiourea ***(11)***. This observation extends to other detergents of the oligo ethylene glycol family, such as the Brij® detergents, i.e., linear alkyl oligo ethylene glycol compounds ***(11)***. However, the most efficient nonionic detergents belong to the glycoside family (e.g., octyl glucoside, dodecylmaltoside), and the latter seems to be efficient for urea alone ***(12)*** as well as for urea-thiourea ***(11,13)***. An example of the variations in protein solubilization induced by the choice of detergent can be seen in **Fig. 1**. The multiple variables playing a role in the solubilization process have also been investigated in **ref. *14***. However, it should not be concluded from the preceding discussion that dodecyl maltoside is the absolute best choice for protein solubilization for 2D electrophoresis. Although the optimal detergent will depend on the sample, a kind of shortlist exists, based on previous work. Thus the best candidates for protein solubilization, at least as a first screen, can be chosen from dodecyl maltoside, ASB14, C7BzO, Brij56, and C13E10, CHAPS being a good choice for soluble proteins.

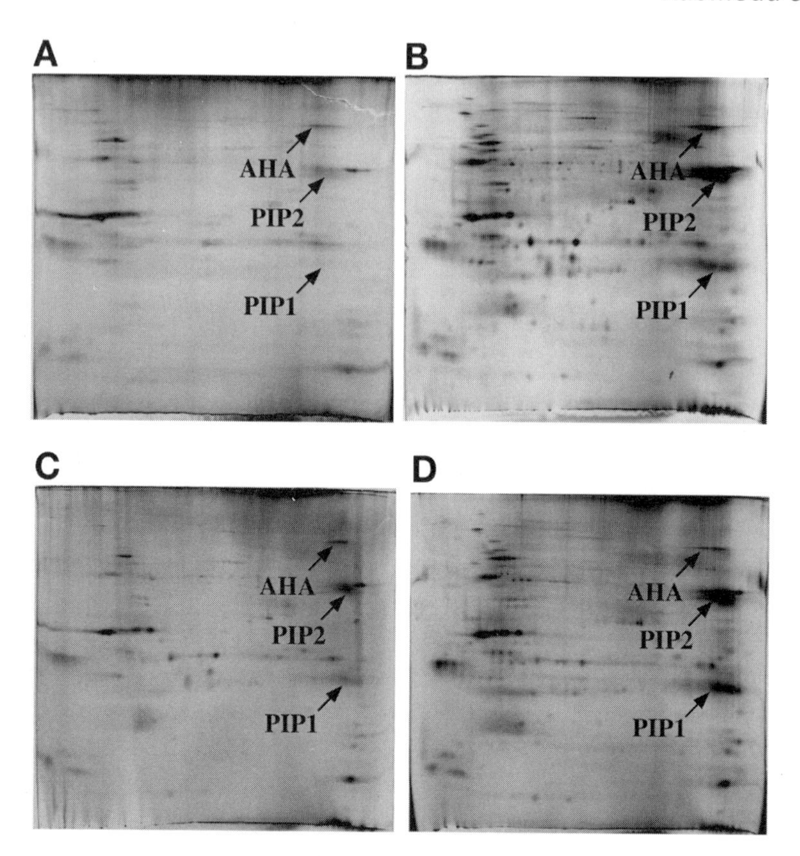

Fig. 1. 2D electrophoretic separation of *A. thaliana* leaf plasma membrane proteins: 60 µg of leaf plasma membrane proteins were loaded on the 2D gels. The first dimension was a 3 to10 linear pH gradient, and the second dimension a 10% acrylamide gel. H+ ATPase (AHA), aquaporin monomer (PIP1), and aquaporin dimer (PIP2) are indicated by arrows. The proteins were extracted and focused in a solution containing 7 *M* urea, 2 *M* thiourea, 20 m*M* DTT, 0.4% carrier ampholytes, and **(A)** 4% CHAPS, **(B)** 2% dodecyl maltoside, **(C)** 2% C7BzO, and **(D)** 2% ASB14.

2. Materials

2.1. Biological Material

Arabidopsis thaliana membrane preparations are obtained according to Santoni (*see* Chapter 11).

2.2. Equipment

1. A tabletop ultracentrifuge, used for membrane preparation and to remove unsolubilized proteins.

2. Immobilized pH gradient (immobilized pH gradient [IPG], linear and nonlinear pH gradient from 3 to 10, 18 cm length; Amersham Pharmacia Biotech).
3. IPGphor apparatus: for isoelectrofocalization of proteins (Amersham Pharmacia Biotech).
4. Tube gels electrophoresis setup (Bio-Rad), for first-dimension gel electrophoresis.
5. Protean II: for SDS-PAGE electrophoresis (Bio-Rad).

2.3. Products and Stock Solutions

1. Dodecyl maltoside, Triton X-100, and CHAPS are best used from 20% (w/v) stock solutions in water. These solutions should be stored at 4°C and show limited conservation (a few weeks).
2. C13E10, Brij 56, and ASB14 are best used from 20% (w/v) stock solutions in ethanol/water (50/50 v/v). These solutions are stable for months at room temperature.
3. Cationic detergents (dodecyl trimethylammonium bromide [DTAB], cetyl trimethyl ammonium bromide [CTAB], benzalkonium chloride) are used as a 20% (w/v) stock solution in water. These solutions are stable at room temperature but are very sensitive to temperature. They sometimes need to be warmed at 37 to 40°C to redissolve the detergent prior to use.
4. Urea stock solution for IEF. It is difficult to go beyond 9 *M* urea at room temperature, which is the concentration used when urea is the sole chaotrope. This means that urea is added as a solid (*see* **Note 1**).
5. Urea-thiourea stock solution: the final chaotrope concentrations are 7 *M* urea and 2 *M* thiourea. This means that a 1.25X concentrated solution can be prepared, which is then simpler to use than reweighing small amounts of solid urea and thiourea for each sample. For 10 mL of this concentrated solution, weigh 5.25 g of urea and 1.9 g of thiourea. Some detergents (e.g., CHAPS or Triton X-100) which are fully compatible with urea, can be added at this stage. Other detergents, which show a more limited urea compatibility (e.g., ASB14), must be added only when the solution is diluted to the final strength. A total volume of 4.2 mL of liquid must be added to the urea and thiourea to make up 10 mL (*see* also **Notes 2** and **3**). This solution is stable for months if stored frozen at –20°C.
6. Urea solution for zone electrophoresis: as urea is used at 4 *M* final concentration, it is quite convenient to prepare an 8 *M* stock solution. For 10 mL, dissolve 4.8 g of urea in 6.4 mL of water. This solution is stable to 2 to 3 d at 4°C.
7. Acidic solubilization buffer for zone electrophoresis: 1 *M* potassium dihydrogen phosphate + 1 µL/mL 85% phosphoric acid.
8. Tri-butylphosphine (TBP) is a liquid (4 *M* when pure). A 40-fold dilution in dimethyl formamide is made just prior to use. This solution is further diluted 50-fold in the sample solution.
9. Tris-carboxyethyl phosphine (TCEP) is a solid. A 1 *M* stock solution in water is made, which is stable for months at –20°C.

3. Methods (*see* Note 4)

3.1 Solubilization in Urea for IEF

3.1.1. Solubilization from a Solid Sample (e.g., Tissue or Cell Pellet)

In this case, the sample volume can often be neglected in the final solubilization volume. A sample solution containing urea (9–9.5 *M* final concentration; *see* **Note 1**), the selected detergent (taken from the list in **Subheading 2.3.1.**) at 2 to 4% (w/v) concentration, carrier ampholytes (0.4% w/v for IPG, 2% w/v for carrier ampholyte [CA]-IEF), and a reducing agent (50 m*M* DTT or 5 m*M* TBP or 5 m*M* TCEP). This solution is added to the solid sample, resulting in a liquid extract. Protein extraction is helped by sonication in a water bath sonicator for approx 30 min. Unsolubilized material is best removed by ultracentrifugation for 30 min at 200,000*g* at room temperature.

3.1.2. Solubilization from a Suspension or Solution

In this case, the volume of the sample must generally be taken into account. It is thus necessary to calculate the final solution. As a rule of thumb, the sample volume can represent up to 35% of the final extraction volume. Solid urea, water, and stock solutions of the detergent, ampholytes, and reducer are used in addition to the liquid sample to build the extraction solution.

3.2. Solubilization in Urea-Thiourea for IEF

With the spreading of IPG using sample application by in-gel rehydration *(15)*, rather large sample volumes can be used. This is especially true when home-made strips are used, as these can be made wider than commercial IPG strips and can thus accommodate a larger volume (up to 1 mL). It is thus often possible to use a dilution approach with the concentrated chaotrope solution and the solid sample resuspended in a minimal volume of water or the liquid sample. If the detergent can be predissolved in the concentrated chaotrope solution, which can also contain the reducer, then the sample volume can represent up to 20% of the total extraction volume. If the detergent must be added only at the end, with a urea concentration not exceeding 8 *M*, then it is more convenient to use 1 vol of sample, 1 vol of detergent stock solution, and to add 8 vol of concentrated chaotrope solution. If this approach leads to too high a volume, two alternate approaches can be considered:

1. Introduce in a sample tube a volume of a stock detergent solution equal to the sample volume. Evaporate the solvent in a SpeedVac. Add the sample and 4 vol of concentrated chaotrope solution.
2. Consider that the sample volume will make 40% of the final sample volume. Weigh the corresponding amounts of urea, thiourea, and solid detergent. Dissolve with the sample in a bath sonicator.

In all cases, an extraction time of 30 to 60 min at room temperature is optimal before centrifugation (200,000*g*, 30 min, room temperature) to remove unsolubilized material.

3.3. Solubilization for Zone Electrophoresis

Solubilization with SDS is not considered in this section, which will deal only with solubilization in urea and cationic detergents prior to off-diagonal electrophoresis ***(16,17)***. The calculations for making the extraction solution are simple, since the sample represents 1/4 of the final extraction volume. To the liquid sample are added (in this order, and expressed as a fraction of the initial sample volume):

0.4 vol of 20% (w/v) cationic detergent stock solution.
0.2 vol of reducer.
0.4 vol of acidic phosphate buffer.
2 vol of 8 *M* urea

The solution is extracted for 30 min in a bath sonicator. Centrifugation at 10,000*g* for 15 min at room temperature may be needed to remove precipitated material.

4. Notes

1. The partial specific volume of urea is a useful number to know to make concentrated urea solution. One gram of urea occupies 0.75 mL in solution. In the same order for most detergents, 1 g occupies 1 mL. This is also true for thiourea. Urea must also never be warmed above 37°C to limit protein carbamylation. As an example, 1 mL of aqueous extract is added to 900 mg urea. This results in 1.675 mL of a 9 *M* urea solution. Otherwise, 1 g urea is added to 1 mL aqueous extract, resulting in 1.75 mL of a 9.5 *M* urea solution.
2. As a matter of fact, most of these extraction solution contain less than 50% water. This means that dissolution of the solids is rather difficult to perform, especially because high temperatures cannot be used (see previous note). The use of a water bath sonicator (marketed for cleaning objects and glassware) is of great help for this difficult solubilization.
3. Many detergents are not fully compatible with urea. Depending on the detergent structure, on the urea concentration, and on the temperature, insoluble detergent-urea complexes can form. Detergents with linear alkyl chains are especially prone to this problem (e.g., ASB 14, Brij 56), which completely prevents the use of commercial linear sulfobetaines, which do not stand more than 4 *M* urea.
4. Many detergents strongly interfere with some popular protein assay methods, whereas others are plagued by reducing compounds (see the corresponding chapter in this book). It is therefore recommended to take this dimension into account when performing protein solubilization. In some cases, one can end with a solubilization cocktail that is incompatible with any protein assay method. In this case, it

is often advisable to determine the protein concentration in the initial sample, prior to extraction, especially if the sample is a suspension. This means in turn that it will not be possible to assess the efficiency of the solubilization process.

References

1. Weber, K. and Kuter, D. J. (1971) Reversible denaturation of enzymes by sodium dodecyl sulfate. *J. Biol. Chem.* **246,** 4504–4509.
2. Ames, G. F. L. and Nikaido, K. (1976) Two-dimensional gel electrophoresis of membrane proteins. *Biochemistry* **15,** 616–623.
3. Remy, R. and Ambard-Bretteville, F. (1987) Two dimensional electrophoresis in the analysis and preparation of cell organelle polypeptides. *Methods Enzymol.* **148,** 623–632.
4. Heizmann, C. W., Arnold, E. M., and Kuenzle, C. C. (1980) Fluctuations of non-histone chromosomal proteins in differentiating brain cortex and cerebellar neurons. *J. Biol. Chem.* **255,** 11,504–11,511.
5. Rabilloud, T., Adessi, C., Giraudel, A., and Lunardi, J. (1987) Improvement of the solubilization of proteins in two-dimensional electrophoresis with immobilized pH gradients. *Electrophoresis* **18,** 307–316.
6. Castellanos-Serra, L. and Paz-Lago, D. (2002) Inhibition of unwanted proteolysis during sample preparation: evaluation of its efficiency in challenge experiments. *Electrophoresis* **23,** 1745–1753.
7. Santoni, V., Molloy, M., and Rabilloud, T. (2000) Membrane proteins and proteomics: un amour impossible? *Electrophoresis* **21,** 1054–1070
8. Chevallet, M., Santoni, V., Poinas, A., et al.(1998) New zwitterionic detergents improve the analysis of membrane proteins by two-dimensional electrophoresis. *Electrophoresis* **19,** 1901–1909.
9. Rabilloud, T., Blisnick, T., Heller, M., et al. (1999) Analysis of membrane proteins by two-dimensional electrophoresis: comparison of the proteins extracted from normal or Plasmodium falciparum-infected erythrocyte ghosts. *Electrophoresis* **20,** 3603–3610
10. Tastet, C., Charmont, S., Chevallet, M., Luche, S., and Rabilloud, T. (2003) Structure-efficiency relationships of zwitterionic detergents as protein solubilizers in two-dimensional electrophoresis. *Proteomics* **3,** 111–121.
11. Luche, S., Santoni, V., and Rabilloud, T. (2003) Evaluation of nonionic and zwitterionic detergents as membrane protein solubilizers in two-dimensional electrophoresis. *Proteomics* **3,** 249–253.
12. Witzmann, F., Jarnot, B., and Parker, D. (1991) Dodecyl maltoside detergent improves resolution of hepatic membrane proteins in two-dimensional gels. *Electrophoresis* **12,** 687–688.
13. Taylor, C. M. and Pfeiffer, S. E. (2003) Enhanced resolution of glycosylphosphatidylinositol-anchored and transmembrane proteins from the lipid-rich myelin membrane by two-dimensional gel electrophoresis. *Proteomics* **3,** 1303–1312.
14. Santoni, V., Kieffer, S., Desclaux, D., Masson, F., and Rabilloud, T. (2000) Membrane proteomics: use of additive main effects with multiplicative interaction model

to classify plasma membrane proteins according to their solubility and electrophoretic properties. *Electrophoresis* **21,** 3329–3344.

15. Sanchez, J.C., Hochstrasser, D., and Rabilloud, T. (1999) In-gel sample rehydration of immobilized pH gradient. *Methods Mol. Biol.* **112,** 221–225.
16. MacFarlane, D.E. (1989) Two dimensional benzyldimethyl-n-hexadecylammonium chloride-sodium dodecyl sulfate preparative polyacrylamide gel electrophoresis: a high capacity high resolution technique for the purification of proteins from complex mixtures. *Anal. Biochem.* **176,** 457–463.
17. Hartinger, J., Stenius, K., Hogemann, D., and Jahn, R. (1996) 16-BAC/SDS-PAGE: a two-dimensional gel electrophoresis system suitable for the separation of integral membrane proteins. *Anal. Biochem.* **240,** 126–133.

13

Two-Dimensional Electrophoresis for Plant Proteomics

Walter Weiss and Angelika Görg

Summary

Two-dimensional gel electrophoresis (2-DE) with immobilized pH gradients (IPGs) combined with protein identification by mass spectrometry (MS) is currently the workhorse for proteome analysis. 2-DE allows separation of highly complex mixtures of proteins according to isoelectric point (p*I*), molecular mass (M_r), solubility, and relative abundance and delivers a map of intact proteins, which reflects changes in protein expression level, isoforms, or posttranslational modifications. 2-DE can resolve more than 5000 proteins simultaneously (approx 2000 proteins routinely) and can detect and quantify <1 ng of protein per spot. Today's 2-DE technology with IPGs has overcome the former limitations of carrier ampholyte-based 2-DE with respect to reproducibility, handling, resolution, and separation of very acidic and/or basic proteins. The development of IPGs between pH 2.5 and 12 has allowed the analysis of very alkaline proteins and the construction of the corresponding databases. Narrow pH range IPGs provide increased resolution ($\Delta pI = 0.001$) and, in combination with prefractionation methods, permit the detection of low abundance proteins. In this article we provide a comprehensive protocol of the current 2-DE technology for plant proteome analysis and describe in detail the individual steps of this technique.

Key Words: Immobilized pH gradient; plant; proteome; two-dimensional gel electrophoresis.

1. Introduction

Since the late 1970s, numerous studies have been performed seeking *changes in genome expression* (note that the term *proteome* was first coined in 1994) in different plant tissues, organs, subcellular compartments, and organelles to analyze the influence of internal or external stimuli, such as response to treatment with hormones, chemicals, or biotic/abiotic stresses (e.g., heat, cold, drought, or ozone) on protein expression patterns, or to monitor

From: *Methods in Molecular Biology, vol. 335: Plant Proteomics: Methods and Protocols*
Edited by: H. Thiellement, M. Zivy, C. Damerval, and V. Méchin © Humana Press Inc., Totowa, NJ

different developmental and growth stages (e.g., seed germination, maturation, or aging) in plants (reviewed in **refs. *1–5***).

Apart from the study of proteome variations within a given genome, proteome variations between different genomes were also widely analyzed *(2)*. For example, phylogenetic relationships and genetic distances were estimated within and between populations to generate genetic maps and to establish dendrograms based on parsimony analysis of qualitative and quantitative variations in protein patterns *(6)*. Other applications included the assessment of genetic diversity/variability for the differentiation and characterization of closely related lines, cultivars *(7)*, hybrids, or mutants. A third field of research was identification of quality-related molecular markers (e.g., for baking, malting, and brewing quality *[8]*), sensitivity/resistance to plant pathogens (such as fungi and viruses), and localization of quantitative trait loci (QTLs) for plant breeding purposes *(9)*. In more recent surveys, the scope has been extended to the study of pathogenic as well as symbiotic plant-microbe interactions (e.g., between legumes and nitrogen-fixing bacteria), identification and characterization of plant allergens in foods (e.g., wheat flour allergens *[10]*), (**Fig. 1**), or assessment of substantial equivalence in genetically modified plants. For a more comprehensive overview, the reader is referred to reviews **refs. *1–5***) on these subjects.

With the exception of a few studies utilizing liquid chromatography/tandem mass spectrometry (LC-MS/MS) in plant proteome analysis, the overwhelming number of the aforementioned surveys was based on two-dimensional gel electrophoresis (2-DE) technology. 2-DE is—and will probably remain for the foreseeable future—the only analytical technique that can be routinely applied for parallel quantitative expression profiling of large sets of complex protein mixtures, such as total cell or tissue extracts. Combined with protein identification by mass spectrometry (MS), 2-DE is currently the workhorse for proteomics. 2-DE couples isoelectric focusing (IEF) and sodium dodecyl sulfate-polyacrylamide gel electrophoresis (SDS-PAGE) to resolve denatured proteins according to two independent parameters, i.e., isoelectric point (p*I*) in the first dimension and molecular mass (M_r) in the second. Depending on the gel size and pH gradient used, 2-DE can resolve more than 5000 proteins simultaneously (approx 2000 proteins routinely) and can visualize and quantify <1 ng of protein per spot, given that a higly sensitive protein detection method has been applied (**Fig. 1**). Equally important, it delivers a map of intact proteins, which reflects changes in protein expression level, isoforms, or posttransla-

Fig. 1. **(A)** IPG–Dalt of Tris-HCl-soluble albumin/globulin wheat seed proteins. First dimension: IPG 4-8 on Multiphor II; sample application by cup loading near the anode (left). Second dimension: vertical SDS-PAGE (13% T). Silver stain. **(B)** Tris-HCl-soluble wheat seed allergens separated by IPG-Dalt and electroblotted onto an

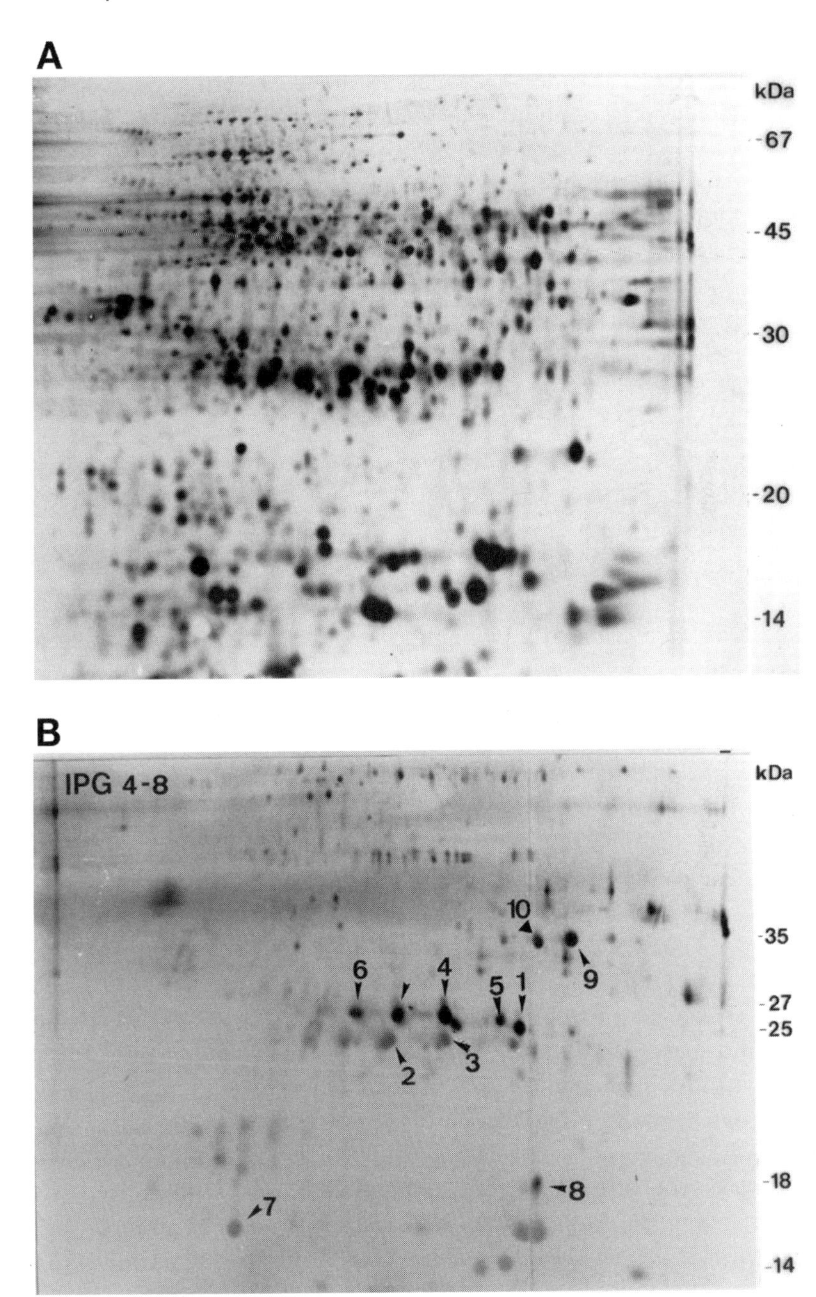

Fig. 1. (*continued from opposite page*) immobilizing PVDF membrane. Immunostaining with a pooled serum of patients suffering from baker's asthma (first antibody), and alkaline phosphatase-labeled antihuman IgE as secondary antibody. Major IgE-binding proteins are labeled with arrows. (Reprinted with permission from **ref. *32***).

tional modifications, which is in contrast to LC-MS/MS-based methods, which perform analysis on peptides, in which M_r and pI information is lost, and in which stable isotope labeling is required for quantitative analysis ***(11)***.

Initial proteome studies were performed by using the 2-DE technology originally described by O'Farrell in 1975 ***(12)***, based on carrier ampholyte-generated pH gradients in the first dimension. The limitations of this 2-DE technology with respect to reproducibility, resolution, separation of very acidic and/or very basic proteins, and sample loading capacity have been largely overcome by the introduction of immobilized pH gradients (IPGs) for the first dimension of 2-DE ***(13)***. IPGs are based on the use of the bifunctional Immobiline^R reagents, a series of 10 chemically well-defined acrylamide derivatives with the general structure CH_2=CH-CO-NH-R, where R contains either a carboxyl or an amino group. These form a series of buffers with different pK values between pK 1 and 13. Since the reactive end is copolymerized with the acrylamide matrix, extremely stable pH gradients are generated, allowing true steady-state IEF with increased reproducibility. Narrow pH range IPGs not only provide increased resolution ($\Delta pI = 0.001$) but also permit detection of lower abundance proteins, whereas alkaline proteins up to pH 12 have been separated under truly steady-state conditions using IPG technology ***(14–17)***.

Other technical improvements, such as more powerful reagents for solubilization of hydrophobic proteins ***(18)***, devices for semiautomatic running of multiple gels in parallel, highly sensitive protein detection procedures based on fluorescent dye technologies for improved reproducibility, more accurate quantitation, and simplified spot pattern comparison, as well as highly sophisticated computer software for the analysis of complex 2-D patterns, have also contributed to the widespread use of 2-DE in proteome analysis (reviewed in **ref. *11***).

Generally speaking, proteome analysis is technically far more challenging than genome analysis owing to the highly diverse physicochemical properties and abundance of proteins. For instance, it has been estimated that >20,000 genes in higher eukaryotic organisms such as plants translate to 50,000 to 100,000 proteins, owing to alternative protein splicing, proteolytic cleavages, phosphorylation, glycosylation, and more than a hundred other possible post-translational modifications. Fortunately, not all these protein variants are expressed in a given tissue at a given time, but nevertheless their number is likely to be in the range of at least 10,000 to 20,000. Moreover, *plant proteome analysis* faces several specific challenges. In particular, sample preparation is difficult because of the rigidity of plant cell walls and because of the accumulation of large quantities of interfering compounds in the central vacuole such as phenolics, pigments, and hydrolytic enzymes, which upon tissue disruption can lead to protein degradation and/or precipitation ***(19,20)***. In addition, owing to the usually low protein content in green plant tissues (typically 2%, compared with approximately 20% in mammalian tissues or microbial cells), methods for enrichment of proteins are usually required.

Despite the aforementioned obstacles, 2-DE-MS has been successfully applied for proteome analysis on the whole plant tissue and subcellular levels. The major steps of the classical 2-DE-MS workflow include: (1) sample preparation/prefractionation and protein solubilization; (2) protein separation by2-DE; (3) protein detection and quantitation; (4) computer-assisted analysis of 2-DE patterns; (5) protein identification and characterization by MS; and (6)2-D protein database construction. The major emphasis of the following sections is on 2-DE technology with IPGs (method 1); the other methods will not be discussed here. The reader is referred to the corresponding chapters of this book and reviews in **refs.** ***1***, ***11***, and ***15*** covering these aspects.

Briefly, the first dimension (IEF) of 2-DE with immobilized pH gradients (IPG-Dalt), according to Görg et al. (***13***; updated in 2000 ***[15]*** and 2004 ***[11]***) is performed in individual, 3-mm-wide and up to 24-cm-long IPG gel strips cast on GelBond PAGfilm (laboratory-made or commercial Immobiline Dry-Strips). Samples can be applied onto the IPG strips either by cup loading or by in-gel rehydration. After completion of IEF, the IPG strips are equilibrated with SDS buffer in the presence of urea, glycerol, dithiothreitol (DTT), and iodoacetamide and applied onto horizontal or vertical SDS gels in the second dimension. 2-DE has been considerably simplified by the use of semiautomated devices such as the *IPGphor* in the first dimension and multiple SDS-PAGE apparatuses for running up to 20 different samples in parallel. After electrophoresis, the separated proteins are visualized by staining with silver nitrate, organic dyes, autoradiography (or phosphor imaging) of radiolabeled samples, or—preferably—by labeling or staining with fluorescent dye molecules ***(11)***.

2. Materials

2.1. Equipment

The following were all procured from GE Healthcare/Amersham Biosciences:

1. Isoelectric focusing device (Multiphor II).
2. IPG DryStrip reswelling tray.
3. IPG DryStrip kit.
4. IPGphor.
5. IPGphor Strip holders.
6. IPGphor Cup loading Strip holders.
7. Multiple vertical SDS electrophoresis apparatus (Ettan DALT).
8. SDS gel casting box.

9. Cassette rack.
10. Glass plates.
11. Thermostatic circulator (Multitemp III).
12. Power supply (Multidrive XL).
13. Laboratory shaker.

2.2. Solutions

1. Urea lysis solution: 9.5 *M* urea, 2% (w/v) 3-[(3-cholamidopropyl)dimethylammonio]-1-propane sulfonate (CHAPS), 2% (v/v) Pharmalyte pH 3 to 10, 1% (w/v) DTT.
   a. To prepare 50 mL of lysis solution, dissolve 30.0 g of urea (Merck, Darmstadt, Germany) in deionized water and make up to 50 mL.
   b. Add 0.5 g of Serdolite MB-1 mixed ion exchange resin (Serva, Heidelberg, Germany; stir for 10 min, and filter.
   c. Add 1.0 g of CHAPS (GE Healthcare, Freiburg, Germany), 0.5 g DTT (Sigma-Aldrich, Taufkirchen, Germany), 1.0 mL Pharmalyte pH 3.0 to 10 (GE Healthcare), and—immediately before use—50 mg of Pefabloc proteinase inhibitor (Merck) to 48 mL of the urea solution (*see* **Notes 1** and **2**).
   d. For solubilization of the more hydrophobic proteins, use thiourea/urea lysis solution (2 *M* thiourea [Sigma-Aldrich], 5–7 *M* urea, 2–4% [w/v] CHAPS, and/or sulfobetaine detergents [e.g., SB 3-10), 1% DTT, 2% [v/v] carrier ampholytes) in combination with IPG strip rehydration solution, consisting of a mixture of urea/thiourea (6 *M* urea, 2 *M* thiourea, 1% [w/v] CHAPS, 15 m*M* DTT, 0.5% [v/v] Pharmalyte, pH 3.0–10.0) ***(18)***.
2. IPG DryStrip rehydration buffer: 8 *M* urea, 0.5% (w/v) CHAPS, 15 m*M* DTT, 0.5% (v/v) Pharmalyte, pH 3.0 to 10.0.
   a. To prepare 50 mL of the solution, dissolve 25.0 g of urea (Merck) in deionized water and complete to 50 mL.
   b. Add 0.5 g of Serdolite MB-1 (Serva), stir for 10 min, and filter.
   c. To 48 mL of this solution add 0.25 g of CHAPS (GE Healthcare), 0.25 mL Pharmalyte, pH 3.0 to 10.0 (40% w/v; GE Healthcare), and 100 mg of DTT (Sigma-Aldrich) and complete to 50 mL with deionized water.
3. IPG strip equilibration buffer: 6 *M* urea, 30% (w/v) glycerol, 2% (w/v) SDS in 0.05 *M* Tris-HCl buffer, pH 8.6.
   a. To make 500 mL, add 180 g of urea (Merck), 150 g of glycerol (Merck), 10 g of SDS (Serva), 16.7 mL of SDS gel buffer (*see* **item 4**), and a few grains of bromophenol blue (Serva).
   b. Dissolve in deionized water and fill up to 500 mL. The buffer can be stored at room temperature up to 2 weeks.
4. SDS gel buffer: 1.5 *M* Tris-HCl, pH 8.6 and 0.4% (w/v) SDS.
   a. To make 500 mL, dissolve 90.85 g of Trizma base (Sigma-Aldrich) and 2.0 g of SDS (Serva) in about 400 mL of deionized water.
   b. Adjust to pH 8.6 with 4 *N* HCl (Merck) and fill up to 500 mL with deionized water.
   c. Add 50 mg of sodium azide (Merck) and filter. The buffer can be stored at 4°C up to 2 wk.

5. Electrode buffer stock solution: to make 5 L of electrode buffer stock solution, dissolve 58.0 g of Trizma base (Sigma-Aldrich), 299.6 g of glycine (Sigma-Aldrich), and 19.9 g of SDS (Serva) in deionized water and complete to 5.0 L.
6. Acrylamide/bisacrylamide solution (30.8% T, 2.6% C): 30% (w/v) acrylamide and 0.8% (w/v) methylenebisacrylamide in deionized water.
   a. To make 500 mL, dissolve 150.0 g of acrylamide (GE Healthcare) and 4.0 g of methylenebisacrylamide (GE Healthcare) in deionized water and fill up to 500 mL.
   b. Add 1 to 2 g of Serdolit MB-1 (Serva), stir for 10 min, and filter.
   c. The solution can be stored up to 2 wk in a refrigerator.
7. Ammonium persulfate solution: 10% (w/v) of ammonium persulfate in deionized water. To prepare 10 mL of the solution, dissolve 1.0 g of ammonium persulfate (GE Healthcare) in 10 mL of deionized water. This solution should be prepared freshly just before use.
8. Displacing solution: 50% (v/v) glycerol in deionized water and 0.01% (w/v) bromophenol blue. To make 500 mL, mix 250 mL of glycerol (100%; Merck) with 250 mL of deionized water, add 50 mg of bromophenol blue (Serva), and stir for a few minutes.
9. Overlay buffer: buffer-saturated 2-butanol. To make 30 mL, mix 20 mL of SDS gel buffer with 30 mL of 2-butanol (Merck) , wait for a few minutes until the two phases have separated, and remove the butanol layer with a pipet.
10. Agarose solution: Suspend 0.5% (w/v) agarose (GE Healthcare) in electrode buffer and melt it in a boiling water bath or in a microwave oven.

3. Methods

3.1. First Dimension: Isoelectric Focusing in IPG Strips (IPG-IEF)

The first dimension of 2-DE with IPGs (IPG-Dalt), isoelectric focusing (IEF), is performed in individual 3-mm-wide IPG gel strips cast on a supporting GelBond PAGfilm™ plastic sheet. A multitude of 7-, 11-, 18-, and/or 24-cm-long ready-made Immobiline DryStrips of almost any desired pH range has been made available, e.g., wide pH ranges of IPG 3 to 10 and IPG 3 to 11, medium pH ranges of IPG 4 to 7 or IPG 6 to 9, and narrow pH ranges of IPG 4 to 5 or IPG 4.5 to 5.5 (e.g., GE Healthcare, Bio-Rad, Sigma-Aldrich, Serva). Alternatively, laboratory-made IPG DryStrips can be used. For details on IPG gel casting, the interested reader is referred to previously published protocols ***(21)***.

IPG DryStrips have to be rehydrated before IEF and are then applied onto the cooling plate of a horizontal isoelectric focusing apparatus ***(13,22)***. More recently, IPG-IEF has been simplified and accelerated by using an integrated system, the *IPGphor*. This instrument features strip holders that provide rehydration of individual IPG DryStrips with or without sample, optional separate sample cup loading, and subsequent IEF with high (8000 V) voltage, without handling the IPG strip after it is placed in a ceramic strip holder.

With IPGs above pH 9, horizontal streaking owing to DTT depletion can occur at the basic end. To avoid streaking, the cysteines should be stabilized as mixed disulfides by using hydroxyethyl-disulfide (HED) reagent (DeStreak™, GE Healthcare) in the IPG strip rehydration solution instead of a reductant such as DTT. Besides the elimination of streaking, the use of HED results in a simplified spot pattern and improved reproducibility ***(23)***. Moreover, with the IPGphor, IEF of strongly alkaline proteins such as ribosomal and nuclear proteins with p*I*s above 10 has been greatly simplified ***(15–17)***.

3.1.1. IPG Strip Rehydration and Sample Application

Prior to IEF, the dried IPG strips must be rehydrated in a reswelling cassette or a reswelling tray to their original thickness of 0.5 mm. IPG DryStrips are rehydrated either with sample already dissolved in rehydration buffer (sample in-gel rehydration) ***(24)*** or with rehydration buffer without sample, followed by sample application by so-called cup-loading. Although sample application by *in-gel rehydration* is more convenient, this procedure is discouraged for samples containing high molecular weight (<100 kDa), very alkaline, and/or very hydrophobic proteins, since these are taken up into the gel with difficulty, because of hydrophobic interactions between proteins and the wall of the reswelling tray, or because of size-exclusion effects of the gel matrix. The latter phenomenon is particularly pronounced if the sample volume significantly exceeds the calculated volume of the IPG strip after reswelling, since higher M_r proteins preferably remain in the excess reswelling solution instead of entering the IEF gel matrix. Cross-contamination is another problem, hence the reswelling tray must be thoroughly cleaned between different experiments. In conclusion, sample in-gel rehydration is less reliable than cup loading, in particular for quantitative analyses.

For *cup loading*, IPG DryStrips are reswollen in rehydration buffer, without sample, however. After IPG strip reydration, samples (20–100 μL) dissolved in lysis solution are put into disposable plastic or silicone rubber cups placed onto the surface of the IPG strip (*see* **Note 3**). The best results are obtained when the samples are applied at the pH extremes, i.e., either near the anode or near the cathode. Sample application near the anode proved to be superior to cathodic application in most cases. When using basic pH gradients such as IPGs 6 to 12 or 9 to 12, anodic application is mandatory for all kinds of samples investigated ***(15–17)***.

The protein amount to be loaded onto a single IPG strip depends on several factors. As a rule of thumb, (1) the longer the separation distance (i.e., the greater the gel size), (2) the narrower the pH gradient, and (3) the less sensitive the protein detection method, the more protein must be applied. Recommended protein amounts range between 50 and 100 μg for analytical (silver-stained)

20×20 cm^2 2-DE gels, and up to 1 mg (or even more) for micropreparative gels. In case of very narrow pH range IPGs, we strongly recommended application of prefractionated samples only ***(25)***.

3.1.1.1. Rehydration of IPG DryStrips Using the IPG DryStrip Reswelling Tray

1. For *sample in-gel rehydration* **(24)**, directly solubilize a cell lysate or tissue sample (5–10 mg protein/mL) in an appropriate quantity of IPG DryStrip rehydration buffer. For 180-mm-long and 3-mm-wide IPG DryStrips, pipet 350 µL of this solution into the grooves of the IPG DryStrip reswelling tray (**Fig. 2**). For longer or shorter IPG strips, the rehydration volume has to be calculated accordingly (e.g., 450 µL for 240-mm-long IPG DryStrips).
2. Remove the protective covers from the surface of the IPG DryStrips and apply the IPG strips, gel side down, into the grooves without trapping air bubbles. Then cover the IPG strip, which must still be moveable and not stick to the tray, with IPG Drystrip cover fluid (which prevents drying out during reswelling), and rehydrate the IPG strips overnight at approximately 20°C. Higher temperatures (>37°C) hold the risk of protein carbamylation, whereas lower temperatures (< 15°C) should be avoided to prevent urea crystallization on the IPG gel.
3. For *cup loading*, the IPG dry strips are rehydrated overnight in rehydration buffer in the reswelling tray as described above in **step 2** but without sample, however.

3.1.1.2. Sample In-Gel Rehydration of IPG DryStrips Using IPGphor Strip Holders

1. Solubilize proteins in sample solubilization buffer (i.e., urea or urea/thiourea lysis solution) and dilute the extract with IPG DryStrip rehydration buffer.
2. Apply the required number of IPGphor strip holders (**Fig. 2**) onto the cooling plate/electrode contact area of the IPGphor.
3. Pipette 350 µL of sample-containing rehydration solution (for 180-mm-long IPG strips) into the strip holder base.
4. Peel off the protective cover sheets from the IPG strip and slowly lower the IPG strip (gel side down) onto the rehydration solution. Avoid trapping air bubbles. The IPG strip must still be moveable and not stick to the tray. Cover the IPG strips with 1 to 2 mL of IPG DryStrip cover fluid and apply the plastic cover. Pressure blocks on the underside of the cover ensure that the IPG strip keeps in good contact with the electrodes as the gel swells. Apply low voltage (30–50 V) during rehydration for improved entry of high M_r proteins ***(15,28)***.

3.1.2. IPG-IEF on a Flat-Bed Apparatus (Multiphor II Unit)

The rehydrated IPG strips may be directly applied onto the cooling plate of the IEF apparatus (1) if running time does not exceed 12 h (which is usually the case for wide and medium pH range IPGs such as IPG 3–10 or 4–7), (2) if the pH gradient does not exceed pH 10.0, and (3) if only small sample volumes (20 µL) are applied by cup loading ***(13,22)***. The application of higher sample volumes (up to 100 µL) is facilitated when the sample cups of the Immobiline

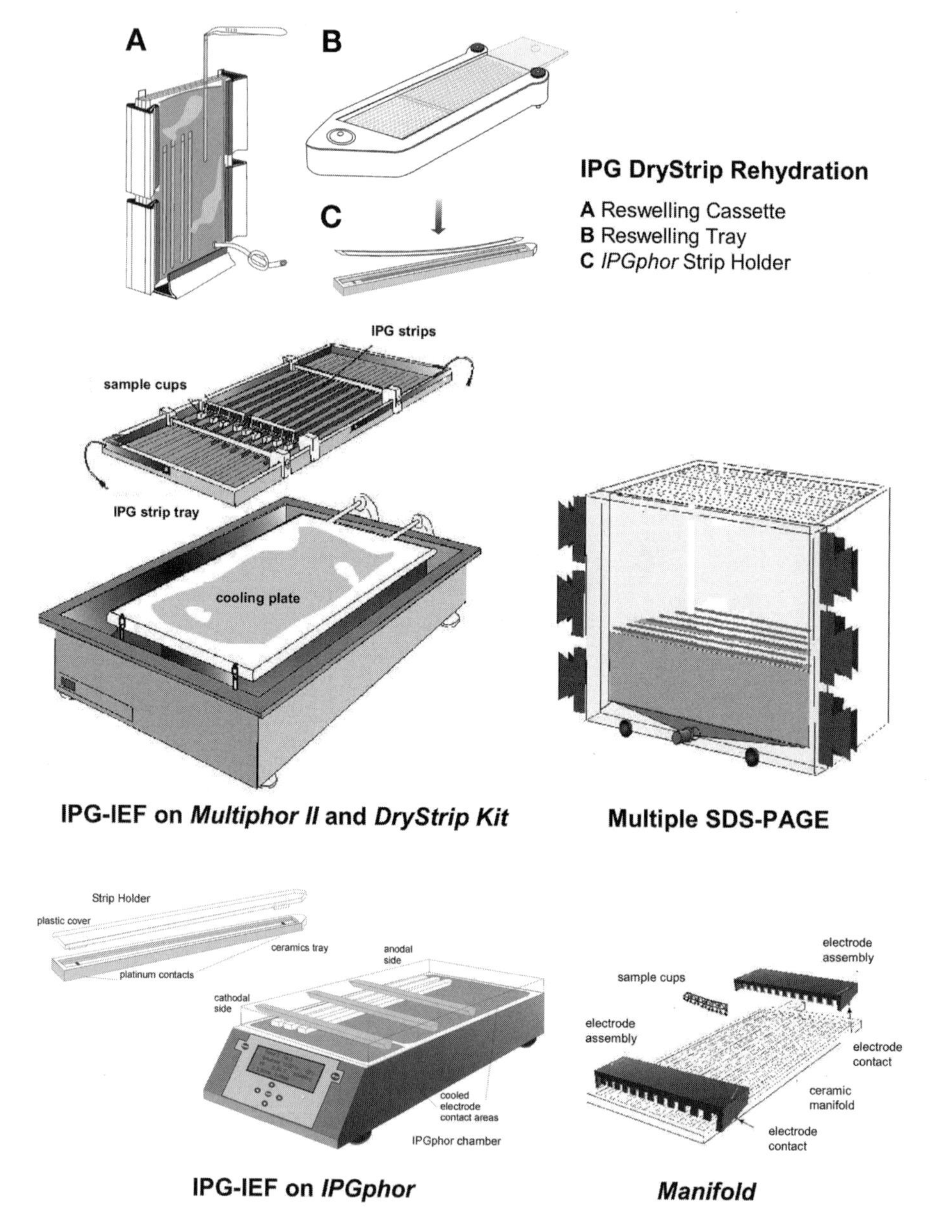

Fig. 2. Procedure of 2-DE with IPGs (IPG-Dalt) based on the protocol of Görg et al. *(**13**,**15**)*.(Reprinted with permission from **ref.** ***33***).

DryStrip Kit are used. When IEF is performed using the DryStrip Kit, the IPG gel strips can be covered by a layer of silicone oil or DryStrip cover fluid. This is mandatory for IEF in very basic (pH > 10.0), or for micropreparative runs in narrow pH gradients (pH range ≤1 unit), whereas in case of broad pH gradients not exceeding pH 10.0 (e.g., IPG 4–7 or 3–10), the DryStrip Kit can be used without oil overlay.

1. Place the cooling plate into the Multiphor II electrophoresis unit. Pipet 3 to 4 mL of kerosene or IPG DryStrip cover fluid onto the cooling plate, and position the Immobiline DryStrip tray on the cooling plate (**Fig. 2**). Avoid trapping air bubbles between the tray and the cooling plate.
2. Connect the electrode leads on the tray to the Multiphor II unit.
3. Pour about 10 mL of IPG DryStrip cover fluid or silicone oil into the tray, and place the corrugated Immobiline strip aligner into the tray on top of the oil.
4. After the IPG strips have been rehydrated (*see* **Subheading 3.1.1.1.**), use clean foreceps to remove the reswollen IPG strips from the reswelling tray. Rinse them with deionized water and blot for a few seconds between two sheets of moist filter paper to remove excess liquid in order to prevent urea crystallization on the surface of the gel during IEF. Transfer the rehydrated IPG gel strips (gel side up and acidic end toward the anode) into adjacent grooves of the aligner in the tray. Align the IPG strips such that the anodic gel edges are lined up.
5. Cut two IEF electrode strips (GE Healthcare) or paper strips prepared from 2-mm-thick filter paper (e.g., MN 440, Macherey & Nagel, Germany) to a length corresponding to the width of all IPG gel strips lying in the tray. Soak the electrode strips with deionized water, remove excessive moisture by blotting with filter paper, and place the moistened IEF electrode strips on top of the aligned strips near the cathode and anode.
6. Position the electrodes and press them gently down on top of the IEF electrode strips.
7. If samples have already been applied by in-gel rehydration, cover the IPG strips with approx 80 mL of DryStrip cover fluid, and continue with **step 12**. In case of sample application by cup loading, continue with **step 8**.
8. Place the sample cups on the sample cup bar, but avoid touching the gel surface with the cups. Moreover, make sure that there is a distance of a few millimeters between the sample cups and the anode (or cathode, in case of cathodic sample application).
9. Move the sample cups into position, one sample cup above each IPG strip, and gently press down the sample cups. The sample cups should form a good seal with the IPG strips but not damage their surface!
10. Once the sample cups are properly positioned, pour about 80 mL of DryStrip cover fluid into the tray so that the IPG gel strips are completely covered. If the oil leaks into the sample cups, suck the oil out, readjust the sample cups, and check for leakage again. Fill up each sample cup with a few drops of DryStrip cover fluid. In case of IEF using wide pH gradients in the range between pH 3 and 10, the oil step can be omitted.
11. Pipet the samples into the cups by underlaying. *Watch again for leakage.*
12. Close the lid of the electrofocusing chamber and start the run according to the parameters given in **Table 1**. For improved sample entry, voltage should be lim-

Table 1
Isoelectric Focusing (IEF) Running Conditions Using the Multiphor II IEF Unit[a]

IPG strip length	180 mm
Temperature	20°C
Current max.	0.05 mA per IPG strip
Power max.	0.2 W per IPG strip
Voltage max.	3500 V
1. Analytical IEF	
Initial IEF:	
Cup loading (20–50 μL)	
150 V, 1–3 h[b]	
300 V, 1–3 h[b]	
600 V, 1 h	
In-gel rehydration (350 μL)	
150 V, 1–3 h[b]	
300 V, 1–3 h[b]	
IEF to the steady state at 3500 V:	
1–1.5 pH units	
e.g., IPG 5–6	24 h
e.g., IPG 4–5.5	20 h
3 pH units	
IPG 4–7	12 h
IPG 6–9	12 h
4 pH units	
IPG 4–8	10 h
IPG 6–10	10 h
5–6 pH units	
IPG 4–9	8 h
IPG 6–12	8 h
7–8 pH units	
IPG 3–10	6 h
IPG 3–11	6 h
8–9 pH units	
IPG 3–12	6 h
IPG 4–12 NL 8 h	

(*continued*)

ited to 150 to 300 to 600 V for the first few hours. Then continue with maximum settings of 3500 V to the steady state (*see* **Note 4**). Current is limited to 0.05 mA/IPG strip. Optimum focusing temperature is 20°C ***(26)***.

13. When the IEF run is completed, remove the electrodes, sample cup bar, and IEF electrode strips from the tray. Use clean forceps and remove the IPG gel strips

Table 1 (*Continued*)
Isoelectric Focusing (IEF) Running Conditions Using the Multiphor II IEF Unit[a]

2. Extended separation distances (240 mm)	
IEF to the steady-state at 3500 V:	
IPG 3–12	8 h
IPG 4–12NL	12 h
IPG 5–6	40 h
3. Micropreparative IEF	
Initial IEF:	
Cup loading (100 μL)	
50 V, 12–16 h	
300 V, 1 h	
In-gel rehydration (350 μL)	
50 V, 12–16 h	
300 V, 1 h	
IEF to the steady state at 3500 V:	
Focusing time of analytical IEF plus approximately 50%	

[a] Procedures are from Görg et al. *(15)*.
[b] Time required for initial IEF (sample entry) depends on salt concentration of the sample. The more salt, the longer.

from the tray. Those IPG gel strips that are not used immediately for a second-dimension run and/or are kept for further reference are stored between two sheets of plastic film at –70°C up to several months.

3.1.3. IPG-IEF Using the IPGphor Unit

IPG-IEF for 2D electrophoresis can be simplified by the use of an integrated instrument, the IPGphor *(27)*. (GE Healthcare; a similar device has recently been developed by Bio-Rad). The IPGphor includes a Peltier element for precise temperature control (between 19.5°C and 20.5°C) and a programmable power supply. The central part of this instrument is the so-called strip holders of different lengths (7, 11, 13, 18, or 24 cm) made from an aluminium oxide ceramic, in which IPG strip rehydration with sample solution and IEF are performed without further handling after the strip is placed into the strip holder (**Fig. 2**). The IPGphor is programmable and can store up to ten different programs. A delayed start is also possible, which allows the user to load the strip holders with sample dissolved in rehydration buffer in the afternoon and automatically start IEF during the night so that IEF is finished the next morning.

When protein separation is performed in alkaline pH ranges (>pH 10.0), much better separations are obtained by applying sample via cup loading on separately rehydrated IPG strips than by sample in-gel rehydration. Sample cup

Table 2
IPGphor Running Conditions
(for Sample In-Gel Rehydration and for Cup Loading)[a]

Gel length	180 mm
Temperature	20°C
Current max	0.05 mA per IPG strip
Voltage max	8000 V
1. Analytical IEF	
Reswelling[b]:	
30 V, 12–16 h[b]	
Initial IEF:	
200 V, 1 h	
500 V, 1 h	
1000 V, 1 h	
IEF to the steady state:	
Gradient from 1000 to 8000 V within 30 min	
8000 V to the steady state, depending on the pH used:	
1–1.5 pH units	
e.g., IPG 5–6	8 h
e.g., IPG 4–5.5	8 h
3 pH units	
IPG 4–7	4 h
4 pH units	
IPG 4–8	4 h
5–6 pH units	
IPG 4–9	4 h
7 pH units	
IPG 3–10 L	3 h
IPG 3–10 NL 3 h	
8–9 pH units	
IPG 3–12	3 h
IPG 4–12	3 h
2. Micropreparative IEF	
Reswelling[b]:	
30 V, 12–16 h*	
IEF to the steady state:	
Focusing time of analytical IEF + additional 50% (approx)	

[a] Procedures from Görg et al. *(15)*.
[b] Omit the low-voltage reswelling step in case of sample cup-loading.

loading is accomplished with special cup loading ("universal") IPGphor strip holders, or with a multiple cup loading strip holder ("Manifold") (**Fig. 2**), which allow(s) the application of quantities up to 100 µL (*see* **Note 3**). The strip holder

Table 3
IPGphor Running Conditions
for Very Alkaline Immobilized pH Gradients (IPGs) (Sample Cup Loading)

Gel length	180 mm
Temperature	20°C
Current max	0.07 mA per IPG strip
Voltage max	8000 V
IPGs 6-12, 9-12, 10-12	
Sample application	anodic
Initial IEF:	
150 V, 1 h	
300 V, 1h	
600 V, 1h	
IEF to the steady-state:	
Gradient from 600 to 8000 V within 30 min	
8000 V to the steady state	
Total volt hours:	
32,000 Vh	

platform regulates temperature and serves as the electrical connector for the strip holders. Besides easier handling, a second advantage of the IPGphor is shorter focusing time, since IEF can be performed at rather high voltage (up to 8000 V).

Typical running conditions for IEF using the IPGphor are given in **Tables 2** and **3**. As indicated earlier, low voltage (30–50 V) should be applied during the rehydration step for improved sample entry of high M_r proteins into the polyacrylamide gel, which otherwise can be a problem with sample in-gel rehydration ***(15,28)***. Then voltage is increased stepwise up to 8000 V. If IPG strips with separation distances ≤11 cm are used, voltage should be limited to 5000 V. For optimum results for samples with high salt concentrations, or when narrow pH intervals are used, it is beneficial to insert moist filter paper pads (size: 4 × 4 mm²) between the electrodes and the IPG strip prior to raising the voltage to 8000 V (*see* **Note 4**). After termination of IEF, the IPG strips are stored as described above in **Subheading 3.1.2., step 13**.

3.1.3.1. IEF of In-Gel Rehydrated Samples

1. Apply the required number of strip holders onto the cooling plate/electrode contact area of the IPGphor (**Fig. 2**). Pipet the appropriate amount (e.g., 350 µL in case of a 180-mm-long IPG strip) of sample-containing IPG DryStrip rehydration buffer into the strip holders, lower the IPG dry strips gel side down into the rehydration buffer, and overlay with IPG DryStrip cover fluid as described in **Subheading 3.1.1.2.**

2. Program the IPGphor (desired rehydration time, volt hours, voltage gradient).
3. After the IPG gel strips have been rehydrated (which requires 6 h at least, but typically overnight), IEF starts according to the programmed parameters listed in **Table 2**.
4. After completion of IEF, store those IPG gel strips that are not used immediately for a second-dimension run between two sheets of plastic film at –70°C.

3.1.3.2. IEF Using Cup Loading Strip Holders

1. Rehydrate IPG DryStrips with rehydration buffer, without sample, however, in a reswelling tray. After the IPG strips have been rehydrated, use clean foreceps to remove the reswollen IPG strips from the reswelling tray or reswelling cassette. Rinse them with deionized water and blot for a few seconds between two sheets of moist filter paper to remove excess liquid in order to prevent urea crystallization on the surface of the gel during IEF, as described in **Subheading 3.1.2.**
2. Apply the required number of the cup loading IPGphor strip holders onto the cooling plate/electrode contact area of the IPGphor instrument, and make sure that the pointed (anodic) ends contact the anodic electrode area. Instead of individual strip holders, the Manifold device may be used.
3. Apply the rehydrated IPG gel strips into the cup loading strip holders (or into the Manifold), gel side upward and pointed (acidic) ends facing toward the anode. Make sure that the cathodic end of the IPG strip is approximately 1.5 cm from the end of the channel and in electrical contact with the electrode rails via the electrode clips.
4. Moisten two filter paper electrode pads (size: 4 × 10 mm^2) with deionzed water, remove excess liquid by blotting with a filter paper, and apply the moistended filter paper pads on the surface IPG gel at the anodic and cathodic ends of the IPG strip between the IPG gel and the electrodes. If necessary (e.g., when the sample contains high amounts of salt), these filter papers should be replaced by fresh ones after several hours.
5. Position the movable electrodes above the electrode filter paper pads. Clip the electrodes firmly onto the electrode paper pads.
6. Position the movable sample cup either near the anode or cathode, and gently press the sample cup onto the surface of the IPG gel strip. The sample cup should form a good seal with the IPG strip but not damage its surface.
7. To confirm that the sample cup does not leak, pipet 100 μL of IPG cover fluid into the cup. If a leak is detected, remove the fluid and use a tissue paper to remove the cover fluid and reposition the sample cup. Check again for leakage. Remove the cover fluid before loading the sample.
8. Overlay each IPG strip with 2 to 4 mL of IPG strip cover fluid (the use of silicone oil or kerosene instead of IPG strip cover fluid is discouraged). In case the cover fluid leaks into the sample cup, rearrange the cup and use tissue paper to remove the cover fluid from the cup. Check again for leakage, and pipet the sample (20–100 μL) in the sample cup.

Table 4
IPG Strip Equilibration Protocol

Reagent	Effect	
50 mM Tris-HCl, pH 8.8		
+2% SDS .		
+6 M urea	Improved protein transfer from IPG strip to SDS gel	
+30% glycerol		
+1% DTT		
+4% iodoacetamide	Removes point streaking Alkylation of SH groups	
	DTT	Iodoacetamide
1. 15 min (10 min)	+	–
2. 15 min (10 min)	–	+

9. Program the instrument (desired volt hours, voltage gradient, temperature, and so on) and run IEF according to the recommended settings in **Tables 2** and **3**. Omit the low-voltage rehydration step recommended for sample in-gel rehydration in **Table 2**.
10. After IEF is complete, proceed with equilibration (*see* **Subheading 3.2.**) and second-dimension IEF (SDS-PAGE) (*see* **Subheading 3.3.**), or store the IPG strips, for up to several months, between two plastic sheets at –70°C.

3.2. IPG Strip Equilibration

Prior to the second-dimension separation (SDS-PAGE), it is essential that the IPG strips be equilibrated to allow the separated proteins to interact fully with SDS. Because the focused proteins bind more strongly to the fixed charged groups of the IPG gel matrix than to carrier ampholyte gels, relatively long equilibration times (10–15 min), as well as urea and glycerol to reduce electroendosmotic effects, are required to improve protein transfer from the first to the second dimension. The by far most popular protocol is to incubate the IPG strips for 10 to 15 min in the buffer originally described by Görg et al. *(13)* (50 mM Tris-HCl [pH 8.8], containing 2% [w/v] SDS, 1% [w/v] DTT, 6 M urea, and 30% [w/v] glycerol; **Table 4**). This is followed by a further 10- to 15-min equilibration in the same solution containing 4% (w/v) iodoacetamide instead of DTT. The latter step is used to alkylate any free DTT, as otherwise it migrates through the second-dimension SDS-PAGE gel, resulting in an artifact known as point streaking that can be observed after silver staining. More importantly, iodoacetamide alkylates sulfhydryl groups and prevents their reoxidation. This two-step reduction/alkylation procedure is highly recommended, since it considerably simplifies downstream sample preparation (protein in-gel digestion) for spot identification by mass spectrometry. After equilibration, the IPG strips are applied onto the surface of the second-dimension horizontal or vertical SDS-PAGE gels.

1. Dissolve 100 mg of DTT (Sigma-Aldrich) in 10 mL of equilibration buffer to make equilibration buffer I.
   a. Make 10 mL per IPG strip.
   b. Place the focused IPG strips into individual test tubes (250 mm long; 20 mm internal diameter), and add 10 mL of equilibration buffer I to each tube.
   c. Seal the tubes with Parafilm, rock them for 15 min on a shaker, and pour off the equilibration buffer. Shorter equilibration times (10 min) can be applied, at the risk, however, that some proteins may not migrate out of the IPG gel strip during sample entry into the SDS-PAGE. In this case it is advisable to check, by staining the IPG strip after removal from the SDS gel, whether all proteins have left the IPG strip.
2. Dissolve 0.4 g of iodoacetamide (Sigma-Aldrich) in 10 mL of equilibration buffer to make equilibration buffer II.
   a. Make 10 mL per IPG strip.
   b. Add this buffer and 50 μL of bromophenol blue (Serva) solution as tracking dye for SDS-PAGE to each tube, and equilibrate for another 15 min with gentle agitation.
3. Pour off equilibration buffer II, and proceed to SDS-PAGE (*see* **Subheading 3.3.**). If SDS-PAGE is performed on a horizontal electrophoresis unit (e.g., Multiphor II), briefly rinse the IPG gel strip with deionized water, and place it on a piece of filter paper at one edge for a few minutes to drain off excess equilibration buffer. If SDS-PAGE is performed on a vertical electrophoresis unit (e.g., Ettan Dalt), briefly rinse the equilibrated IPG strip with electrode buffer.

3.3. Second Dimension: Multiple Vertical SDS-PAGE

SDS-PAGE can be performed on horizontal or vertical systems ***(29)***. Horizontal setups ***(30)*** are ideally suited for ready-made gels (e.g., ExcelGel SDS; Amersham Biosciences/GE Healthcare), whereas vertical systems are preferred for multiple runs in parallel, in particular for large-scale proteome analysis, which usually requires simultaneous electrophoresis of batches of second-dimension SDS-PAGE gels for higher throughput and maximal reproducibility ***(31)***.

3.3.1 SDS Gel Casting

1. The gel casting cassettes (200 × 250 mm^2) are made in the shape of books consisting of two 3-mm-thick glass plates, connected by a hinge strip, and two 1.0-mm-thick spacers in between them. Stack 14 cassettes vertically into the gel

Table 5
Recipes for Casting Vertical SDS Gels (7.5, 10, 12.5, or 15% T)

	7.5% 2.6% C	T 10% T 2.6% C	12.5% T 2.6% C	15% T 2.6% C
Acrylamide/bisacrylamide (30.8% T, 2.6% C)	244 mL	325 mL	406 mL	487 mL
Gel buffer	250 mL	250 mL	250 mL	250 mL
Glycerol (100%)	50.0 g	50.0 g	50.0 g	50.0 g
Deionized water	461 mL	380 mL	299 mL	218 mL
TEMED (100%)	50 µL	50 µL	50 µL	50 µL
Ammonium persulfate (10%)	7.0 mL	7.0 mL	7.0 mL	7.0 mL
Final volume	1000 mL	1000 mL	1000 mL	1000 mL

casting box of the Ettan Dalt II apparatus with the hinge strips to the right, interspersed with plastic sheets (e.g., 0.05-mm-thick polyester sheets).

2. Place the front plate of the casting box in place and screw on the nuts (hand tight) (**Fig. 2**).
3. Connect a polyethylene tube (i.d. 5 mm) to a funnel held in a ring-stand at a level of about 30 cm above the top of the casting box. The other end of the tube is placed in the grommet in the casting box side chamber.
4. Fill the side chamber with 100 mL of displacing solution.
5. Immediately before gel casting, add TEMED and ammonium persulfate solutions to the gel solution (**Table 5**). To cast the gels, the gel solution (830 mL) is poured into the funnel. *Avoid introduction of any air bubbles into the tube*. Do not fill the cassettes with acrylamide solution completely, as some space at the top (approx 10 mm) is needed to fix the IPG strip to the SDS gel with hot agarose.
6. When pouring is complete, the tube is removed from the side chamber grommet so that the level of the displacing solution in the side chamber falls.
7. Very carefully pipet about 1 mL of overlay buffer onto the top of each gel to obtain a smooth, flat gel top surface.
8. Allow the gels to polymerize at approximately 20°C for at least 3 h, but preferably overnight for higher reproducibility.
9. After gel polymerization, remove the front of the casting box, and carefully unload the gel cassettes from the box, using a blade to separate the cassettes. Remove the polyester sheets that had been placed between the individual cassettes.
10. Wash each cassette with water to remove any acrylamide on the outer surface, and drain excess liquid off the top surface. Since only 12 gel cassettes fit into the electrophoresis unit, unsatisfactory gels should be discarded, in particular gels with uneven thickness, i.e., usually those at the outer edges of the gel casting cassette.
11. Gels that are not needed immediately can be wrapped in plastic wrap and stored in a refrigerator (4°C) for up to 2 d.

3.3.2. Multiple SDS-PAGE Using the Ettan Dalt II Vertical Electrophoresis Unit

1. Add 1875 mL of electrode buffer stock solution and 5625 mL of deionized H_2O to the lower electrophoresis buffer tank of the Ettan DALT II unit. Mix, and turn on cooling (25°C).
2. Support the DALT gel cassettes (containing the SDS gels) in a vertical position on the cassette rack to facilitate the application of the IPG gel strips.
3. Briefly rinse the equlibrated IPG strip with electrode buffer (diluted 1:1 with H_2O) and place the IPG strip on top of the DALT gel cassette.
   a. Use a thin spatula or a ruler to push against the plastic backing of the IPG strip and slide it into the gap between the two glass plates.
   b. Add 2 mL of hot (75°C) agarose solution, and continue to slide the strip down onto the surface of the SDS gel until good contact is achieved. Avoid trapping air bubbles between the IPG strip and the SDS gel surface.
   c. For co-electrophoresis of molecular weight marker proteins, soak a filter paper pad ($2 \leftrightarrow 4$ mm^2) with 5 μL of SDS marker proteins dissolved in electrophoresis buffer, let it dry, and apply it to the left or right of the IPG strip.
   d. Dried filter paper pads soaked with molecular weight marker proteins can be stored in microfuge tubes at –70°C.
4. Allow the agarose to solidify for at least 5 min before placing the slab gel into the electrophoresis apparatus (*see* **step 5**). Repeat this procedure for the remaining IPG strips. Although embedding in agarose is not absolutely necessary, it ensures much better contact between the IPG gel strip and the top of the SDS gel.
5. Wet the outside of the gel cassettes by dipping them into electrode buffer to make them fit more easily into the electrophoresis unit. Insert them in the electrophoresis apparatus. If necessary, push blank cassette inserts into any unoccupied slots. Seat the upper buffer chamber over the gels, and fill it with 2.5 L of electrode buffer (1250 mL of buffer stock solution + 1250 mL of deionized H_2O).
6. Place the safety lid on the electrophoresis unit and start SDS-PAGE with 5 mA per SDS gel (100 V maximum setting) for approx 2 h. Continue with 15 mA per SDS gel (200 V maximum setting) for approx 16 h overnight, or higher current for faster runs (30 mA per SDS gel for approx 8 h).
7. Terminate the run after the bromophenol blue tracking dye has migrated off the lower end of the gel.
8. After electrophoresis, carefully open the cassettes with a plastic spatula. Use the spatula to separate the agarose overlay from the polyacrylamide gel. Carefully peel the gel from the glass plate, lifting the gel by its lower edge, and place it in a box of fixing or staining solution.

4. Notes

1. Lysis solution and IPG strip rehydration buffer should always be prepared freshly. Alternatively, small aliquots (1 mL) can be stored at –70°C. Lysis and rehydration solution thawed once should not be refrozen again.

2. Never heat urea solutions above 37°C, to avoid carbamylation of protein amino groups. It is important that the urea solution be deionized with an ion exchanger prior to adding the other chemicals, because urea in aqueous solution exists in equilibrium with ammonium cyanate, which can react with the NH_3^+ of protein side chains (e.g,. lysine) and introduce charge artifacts (= additional spots) on the 2-D gel. In addition, always include carrier ampholytes in the lysis solution and rehydration buffer as cyanate scavenger.
3. If sample volumes exceeding 100 µL are to be applied onto IPG strips by cup loading, pipet in 100 µL, and run IEF with limited voltage (max. 300 V) until the sample has migrated out of the cup. Then apply another 100 µL and repeat the procedure until the whole sample has been applied.
4. Theoretically, no further intervention is required after the start of IPG-IEF until IEF has been completed. In practice, however, superior results are obtained if the electrode filter pads between the IPG strip and the electrodes are replaced by new ones after the sample has entered the IPG strip. This is particularly important for samples that contain high amounts of salts and/or protein, because salt contaminants have quickly moved through the gel and have now collected in the electrode papers, or when very alkaline IPGs (e.g., IPG 6–12 or IPG 9–12) are used. In these cases, the filter paper pads should be replaced every 2 h. For IEF with very alkaline, narrow range IPGs, such as IPG 10 to 12, this procedure should be repeated once an hour.
5. When basic IPG gradients exceeding pH 10.0 are used for the first dimension (e.g., IPG 6–12 or IPG 9–12), horizontal streaking can often be observed at the basic end of 2-D protein profiles. This problem may be resolved substituting DTT in the rehydration buffer of the IPG strip by a disulfide such as hydroxyethyl disulfide ***(23)*** and by application of high voltages (8000 V) for short running times ***(15–17)***.

References

1. Damerval, C., Zivy, M., Granier, F., and de Vienne, D. (1989) Two-dimensional electrophoresis in plant biology. *Adv. Electrophoresis* **2,** 263–340.
2. Thiellement, H., Bahrman, N., Damerval, C., et al. (1999) Proteomics for genetic and physiological studies in plants. *Electrophoresis* **20**, 2013–2026.
3. Thiellement, H., Zivy, M., and Plomion, C. (2002) Combining proteomic and genetic studies in plants. *J. Chromatogr. B*. **782,** 137–149.
4. Cánovas, F. M., Dumas-Gaudot, E., Recorbet, G., Jorrin, J., Mock, H.-P., and Rossignol, M. (2004) Plant proteome analysis. *Proteomics* **4,** 285–298.
5. Bolwell, G. P., Slabas, A.R., and Whithelegge, J. P. (2004) Proteomics: empowering systems biology in plants. *Phytochemistry* **65,** 1665–1669.
6. Posch, A., van den Berg, B., Duranton, C., and Görg, A. (1994) Polymorphism of pepper (*Capsicum annuum* L.) seed proteins studied by two-dimensional electrophoresis with immobilized pH gradients: methodical and genetic aspects. *Electrophoresis* **15,** 297–304.

7. Görg, A., Postel, W., Baumer, M., and Weiss, W. (1992) Two-dimensional polyacrylamide gel electrophoresis, with immobilized pH gradients in the first dimension, of barley (*Hordeum vulgare* L.) seed proteins: discrimination of cultivars with different malting grades. *Electrophoresis* **13,** 192–203.
8. Görg, A., Postel, W., and Weiss, W. (1992) Detection of polypeptides and amylase isoenzyme modifications related to malting quality of barley (*Hordeum vulgare* L.) by 2-D electrophoresis and isoelectric focusing with immobilized pH gradients. *Electrophoresis* **13,** 759–770.
9. Consoli, L., Lefèvre, A., Zivy, M., de Vienne, D., and Damerval, C (2002) QTL analysis of proteome and transcriptome variations for dissecting the genetic architecture of complex traits in maize. *Plant Mol. Biol.* **48,** 575–581.
10. Weiss, W., Huber, G., Engel, K. H., Pethran A., Dunn, M. J., Gooley, A.A., and Görg, A. (1997) Identification and characterization of wheat grain albumin/globulin allergens. *Electrophoresis* **18,** 826–833.
11. Görg, A., Weiss, W., and Dunn, M. J. (2004) Current two-dimensional electrophoresis technology for proteomics. *Proteomics* **4,** 3665–3685.
12. O' Farrell, P. H. (1975) High resolution two-dimensional electrophoresis of proteins. *J. Biol. Chem.* **250,** 4007–4021.
13. Görg, A., Postel, W., and Günther, S. (1988) The current state of two-dimensional electrophoresis with immobilized pH gradients. *Electrophoresis* **9,** 531–546.
14. Görg, A., Obermaier, C., Boguth, G., Csordas, A., Diaz, J. J. and Madjar, J. J. (1997) Very alkaline immobilized pH gradients for two-dimensional electrophoresis of ribosomal and nuclear proteins. *Electrophoresis* **18,** 328–337.
15. Görg, A., Obermaier, C., Boguth, G., Harder, A., Scheibe, B., Wildgruber, R. and Weiss, W. (2000) The current state of two-dimensional electrophoresis with immobilized pH gradients. *Electrophoresis* **21,** 1037–1053.
16. Wildgruber, R., Reil, G., Drews, O., Parlar, H. and Görg, A. (2002) Web-based two-dimensional database of *Saccharomyces cerevisiae* proteins using immobilized pH gradients from pH 6 to pH 12 and matrix-assisted laser desorption/ionization-time of flight mass spectrometry. *Proteomics* **2,** 727–732.
17. Drews, O., Reil, G., Parlar, H., and Görg, A. (2004) Setting up standards and a reference map for the alkaline proteome of the Gram-positive bacterium *Lactococcus lactis*. *Proteomics* **4,** 1293–1304.
18. Rabilloud, T., Adessi, C., Giraudel, A., and Lunardi, J. (1997) Improvement of the solubilization of proteins in two-dimensional electrophoresis with immobilized pH gradients. *Electrophoresis* **18,** 307–316
19. Hirano, H., Islam, N., and Kawasaki, H. (2004) Technical aspects of functional proteomics in plants. *Phytochemistry* **65,** 1487–1498.
20. Méchin, V., Consoli, L., Le Guilloux, M., and Damerval, C. (2003) An efficient solubilization buffer for plant proteins in immobilized pH gradients. *Proteomics* **3,** 1299–1302.
21. Görg, A. and Weiss, W. (2000) 2D electrophoresis with immobilized pH gradients, in *Proteome Research: Two Dimensional Electrophoresis and Identification methods* (Rabilloud, T., ed.), Springer, New York, pp. 57–106.

22. Görg, A. and Weiss, W. (1999) Analytical IPG-Dalt. *Methods Mol. Biol.* **112,** 189–195.
23. Olsson, I., Larsson, K., Palmgren, R., and Bjellqvist, B. (2002) Organic disulfides as a means to generate streak-free two-dimensional maps with narrow range IPG strips as first dimension. *Proteomics* **2,** 1630–1632.
24. Rabilloud, T., Valette, C., and Lawrence, J. J. (1994) Sample application by in-gel rehydration improves the resolution of two-dimensional electrophoresis with immobilized pH-gradients in the first dimension. *Electrophoresis* **15,** 1552–1558.
25. Görg, A., Boguth, G., Köpf, A., Reil, G., Parlar, H., and Weiss, W. (2002) Sample prefractionation with Sephadex isoelectric focusing prior to narrow pH range two-dimensional gels. *Proteomics* **2,** 1652–1657.
26. Görg, A., Postel, W., Friedrich, C., Kuick, R., Strahler, J. R., and Hanash, S. M. (1991) Temperature-dependent spot positional variability in two-dimensional polypeptide patterns. *Electrophoresis* **12,** 653–658.
27. Islam, R., Ko, C., and Landers, T. (1998) A new approach to rapid immobilised pH gradient IEF for 2-D electrophoresis. *Science Tools* **3,** 14–15.
28. Görg, A., Obermaier, C., Boguth, G., and Weiss, W. (1999) Recent developments in 2-D gel electrophoresis with immobilized pH gradients: wide pH gradients up to pH 12, longer separation distances and simplified procedures. *Electrophoresis* **20,** 712–717.
29. Görg, A., Boguth, G., Obermaier, C., Posch, A., and Weiss, W. (1995) Two-dimensional polyacrylamide gel electrophoresis with immobilized pH gradients in the first dimension (IPG-Dalt): the state of the art and the controversy of vertical versus horizontal systems. *Electrophoresis* **16,** 1079–1086.
30. Görg, A. and Weiss, W. (1999) Horizontal SDS-PAGE for IPG-Dalt. *Methods Mol. Biol.* **112,** 235-244.
31. Anderson, N. L. and Anderson, N. G. (1978) Analytical techniques for cell fractions: multiple gradient-slab gel electrophoresis. *Anal. Biochem.* **85,** 341–354.
32. Posch, A., Weiss, W., Wheeler, C., Dunn, M. J., and Görg, A. (1995) Sequence analysis of wheat grain allergens separated by 2-D electrophoresis with immobilized pH gradients. *Electrophoresis* **16,** 1115–1119.
33. Westermeier, R. and Görg, A. (2005) in: *Protein Purification*, 3rd ed., (Janson, J. C., ed.), John Wiley & Sons, New York.

14

Visible and Fluorescent Staining of Two-Dimensional Gels

François Chevalier, Valérie Rofidal, and Michel Rossignol

Summary

Staining of two-dimensional gel constitutes a crucial step in comparative proteome analysis with respect to the number of proteins analyzed, the accuracy of spot quantification, and compatibility with mass spectrometry. The present chapter describes procedures for several visible and fluorescent dyes compatible with mass spectrometry: colloidal Coomassie blue, silver nitrate, Sypro Ruby®, Deep Purple®, and 5-hexadecanoylamino-fluorescein.

Key Words: Staining protocol; two-dimensional electrophoresis; silver nitrate; colloidal Coomassie blue; Sypro Ruby; Deep Purple; C16 fluorescein.

1. Introduction

Protein separation by electrophoresis is largely used in proteomic approaches because of high resolution, availability of powerful image analysis software for gel comparison, and compatibility with subsequent protein characterization by mass spectrometry (MS) *(1,2)*. For these various aspects, the selection of the protein staining procedure is of major importance.

1.1. Dyes Available for Gel Staining

Classically, *Coomassie blue* was the most widely used dye. However, it suffers from a low sensitivity in protein detection (a few tens of nanograms), which can be improved, however, by using the colloidal version (CCB) *(3,4)*. The binding behavior is attributed to van der Waals forces and hydrophobic interactions. Such noncovalent binding of the dye allowed an excellent compatibility with matrix-assisted laser desorption ionization-time of flight (MALDI-TOF) MS *(5)*. CCB has been considered for long time as a convenient dye for large-scale proteomic analysis.

From: *Methods in Molecular Biology, vol. 335: Plant Proteomics: Methods and Protocols*
Edited by: H. Thiellement, M. Zivy, C. Damerval, and V. Méchin © Humana Press Inc., Totowa, NJ

The other classical protein stain, *silver nitrate* (SN), displays an excellent sensitivity (approx1 ng) but shows staining saturation and could interfere with protein analysis by MS. Two categories of silver staining were used to visualize proteins in gel: the acidic silver nitrate and the alkaline silver diamine procedure, which differ in binding specificity, sensitivity, cost, and safety risk ***(6,7)***. Proteins bind silver by salt formation on an acidic group of glutamic and aspartic acid residues or by complex formation through nucleophilic groups of histidyl, cysteyl, methionyl, or lysyl residues ***(6)***. Several modified protocols have been proposed since the first descriptions 25 yr ago. Owing to the complex chemistry involved ***(6)***, many modifications have been proposed to decrease background and increase sensitivity ***(8–10)***. More recently, efforts were oriented toward the compatibility of silver stains with MS ***(11–16)***. Because of the oxidative attacks of silver ions on the proteins and the use of various sensitizing pretreatments of gels ***(6)***, irreversible modifications of amino acids have limited peptide mass fingerprint analysis or other MS analysis. Most adaptations consisted of omitting crosslinking and sensitizing agents such as glutaraldehyde and formaldehyde ***(5,11,14–16)*** and to adapt the destaining method before enzymatic digestion ***(13)***. In this chapter, we describe an acidic procedure that displays a good sensitivity and a clear background ***(15)*** and is compatible with MS.

In the last few years, different fluorescent dyes have been introduced and have proved to combine high sensitivity and compatibility with MS. These include both commercially available stains, such as the series of Sypros and ruthenium red-based dyes for which synthesis procedures have been published ***(17–21)***. *Sypro Ruby* (SR) apparently includes the favorable features of CCB ***(17,18)***. SR is a luminescent ruthenium complex that interacts noncovalently with proteins. As no irreversible modification of amino acids occurs during staining, satisfactory MS compatibility is expected ***(17,19)***. Additionally, SR was shown to have a broader linear dynamic range ***(22)*** and a higher sensitivity than SN ***(23)***, suggesting a profitable use of this dye for large-scale proteomic analysis despite some propensity to saturation ***(23)***. Nevertheless, the necessity for a fluorescent scanner, added to the cost of the dye itself, even if a ruthenium copy existed ***(21,24)***, has limited the use of SR up to now.

Very recently, alternative molecules have been proposed. *Deep Purple* (DP) initially named "Lightning Fast," a sensitive fluorescent-based stain ***(25)*** based on a natural compound extracted from the fungus *Epicoccum nigrum,* is now commercially available (Amersham Biosciences). The fluorescent polyketide is able to bind with proteins and possibly to react on lysyl residues for fluorescence emission ***(26)***. It was described as more sensitive than SR and compatible with MALDI-TOF mass spectrometry ***(25)***. However, staining saturation was also demonstrated ***(23)***.

Fluorescein derivatives constitute another recent fluorescent alternative. Three derivative with various lengths of hydrocarbon tails were examined as fluorescent dyes for gel staining *(27)*. In this chapter, we present the *C16 fluorescein* (C16-F) procedure, which exhibits good sensitivity and clear background, with no propensity to saturation *(23)*.

1.2. Importance of Dyes Selection in a Proteomic Analysis

Large-scale proteome comparison from 2D gels is now widely used for various goals, such as the characterization of genetic diversity *(28,29)*, the analysis of developmental processes *(30,31)*, or the study of interactions with both abiotic and biotic environments *(32,33)*. The selection of dyes in a proteomic analysis is of crucial importance, and time of manipulation, cost of chemical products, equipment required, sensitivity of dyes, and compatibility with MS constitute determinant parameters for the choice. In a recent work, five procedures were compared to help in the selection of a protein stain convenient for large-scale comparison of 2D gel electrophoresis patterns by image analysis *(23)*. It was shown that, for the same protein amount, the number of detected spots increased by a factor of 3 in a sequence CCB < C16-F < SN DP < SR. This was correlated with differences in sensitivity between dyes that led to the detection of additional spots belonging to classes of lower abundance. Analysis of the distribution of variation coefficients for spots from replicates also showed differences in staining reproducibility between dyes that in the order SR > C16-F > DP > SN > CCB.

The present chapter details the gel staining protocols for the five dyes. To help readers in the selection of a procedure, results obtained with a same 100 µg protein sample from an *Arabidopsis* cell suspension are shown.

2. Materials

Use only ultrapure water for all solutions and washing steps.

2.1. Protein Extraction and 2DE

1. Extraction solution: 90% cold (–20°C) acetone (v/v), 10% trichloroacetic acid (TCA) solution (v/v), 0.07% 2-mercaptoethanol (v/v).
2. Washing solution: 90% cold (–20°C) acetone (v/v), 0.07% 2-mercaptoethanol (v/v).
3. Solubilization solution: 9 *M* urea, 4% CHAPS (w/v), 0.05% Triton X-100 (v/v), 65 m*M* dithiothreitol (DTT).
4. First-dimension solution: 0.5% immobilized pH gradient (IPG) buffer 4 to 7 (v/v), 0.002% (w/v) bromophenol blue.
5. Reduction solution: 50 m*M* Tris-HCl, pH 8.8, 6 *M* urea, 30% (v/v) glycerol, 2% (w/v) SDS, 130 m*M* DTT (*see* **Note 1**).
6. Alkylation solution: 50 m*M* Tris-HCl, pH 8.8, 6 *M* urea, 30% (v/v) glycerol, 2% (w/v) SDS, 130 m*M* iodoacetamide (*see* **Note 1**).

7. Agarose solution: 0.6% (w/v) low-melt agarose in running buffer (*see* **Subheading 3.1.9.**) with traces of bromophenol blue.
8. Acrylamide gel solution: 11% acrylamide, 0.375 *M* Tris-HCl, pH 8.8, 0.1% SDS, 0.05% ammonium persulfate, 0.0003% TEMED (*see* **Note 1**).
9. Running buffer: 25 m*M* Tris, 192 m*M* glycine, 0.1% SDS (w/v) (*see* **Note 1**).

2.2. Gel Staining with CCB (see Note 2)

1. Fixation solution: 50% (v/v) ethanol, 2% (v/v) phosphoric acid.
2. Incubation solution: 34% (v/v) methanol, 2% (v/v) phosphoric acid, 17% (w/v) ammonium sulfate. First mix ammonium sulfate, phosphoric acid, and water (180 mL of water for a 300 mL total volume), and then gently add methanol under agitation.
3. Staining solution: 34% (v/v) methanol, 2% (v/v) phosphoric acid, 17% (w/v) ammonium sulfate, 0.05% (w/v) Coomassie brillant blue G-250. First mix ammonium sulfate, phosphoric acid, Coomassie brillant blue, and water (180 mL of water for a 300 mL total volume), and then gently add methanol under agitation.

2.3. Gel Staining with SN (see Note 3)

1. Fixation solution: 50% (v/v) methanol, 12% (v/v) acetic acid, 0.05% (v/v) formaldehyde.
2. Washing solution: 35% (v/v) ethanol.
3. Sensitization solution: 0.02% (w/v) $Na_2S_2O_3$.
4. Staining solution: 0.2% (w/v) silver nitrate, 0.076% (v/v) formaldehyde.
5. Development solution: 6% (w/v) Na_2CO_3, 0.05% (v/v) formaldehyde, 0.0004% (w/v) $Na_2S_2O_3$.
6. Stop solution: 50% (v/v) methanol, 12% (v/v) acetic acid.
7. Storage solution: 1% (v/v) acetic acid.

2.4. Gel Staining with SR (see Note 4)

1. Fixation solution: 30% (v/v) ethanol, 10% (v/v) acetic acid.
2. Staining solution: Sypro Ruby ready to use commercial solution.
3. Washing solution: 10% (v/v) ethanol, 7% acetic acid.

2.5. Gel Staining with DP (see Note 5)

1. Fixation solution: 10% (v/v) methanol, 7.5% (v/v) acetic acid.
2. Staining solution: 200X Deep Purple commercial solution diluted 200 times in water.
3. Washing solution 1: 0.1% (v/v) NH_4OH.
4. Washing solution 2: 0.75% (v/v) acetic acid.

2.6. Gel Staining with C16-F (see Note 6)

1. Fixation and staining solution: 30% (v/v) ethanol, 7.5% (v/v) acetic acid, 1 μ*M* 5-hexadecanoylamino-fluorescein.
2. Washing solution: 7.5% acetic acid.

2.7. Equipment

1. 2DE: standard modern equipment for two-dimensional gel electrophoresis using IPG strips in the first dimension (like the IPG-Phor and the Dalt systems [Amersham Bioscience, Buckinghamshire, UK] for the gels shown here).
2. Image acquisition: visible scanner or densitometer (like the GS-710 [Bio-Rad, Hercules, CA] for gels shown here), multiwavelength fluorescent scanner (like a FLA-5000 [Fuji Photo Film Company, Tokyo, Japan] for the gels shown here).

3. Methods

3.1. Preparation of Total Soluble Protein Extracts (TCA/Acetone Protocol)

1. Collect *Arabidopsis thaliana* cells from a cell suspension.
2. Grind in liquid nitrogen.
3. Mix the fine powder with extraction solution and incubate at –20°C for at least 30 min (*see* **Note 7**).
4. Centrifuge insoluble material at 42,000*g* and wash pellets three times (*see* **Note 8**) with washing solution (*see* **Note 8**).
5. Allows pellets to dry in air.
6. Solubilize proteins by shacking pellets for 2 h (*see* **Note 9**) in solubilization solution.
7. Estimate the protein amount according to the method of Bradford *(34)* using an aliquot sample and serum albumin as standard (in the presence of the same proportion of solubilization solution as introduced by aliquot samples).

3.2. Two-Dimensional Gel Electrophoresis

1. Hydrate IPG strips (18 cm, pH 4.0–7.0) directly with 100 μg protein solution complemented with first-dimension solution.
2. Perform isoelectric focusing until 100 kV/h.
3. Before the second dimension, reduce and then alkylate proteins in reduction solution and then in alkylation solution for 15 min each.
4. Embed strips using agarose solution on the top of a 11% acrylamide gel.
5. Perform SDS-PAGE in running buffer, at 15 mA per gel, overnight and at 10°C (*see* **Note 10**).

3.3. Staining and Imaging with CCB (see Notes 11 and 12)

Gels are stained according to the method of Neuhoff et al. *(3)*. For each step, use 300 mL vol per 2D gel. Perform all steps under shaking.

1. Fix proteins by immerging gels in fixation solution for at least 2 h and up to overnight.
2. Wash gels three times for 30 min with water.
3. Transfer gels to incubation solution for 1 h (*see* **Note 13**) and then to staining solution (*see* **Notes 13** and **14**).
4. Allow staining to develop for 5 d under mild agitation.
5. Wash gels in water and acquire an image using a densitometer at 300 dpi (**Fig. 1**).

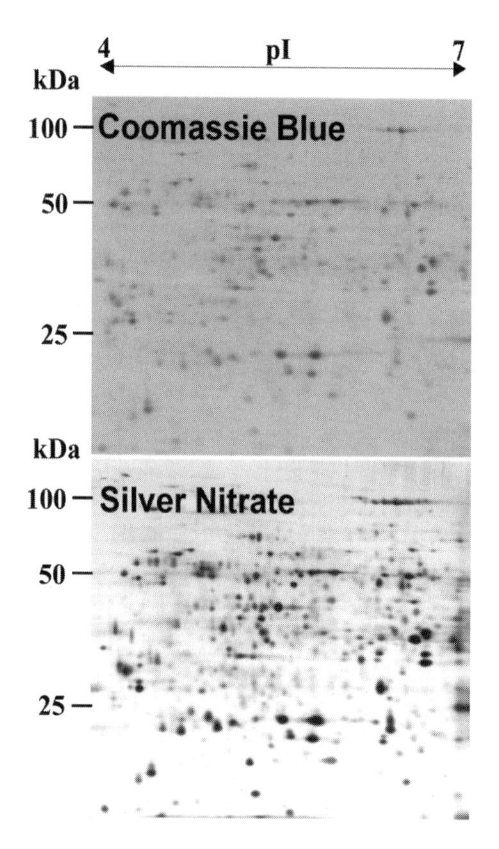

Fig. 1. Images of gels obtained using the visible dyes colloidal Coomassie blue and silver nitrate. Gels were loaded with 100 μg total soluble protein from *Arabidopsis* cultured cells.

3.4. Staining and Imaging with SN (see Notes 11 and 12)

Gels are stained according to the "Vorum" method of Mortz et al. *(15)*. For each step, use 300 mL vol per 2D gel. Perform all steps under shaking.

1. Fix proteins by immerging gels in fixation solution for at least 2 h or overnight.
2. Wash gels three times for 20 min in washing solution.
3. Briefly incubate gels (2 min) in sensitization solution (*see* **Note 15**).
4. Wash gels three times for 5 min with water.
5. Incubate gels for 20 min in staining solution.
6. Quickly wash gels two times for 1 min in water.
7. Allow color to develop for 5 to 10 min in development solution (*see* and **Note 16**).

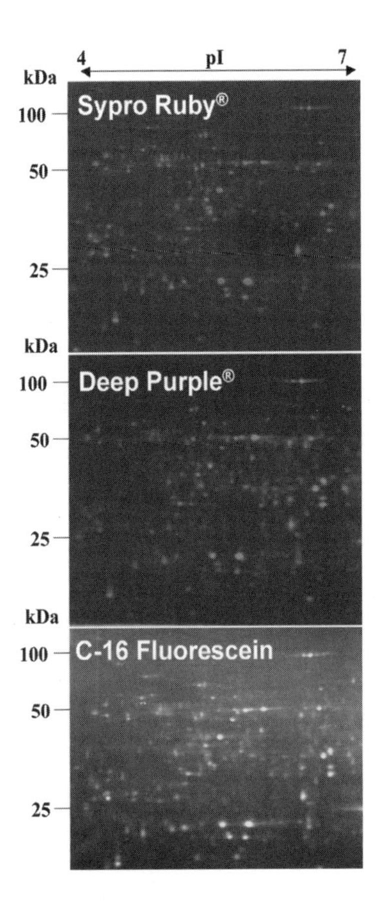

Fig. 2. Images of gels obtained using the fluorescent dyes Sypro Ruby, Deep Purple, and C16 Fluorescein. Gels were loaded with 100 μg total soluble protein from *Arabidopsis* cultured cells.

8. Stop development for 5 min in the stop solution and transfer to the storage solution.
9. Acquire an image using a densitometer at 300 dpi (**Fig. 1**).

3.5. Staining and Imaging with SR (see Notes 11, 12, and 17)

For each step, use 300 mL vol per 2D gel. Perform all steps under shaking.

1. Fix proteins by immerging gels in fixation solution for 30 min.
2. Incubate gels for least 30 min (up to 1 d) in staining solution.
3. Wash gels for 30 min in washing solution.
4. Acquire an image using a FLA-5000 analyzer with 473-nm laser excitation and a long-pass filter, Y510. Select 100 μm resolution with a 16-bit gray scale level, and apply 700 V to the photomultiplier tube (**Fig. 2**).

3.6. Staining and Imaging with DP (see *Notes 11, 12, and 17)*

For each step, use 300 mL vol per 2D gel. Perform all steps under shaking.

1. Fix proteins by immerging gels in fixation solution for 1 h.
2. Wash gels four times for 10 min each in water.
3. Incubate gels for 1 h in staining solution.
4. Wash gels two times for 10 min in washing solution 1 and then two times for 10 min in washing solution 2.
5. Acquire an image using an FLA-5000 analyzer with 532-nm laser excitation and a long-pass filter, O575. Select 100 μm resolution with a 16-bit gray scale level and apply 700 V to the photomultiplier tube (**Fig. 2**).

3.7. Staining and Imaging with C16-F (see *Notes 10, 11, 17, and 18)*

Gels are stained according to Kang et al. *(27)*. For each step, use 300 mL vol per 2D gel. Perform all steps under shaking.

1. Simultaneously fix and stain proteins by immerging gels in the fixation and staining solution for 1 h.
2. Wash gels at least two times for 5 min each in washing solution.
3. Acquire an image using an FLA-5000 analyzer with 473-nm laser excitation and a long-pass filter, Y510. Select 100 μm resolution with a 16-bit gray scale level and apply 700 V to the photomultiplier tube (**Fig. 2**).

3.8 Comparison Between Dyes

When using 100 μg total soluble proteins from *Arabidopsis*, the staining procedures above allow for the routine detection of 250 spots with CCB, 450 spots with C16-F, 550 spots with DP, 600 spots with SN, and 800 spots with SR (**Figs. 1** and **2**). The main features of the procedures are summarized in **Table 1**. Protocols are compared according to the type of dye, number of steps, and duration of manipulation. The quality of images is estimated as a function of the background, spot saturation, and sensitivity of the dyes.

4. Notes

1. For gels to be stained using Deep Purple, high-quality SDS is recommended to avoid nonspecific stain and background (such as that from USB, Cleveland, OH).
2. Different sources of CCB are available. In this work, Coomassie brillant blue G-250 (Ref 161-0406, Biorad, Hercules, CA, USA) was used.
3. Different sources of SN are available. In this work, silver nitrate (cat. no. S-0139, Sigma-Aldrich, St. Louis, MO) was used.
4. Sypro Ruby is distributed by different purchasers. In this work, SR (cat. no. 1703125, Bio-Rad) was used.
5. Deep Purple (cat. no. RPN6305, Amersham Biosciences, Buckinghamshire, UK).
6. 5-Hexadecanoylamino-fluorescein (cat. no. H-110, Molecular Probes, Eugene, OR).

Table 1
Comparison of Practical Features of the Staining Procedures

	CCB	SN	SR	DP	C16-F
Type of stain	Visible	Visible	Fluorescent	Fluorescent	Fluorescent
Steps to stain	4	8	3	5	2
Duration	5 d	4 h	3 h	3 h	3 h
Quality of background	+/–	++	+	+/–	–
Absence of spot saturation	++	+/–	+/–	+/–	+
Sensitivity	+/–	+	++	++	+
Cost	++	++	–	–	++

++, very good/cheap; +, good; +/–, medium; –, bad/expensive.

7. Protein precipitation is usually sufficient after 30 min at –20°C. Prolonged storage at –20°C (overnight) is recommended to improve precipitation, especially for small and/or hydrophilic molecules.
8. Pellets (without supernatant) can be stored at –20°C.
9. Solubilization can be improved by a prolonged agitation step (overnight) and/or by a short sonication step using a bath sonicator.
10. After the first 2 h, in which proteins are electroeluted from the strip into the second dimension gel at 15 mA per gel, the current can be increased to 30 mA to accelerate migration.
11. After migration (when the bromophenol blue front arrives at the bottom of gel), gels need to be quickly transferred to fixation solution to limit passive protein diffusion.
12. To limit manipulation of large gels (and to avoid gel breaking), it is recommended to change solution between consecutive steps by aspiration using a vacuum pump apparatus.
13. To prepare CCB incubation solution and staining solution, be careful to add methanol solution slowly (with Coomassie blue for staining solution) to the ammonium sulfate solution under agitation. When mixing methanol solution and ammonium sulfate solution, ammonium sulfate can precipitate, resulting in the formation of a compact block. To avoid such problems, always add methanol very carefully to the ammonium sulfate solution (dissolve ammonium sulfate in about 180 mL water for a 300 mL final volume), and never add ammonium sulfate solution to methanol solution. During the mixing step, a white ammonium sulfate precipitate can occur. To help ammonium sulfate resolubilization, just add a small volume of water.
14. To avoid the formation of Coomassie blue crystals, prepare staining solution 2 h before use. Coomassie blue needs to be first dissolved in methanol for 1 h. Methanol with Coomassie blue needs to be mixed with ammonium sulfate at least 1 h to obtain an homogeneous blue solution.
15. Sensitization is a quick but very important step allowing further silver fixation to proteins. This step is done in an alcohol-free solution; as the preceding steps

involve a 35% alcohol solution, gels tend to float. To avoid such problems, increase agitation during this step, and use a two times greater volume.

16. Convenient development is obtained within 5 to 10 min. As the chemical reaction does not stop immediately when development solution is removed to the stop solution, an anticipation of the final staining is needed to limit overstaining of protein spots.
17. In the case of fluorescent dyes, all staining steps must be carried out in the dark. Aluminum foil can be used to envelop staining containers.
18. For the C16 fluorescein stain, a glass container must be used to avoid dye adsorption to plastic.

References

1. Van Wijk, K. J. (2001) Challenges and prospects of plant proteomics. *Plant Physiol.* **126,** 501–508.
2. Patton, W. F. (2002) Detection technologies in proteome analysis. *J. Chromatogr. B.* **771,** 3–31.
3. Neuhoff, V., Arold, N., Taube, D., and Ehrhardt, W. (1988) Improved staining of proteins in polyacrylamide gels including isoelectric focusing gels with clear background at nanogram sensitivity using Coomassie Brilliant Blue G-250 and R-250. *Electrophoresis* **9,** 255–262.
4. Neuhoff, V., Stamm, R., Pardowitz, I., Arold, N., Ehrhardt, W., and Taube, D. (1990) Essential problems in quantification of proteins following colloidal staining with Coomassie brilliant blue dyes in polyacrylamide gels, and their solution. *Electrophoresis* **11,** 101–117.
5. Scheler, C., Lamer, S., Pan, Z., Li, X. P., Salnikow, J., and Jungblut, P. (1998) Peptide mass fingerprint sequence coverage from differently stained proteins on two-dimensional electrophoresis patterns by matrix assisted laser desorption/ionization-mass spectrometry (MALDI-MS). *Electrophoresis* **19,** 918–927.
6. Rabilloud, T. (1990) Mechanisms of protein silver staining in polyacrylamide gels: a 10-year synthesis. *Electrophoresis* **11,** 785–794.
7. Patton, W. F. (2000) A thousand points of light: the application of fluorescence detection technologies to two-dimensional gel electrophoresis and proteomics. *Electrophoresis* **21,** 1123–1144.
8. Rabilloud, T., Carpentier, G., and Tarroux, P. (1988) Improvement and simplification of low-background silver staining of proteins by using sodium dithionite. *Electrophoresis* **9,** 288–291.
9. Swain, M. and Ross, N. W. (1995) A silver stain protocol for proteins yielding high resolution and transparent background in sodium dodecyl sulphate-polyacrylamide gels. *Electrophoresis* **16,** 948–951.
10. Rabilloud, T. (1992) A comparison between low background silver diammine and silver nitrate protein stains. *Electrophoresis* **13,** 429–439.
11. Shevchenko, A., Wilm, M., Vorm, O., and Mann, M. (1996) Mass spectrometric sequencing of proteins silver-stained polyacrylamide gels. *Anal Chem.* **68,** 850–858.

12. Larsson, T., Norbeck, J., Karlsson, H., Karlsson, K. A., and Blomberg, A. (1997) Identification of two-dimensional gel electrophoresis resolved yeast proteins by matrix-assisted laser desorption ionization mass spectrometry. *Electrophoresis* **18,** 418–423.
13. Gharahdaghi, F., Weinberg, C. R., Meagher, D. A., Imai, B. S., and Mische, S. M. (1999) Mass spectrometric identification of proteins from silver-stained polyacrylamide gel: a method for the removal of silver ions to enhance sensitivity. *Electrophoresis* **20,** 601–605.
14. Yan, J. X., Wait, R., Berkelman, T., et al. (2000) A modified silver staining protocol for visualization of proteins compatible with matrix-assisted laser desorption/ionization and electrospray ionization-mass spectrometry. *Electrophoresis* **21,** 3666–3672.
15. Mortz, E., Krogh, T.N., Vorum, H., and Gorg, A. (2001) Improved silver staining protocols for high sensitivity protein identification using matrix-assisted laser desorption/ionization-time of flight analysis. *Proteomics* **1,** 1359–1363.
16. Richert, S., Luche, S., Chevallet, M., Van Dorsselaer, A., Leize-Wagner, E., and Rabilloud, T. (2004) About the mechanism of interference of silver staining with peptide mass spectrometry. *Proteomics* **4,** 909–916.
17. Berggren, K., Steinberg, T. H., Lauber, W. M., et al. (1999) A luminescent ruthenium complex for ultrasensitive detection of proteins immobilized on membrane supports. *Anal. Biochem.* **276,** 129—143.
18. Berggren, K., Chernokalskaya, E., Steinberg, T. H., et al. (2000) Background-free, high sensitivity staining of proteins in one- and two-dimensional sodium dodecyl sulphate-polyacrylamide gels using a luminescent ruthenium complex. *Electrophoresis* **12,** 2509–2521.
19. Berggren, K. N., Schulenberg, B., Lopez, M. F., et al. (2002) An improved formulation of SYPRO Ruby protein gel stain: comparison with the original formulation and with a ruthenium II tris (bathophenanthroline disulfonate) formulation. *Proteomics* **2,** 486–498.
20. Steinberg, T. H., Chernokalskaya, E., Berggren, K., et al. (2000). Ultrasensitive fluorescence protein detection in isoelectric focusing gels using a ruthenium metal chelate stain. *Electrophoresis* **21,** 486–496.
21. Rabilloud, T., Strub, J. M., Luche, S., van Dorsselaer, A., and Lunardi, J. (2001) A comparison between Sypro Ruby and ruthenium II tris (bathophenanthroline disulfonate) as fluorescent stains for protein detection in gels. *Proteomics* **1,** 699–704.
22. Lopez, M. F, Berggren, K, Chernokalskaya, E, Lazarev, A, Robinson, M, and Patton, W. F. (2000) A comparison of silver stain and SYPRO Ruby Protein Gel Stain with respect to protein detection in two-dimensional gels and identification by peptide mass profiling. *Electrophoresis* **21,** 3673–3683.
23. Chevalier, F., Rofidal, V., Vanova, P., Bergoin, A., and Rossignol, M. (2004) Proteomic capacity of recent fluorescent dyes for protein staining. *Phytochemistry* **65,** 1499–1506.
24. Lamanda, A., Zahn, A., Roder, D., and Langen, H. (2004) Improved ruthenium II tris (bathophenantroline disulfonate) staining and destaining protocol for a better

signal-to-background ratio and improved baseline resolution. *Proteomics* **4,** 599–608.

25. Mackintosh, J. A., Choi, H. Y., Bae, S. H., et al. (2003) A fluorescent natural product for ultra sensitive detection of proteins in one-dimensional and two-dimensional gel electrophoresis. *Proteomics* **3,** 2273–2288.
26. Bell, P. J. and Karuso, P. (2003) Epicocconone, a novel fluorescent compound from the fungus *Epicoccum nigrum*. *J. Am. Chem. Soc.* **125,** 9304–9305.
27. Kang, C., Kim, H. J., Kang, D., Jung, D. Y., and Suh, M. (2003) Highly sensitive and simple fluorescence staining of proteins in sodium dodecyl sulphate-polyacrylamide-based gels by using hydrophobic tail-mediated enhancement of fluorescein luminescence. *Electrophoresis* **24,** 3297–3304.
28. Thiellement, H., Bahrman, N., Damerval, C., et al. (1999) Proteomics for genetic and physiological studies in plants. *Electrophoresis* **20,** 2013–2026.
29. Marquès, K., Sarazin, B., Chané-Favre, L., Zivy, M., and Thiellement, H. (2001) Comparative proteomics to establish genetic relationships in the Brassicaceae family. *Proteomics* **1,** 1457–1462.
30. Santoni, V., Delarue, M., Caboche, M., and Bellini, C. (1997) A comparison of two-dimensional electrophoresis data with phenotypical traits in *Arabidopsis* leads to the identification of a mutant (cri1) that accumulates cytokinins. *Planta* **202,** 62–69.
31. Gallardo, K., Job, C., Groot, S. P., et al. (2002) Proteomics of *Arabidopsis* seed. *Plant Physiol.* **129,** 823–837.
32. Costa, P., Bahrman, N., Frigerio, J. M., Kremer, A., and Plomion, C. (1998) Water-deficit-responsive proteins in maritime pine. *Plant Mol. Biol.* **38,** 587–596.
33. Bestel-Corre, G., Dumas-Gaudot, E., Poinsot, V., et al. (2002) Proteome analysis and identification of symbiosis-related proteins from *Medicago truncatula* Gaertn. by two-dimensional electrophoresis and mass spectrometry. *Electrophoresis* **23,** 122–137.
34. Bradford, M. M. (1976) A rapid and sensitive method for the quantitation of microgram quantities of protein utilizing the principle of protein-dye binding. *Anal. Biochem.* **72,** 248–254.

15

Two-Dimensional Differential In-Gel Electrophoresis (DIGE) of Leaf and Roots of *Lycopersicon esculentum*

Matthew Keeler, Jessica Letarte, Emily Hattrup, Fatimah Hickman, and Paul A. Haynes

Summary

In this report we present a detailed protocol for the analysis of differential protein expression between two plant tissue samples. The protocol involves harvesting of leaves and roots from mature tomato plants, preparing protein extracts from the harvested tissues, fluorescent labeling of each sample prior to differential in-gel electrophoresis (DIGE), first- and second-dimension electrophoretic separations, and image analysis to visualize and quantify differential protein expression. This protocol is adaptable for use with a wide variety of plant materials and can be used to measure protein expression changes occurring in response to abiotic stress, biotic stress, genetic manipulation, selective breeding, and many other conditions. In addition to the detailed protocol, we also present the results of a representative experiment analyzing subtle changes in protein expression in the roots of tomato plants grown under control and salt-stress conditions.

Key Words: 2D protein gel electrophoresis; differential in-gel electrophoresis; DIGE; plant protein; tomato; CyDye protein staining; differential wavelength fluorescence imaging; *Lycopersicon esculentum*.

1. Introduction

One mainstay technology of proteomics is high-resolution two-dimensional gel electrophoresis (2-DE) *(1,2)* used in conjunction with protein identification by mass spectrometry *(3–6)*. The most common implementation of this technique is to couple a first dimension of protein separation by charge (isoelectric focusing) with a second dimension of protein separation by apparent size (sodium dodecyl sulfate-polyacrylamide gel electrophoresis [SDS-PAGE]). A 2D gel system can be loaded with amounts of up to milligrams of protein and can separate and visualize thousands of protein spots. Although this technique

From: *Methods in Molecular Biology, vol. 335: Plant Proteomics: Methods and Protocols*
Edited by: H. Thiellement, M. Zivy, C. Damerval, and V. Méchin © Humana Press Inc., Totowa, NJ

has been very widely applied, several technical limitations still exist. The protein expression patterns on a given 2D gel can usually not be exactly replicated, because of subtle changes in experimental conditions, principally involving protein solubility. This makes it difficult to pinpoint subtle changes in protein expression level between gels and even more difficult to quantitate such changes accurately. A comparison of protein expression profiles from two different samples run on parallel gels can be performed using various software programs, but these analyses typically require a significant amount of image manipulation in order to align protein spots exactly. These and other difficulties, such as the limited useful dynamic range of many protein staining techniques, limit both the speed and accuracy of quantitation of protein spots in 2D gel electrophoresis.

Two-dimensional differential gel electrophoresis technology (2D-DIGE) *(7)* adds an accurate quantitative component to comparative 2D gel analysis. It can also be used to compare protein abundance changes across multiple samples simultaneously with concurrent statistical measurements of confidence *(8,9)*. Protein samples to be compared are first labeled with high-sensitivity cyanine fluorescent dyes and then mixed together and run on the same 2D gel. This removes any gel-to-gel variability, thus allowing for much higher accuracy of relative quantitation. Protein abundance changes can be quantified using Cy dyes over approximately 4 orders of magnitude *(8)*. The 2-DE gel pattern is visualized by fluorescent image analysis using sequential acquisitions at excitation wavelengths appropriate for the dyes used. The protein/dye ratio is kept deliberately high (>20:1), to try and limit the labeling of proteins to one molecule of dye per protein. The charge and mass of the fluorescence dyes used are carefully matched to reduce protein migration shifts during electrophoresis that are caused by the dye group. A numerical comparison of the images generated by scanning of the DIGE gel at two wavelengths allows for the relative quantitation of protein, or proteins, present in the spot. The use of a third dye permits the inclusion of a pooled internal standard, which can then be used to standardize quantitation across a series of comparative gels of different samples.

There are still, however, some technical limitations inherent in the DIGE approach. The fact that only a small portion of the protein present in the gel is actually labeled means that if the fluorescent spot is excised from the gel for further analysis, the majority of the protein can be left behind in the gel, with obvious adverse effects. This is especially true of very small proteins: even a single dye molecule can make an observable difference in protein migration. To avoid this problem, most users poststain the gel after fluorescent image analysis with Sypro ruby *(10)* or silver (as we describe in this report). This in turn can lead to some problems with the alignment of fluorescent and poststaining images to make sure the correct spots are excised, especially in very crowded gel regions, but it is preferable to the alternative. An alternative DIGE strategy using different reagents that label proteins to saturation, avoiding these problems, has been reported in an evaluation study ***(11)*** but is not yet widely available.

The significance of quantitative analysis results from DIGE is another area of some contention. The system is currently available only from Amersham Biosciences (now part of GE Healthcare) and is thus very expensive owing to lack of competition. It is possible to circumvent this problem by synthesizing and using the *N*-hydroxysuccinimide esters of commercially available cyanine dyes ***(12)***. The accompanying deCyder software (also from Amersham Biosciences) distinguishes protein expression changes that are statistically significant. The comparison of fluorescent spot intensities using the software module is relatively more objective and accurate than conventional approaches involving manual adjustment of the brightness and contrast of two side-by-side images. It is essential to try and match protein loading of the two samples for comparison as closely as possible, to avoid skewing the results. Most users also report at least the results of duplicate experiments, since there is still some inherent variation, probably owing to small protein solubility differences rather than electrophoretic parameters. Attempts have been made to overcome these problems using both integration of DNA expression results with proteomics data ***(13)*** and an alternative normalization algorithm ***(14)***. More accurate quantitation, especially among sets of samples run on different gels, can be achieved by using all three cyanine dyes in a more complicated protocol involving the use of a pooled internal standard ***(15)***.

The utility of 2D-DIGE has already been demonstrated for a number of different biological applications, including, for example, analysis of protein changes in esophageal ***(10)***, gastric ***(16)***, colon ***(17)***, and breast ***(18)*** cancers, leukemia cells ***(19)***, mouse liver toxicity ***(20,21)***, brain tissue ***(22,23)***, and *E. coli* ***(24)*** and yeast ***(12)*** grown under stress conditions. To the best of our knowledge, however, there is as yet no published report showing the use of 2D-DIGE analysis on plant protein preparations. This is most likely because 2D gels of plant tissue protein extract preparations are somewhat problematic in themselves, owing to the high levels of other organic compounds found in such preparations. These include complex carbohydrates, DNA, organic acids, polyphenols, and other structural components of mature adult plants. We have adapted the 2D-DIGE technology for use in our laboratory, in studies of protein expression in tomato plants subjected to heat, light, and salt stress conditions. We first prepare protein extracts from leaf or root tissue using the trichloroacetic acid (TCA)/acetone precipitation procedure used in many previous studies

(25–28). The TCA/acetone powder is then washed briefly to remove excess acid and extracted in 2D gel sample rehydration buffer containing thiourea. Following quantitation of each of a pair of samples and adjustment to equal concentrations, the protein extracts are labeled with cyanine dyes and then mixed together prior to first-dimension isoelectric focusing.

Using this approach we have been able to quantify accurately a large number of protein expression changes induced in the leaves and roots of tomato plants by the imposition of different abiotic stresses. We have also further characterized these protein expression changes by identifying proteins contained in differentially expressed 2D gel spots using nano liquid chromatography (nanoLC)-tandem mass spectrometry (nanoLC-MS/MS) and database searching. The following detailed protocol describes the methods used for our experiments on tomato leaf and root tissue and includes protein extraction, CyDye labeling, 2-DE, fluorescent image analysis, and poststaining. Results are also presented from a representative experiment showing quantitation of differential protein expression in tomato roots grown under salt stress conditions.

2. Materials

2.1. Instrumentation

1. Bio-Rad PROTEAN IEF Cell.
2. Bio-Rad 24-cm Isoelectric Focusing Tray with Lid.
3. Amersham Biosciences Ettan DALTsix Electrophoresis Unit.
4. Amersham Biosciences Ettan DALT casting chamber.
5. Amersham Biosciences Electrophoresis Power Supply EPS 3501 XL.
6. Amersham Biosciences Typhoon multiwavelength fluorescent scanner.
7. Thermolyne Maxi Mix II vortexer.
8. Eppendorf Centrifuge 5415 D.
9. Beckman Coulter Avanti Centrifuge J-20.
10. Sigma Aldrich MicroCentrifuge SD.
11. Fisher Scientific Isotemp Incubator.
12. Ceramic mortar and pestle.
13. Low-fluorescence glass plates for casting 24 × 20-cm polyacrylamide gels (The Gel Company, San Francisco, CA).

2.2. Consumables

1. Amersham Biosciences Immobiline DryStrip pH 3.0–10.0 NL, 24 cm.

2.3. Chemicals, Buffers, and Solutions

1. Milli-Q water (18 MΩ resistance).
2. Leaf resuspension buffer: 10% TCA (EMD Chemical, Gibbstown, NJ), 0.07% (v/v) mercaptoethanol (EMD Chemical) in acetone (EMD Chemical).
3. EDTA wash solution: 0.07% (v/v) mercaptoethanol, 2 m*M* EDTA in acetone.

4. 100 m*M* Tris-HCl (Sigma, St. Louis, MO), pH 8.5.
5. Thiourea sample buffer: 2 *M* thiourea (Bio-Rad, Hercules, CA), 7 *M* urea (Bio-Rad), 2% (w/v) dithiothreitol (DTT) (Bio-Rad), 4% (w/v) 3[3-cholaminopropyl diethylammonio]-1-propane sulfonate (CHAPS; Amersham Biosciences, Piscataway, NJ).
6. *N,N*-dimethylformamide (DMF; DriSolv, EMD Chemical).
7. CyDye DIGE Fluor Cy2, Cy3, Cy5 (Amersham Biosciences).
8. 10 m*M* Lysine (Sigma).
9. Immobilized pH gradient (IPG) rehydration buffer: 0.012% Amersham Biosciences DeStreak Reagent, 2 *M* thiourea, 7 *M* urea, 2% (w/v) DTT, 4% (w/v) CHAPS, 2% (v/v) Amersham Biosciences IPG buffer, pH 3–10 pH NL, trace Bromophenol Blue.
10. Mineral oil (Bio-Rad).
11. Ethanol (EMD Chemical).
12. Glacial acetic acid (J. T. Baker, Phillipsburg, NJ).
13. Bind-Silane and Repel-Silane (Amersham Biosciences).
14. Bind-Silane working solution: 8 mL ethanol, 200 μL glacial acetic acid, 1.8 mL Milli-Q H_2O, 10 μL Bind-Silane.
15. 40% Acrylamide stock solution (40% acrylamide/bisacrylamide 37.5:1, 2.6% C; EMD Chemical).
16. 1.5 *M* Tris-HCl, pH 8.8.
17. 10% (w/v) SDS (Bio-Rad) solution.
18. 10% (w/v) Ammonium persulfate (J.T. Baker) solution.
19. TEMED (*N,N,N,N*-tetramethylethylenediamine; Bio-Rad).
20. Reduction reequilibration buffer: 50 m*M* Tris-HCl, 6 *M* urea, 30% (v/v) 87% glycerol (EMD Chemical), 2% (w/v) SDS, 0.5% (w/v) DTT.
21. Alkylation reequilibration buffer: 50 m*M* Tris-HCl, 6 *M* urea, 30% (v/v) 87% glycerol, 2% (w/v) SDS, 4.5% (w/v) iodoacetamide (Bio-Rad).
22. SDS electrophoresis running buffer: 25 m*M* Tris-HCl, 192 m*M* glycine, 0.2% (w/v) SDS.
23. 0.5% (w/v) Agarose sealing solution: 0.5% (w/v) agarose (Bio-Rad), 25 m*M* Tris-HCl, 192 m*M* glycine (Bio-Rad), 0.2% (w/v) SDS, trace Bromophenol Blue. Heat up in a microwave immediately prior to use to liquefy the agarose.

3. Methods

3.1. Leaf and Root Tissue Harvesting and Protein Precipitation

1. Cut green nonsenescent leaves from the middle of the plant, and immediately place in a Ziploc plastic bag and freeze. If there is no freezer readily available, place on ice until leaves can be stored at –20°C.
2. For root tissue, dig up plants and cut the roots off at least 1 in. below the start of the green stem. Shake the roots and dip rapidly into three successive large beakers of water to remove soil; then immediately place in a Ziploc plastic bag and freeze, or temporarily store on ice.

3. Weigh out 2.5 g of frozen leaf or root tissue with stems and other extraneous materials removed. Cut the tissue into small pieces with chilled scissors into a chilled ceramic mortar and pestle. Grind to a fine powder in the presence of liquid nitrogen. More than one application of liquid nitrogen may be necessary to keep the tissue frozen.
4. To begin protein extraction, transfer the frozen leaf or root tissue powder to a 40-mL centrifuge tube, and resuspend the powder in 25 mL leaf resuspension buffer. Shake and mix thoroughly to ensure that all the powder has been resuspended. Let stand at –20°C for 45 min and shake again. Centrifuge the sample for 15 min at 35,000*g*.
5. Remove the supernatant with a glass pipette, trying not to disturb the pellet. Wash the pellet with the same volume of EDTA wash solution (*see* **Note 1**). Shake to disperse the pellet, let sit for 3–4 min on ice, and centrifuge again for 15 minutes at 35,000g. Repeat washings at least twice, or more if needed until leaf tissue protein pellet is no longer green.
6. Lyophilize the pellet. The resulting dry powder should be pale brown; it contains protein as well as other cell wall and fibrous materials.

3.2. Preparing Protein Sample from TCA/Acetone Powder

1. Weigh out 15 mg of the TCA/acetone powder (prepared as in **Subheading 3.1.**) from two plant tissue samples you wish to compare, and place in 1.5-mL microcentrifuge tubes with screw caps.
2. Pipet 1mL of 100 m*M* Tris-HCl (pH 8.5) into each tube, vortex for 1 min, and spin at 18,000*g* for 10 min. Remove and discard supernatant (*see* **Note 2**).
3. Pipet 1 mL of thiourea sample buffer into each tube. To extract protein from the powder, vortex for 5 min, place in a sonicating water bath for 10 min, and then place on a rotating shaker for 30 min.
4. Spin the pellet down at 18,000*g* for 10 min. The supernatant now contains protein extracted from the leaf tissue, usually in the range 0.5 to 1 μg/μL.

3.3. Preparing CyDyes and Fluorescent Protein Labeling

1. Obtain fresh DMF to reconstitute the CyDyes (*see* **Note 3**). DMF is degraded by oxygen, so always use fresh when preparing CyDyes.
2. To create a working solution of CyDye, add 1.5 μL of DMF to every 1 μL of CyDye. One microliter of working solution is intended to label 50 μg of protein (*see* **Note 4**). Owing to the light sensitivity of the CyDyes, keep them in the dark by wrapping the tubes with aluminum foil.
3. Pipet out 100 μL of each sample, containing 50 μg of protein (*see* **Note 5**), into 0.6-mL microcentrifuge tubes. (The samples should still be separate at this time.) Pipet 1 μL of the working solution of Cy3 into one sample and 1 μL of the working solution of Cy5 into the other sample (*see* **Note 6**).
4. Vortex for 10 s and spin briefly in a countertop centrifuge. Repeat twice.
5. Incubate on ice for 30 min, making sure to keep the tubes away from light.
6. To quench the labeling reaction, add 1 μL of 10 m*M* lysine to each sample. Incubate on ice (covered) for 10 more min.

7. Combine 250 µL of the IPG rehydration buffer with the 100 µL fluorescently labeled aliquot of both samples for a final volume of 450 µL in a solution ready for isoelectric focusing (*see* **Note 7**).

3.4. First-Dimension Protein Separation: Isoelectric Focusing

1. Place the Bio-Rad focusing tray in the Bio-Rad focusing unit. Pipet the sample into the tray, aiming for the middle of the lane, and avoiding bubbles.
2. Peel away the plastic cover from the 24-cm 3 to 10 IPG dry strip using tweezers. Place it into the tray, gel side down, ensuring that the positive side is lined up with positive end of the tray and the negative side at the negative end of the tray. By carefully raising and lowering one end of the strip, make sure the entire length of the strip is covering the sample and there are no bubbles remaining.
3. Cover the strip with mineral oil to prevent it from drying out when voltage is applied.
4. A minimum of 67,000 volt-hours is usually needed for complete focusing.
   a. A recommended program is as follows: 50 V active rehydration for 12 h; 500 V for 1 h; 1000 V for 1 h; 8000 V for 9 h, 100 V for 5 h.
   b. The first step is the rehydration step and should not be less than 12 h. The CyDyes are light sensitive, so the focusing tray should be covered with something that will block out the light, or the whole apparatus should be used in a darkened room.
   c. The last step in the program is the removal window. The strip can be removed at any point during this time. The low voltage ensures that the strip remains completely focused until removed from the tray.

3.5 Preparing SDS-PAGE Second-Dimension Gels

1. Low-fluorescence glass plates must be used to minimize background interference.
   a. For ease of handling during scanning, the gels are bound to the spacer side glass plate using Bind-Silane, and the opposing plates are treated with Repel-Silane to facilitate removal.
   b. Wipe all the nonspacer side low-fluorescence glass plates with Milli-Q H_2O and paper towels, and wipe again with ethanol and Kimwipes.
   c. Apply Bind-Silane working solution to the nonspacer side plates in 0.5-mL increments, spreading evenly over plate with a crew wipe, until each plate has received a total of 1 mL of working solution
2. Wipe spacer-side plates with ethanol first, followed by water, using a Kimwipe. Apply Repel-Silane to the spacer side opposing plates in 0.5-mL increments, spreading evenly over plate with a crew wipe, until each plate has received a total of 1 mL. Let dry for 5–10 min, remove excess Repel-Silane with a crew wipe, wipe the plates with ethanol and milli-Q water, and remove excess liquid.
3. Allow plates to dry in a dust-free environment for a minimum of 3 h, noting that Bind-Silane- and Repel-Silane-treated plates should be stored in physically separate areas. Use nitrogen gas to spray off any dust from plates prior to assembly in gel casting apparatus.
4. Apply paper dot markers to the treated side of each Bind-Silane-treated (nonspacer) plate. One marker should be placed at the midpoint of each of the

shorter edges of the plate approximately 1.5 cm from the edge of the plate, and markers should be placed at the bottom of the plate about ¬ cm from the edge according to a numbering system that will allow you to distinguish the gels from each other during subsequent handling steps.

5. Prepare 12.5% gels using reagents as follows. This makes 900 mL, which is enough for 12 gels: 281 mL acrylamide 40% stock; 225 mL 1.5 *M* Tris-HCl, pH 8.8; 376 mL Milli-Q H_2O; 9 mL 10% SDS solution; 9 mL 10% ammonium persulfate solution; 125 μL TEMED. Add acrylamide, Milli-Q H_2O, Tris-HCl, and SDS to a gel beaker, mix thoroughly, filter the solution (0.2 μ*M*), and place in a reservoir attached to the gel casting chamber by a peristaltic pump.
6. Turn stirrers on. Add ammonium persulfate and TEMED and quickly (within 15 s) turn on peristaltic pump.
   a. When almost all the solution has been pumped in, turn the mixer on its side to get as much of the solution in as possible.
   b. Before air reaches the tubing, add displacement solution to the hose side chamber.
   c. Continue pumping slowly until 12.5% solution is just below the top of the no spacer plates.
7. Turn off the pump and add enough water-saturated butanol to the top and back of the gels to cover them completely. (Butanol is on top of the partitioned solution.) Allow gels to set for 1 h, then pour off butanol, and cover with Milli-Q H_2O.
8. Ideally gels should be left to sit overnight at room temperature, or they can be run after they have completely solidified. Disassemble the casting apparatus and rinse acrylamide from outside of plates. Use gels immediately or wrap in plastic wrap with water and refrigerate for later use (typically 1-wk storage maximum).

3.6. Second-Dimension Protein Separation: SDS-PAGE

1. Remove the isoelectric focusing strip from the tray, and gently wipe the front and back with a Kimwipe to remove excess mineral oil. Take care not to press the Kimwipe into the gel. The strip can be frozen at this point at –80°C if desired.
2. Place the strip in a reequilibration tray, gel side up.
   a. Wash in 2 mL of freshly prepared reduction reequilibration buffer for 30 min on a shaker.
   b. Decant the solution and add 2 mL of freshly prepared alkylation reequilibration buffer.
   c. Place on the shaker for another 30 min. Decant the alkylation buffer.
   d. Add 2 mL of SDS electrophoresis running buffer to the strip, and let sit on shaker for an additional 5 min.
3. Carefully place the strip between the plates of a previously prepared 24-cm gel (*see* **Subheading 3.5.**). Placing the plastic side of the strip against the back plate will make it easier to slide the gel strip down onto the gel so it lays flat. There should be no bubbles between the edge of the strip and the surface of the gel. Seal the strip in place using the 0.5% (w/v) agarose sealing solution. Pipet the solution on top of the gel and allow it to solidify at room temperature.

4. Place the gel cassette rack inside the Ettan DALTsix Electrophoresis Unit, which is connected to a refrigerated recirculating water cooler.
   a. Place the gel in the outermost slot. When running multiple gels, place them evenly on both sides, all facing the same direction.
   b. Blank cassette inserts should be placed in the tank until each slot is filled.
   c. Fill the tank running buffer until it has reached the maximum fill line.
   d. Connect the lid to the power supply, and start electrophoresis.
   e. If the unit is running overnight, set the power supply to 2 W/gel. In the morning, or when running the gel during the day, the wattage can be increased but should not exceed 8 W/gel.

3.7. Scanning Fluorescent Image

1. Turn on the Typhoon scanner, and wait approx 30 min for the instrument to warm up before you start the first scan. Scanning before the instrument is warmed up can affect the accuracy of a scan.
2. Clean the Typhoon platen before and after using the Typhoon; use ethanol and crew wipes. It is easy to scratch the glass surface, so do not use paper towels.
3. Remove glass cassettes from the gel tank and wipe outside surfaces clean. Place fluorescently labeled gel, still inside the glass cassette, on the scanner platen in the correct orientation.
4. Scan image at the appropriate emissions filters and wavelengths for each label used: Cy 3 with 580 BP emission filter, green laser 532 nm; Cy 5 with 670 BP emission filter, red laser 633 nm; and Cy2 with 520 BP emission filter, blue laser, 488 nm. Initial scan parameters are press sample, depth +3 mm, 600 V photomultiplier tube setting and 500 μm pixel size, before repeating at 50 to 100 μm pixel size for a high-resolution scan.
5. Images are acquired using the Typhoon scanner control software, saved in Data Set format with an associated folder containing gel images, which can be opened in ImageQuant, and then resaved as either Data Set format for further processing in DeCyder (Amersham Biosciences), or Tiff format for further processing by other image analysis packages, such as Progenesis (Nonlinear Dynamics).

3.8. Image Analysis

The acquired images from the different wavelengths scanned are then compared and analyzed using DeCyder (*see* **Note 8**). The programs are too complicated to describe in detail here, but the basic steps to follow in DeCyder are as follows.

1. Ensure that images are cropped to match each other exactly in size, then use the "Process gel image" command to measure and normalize spot volume intensity ratios for each spot, and generate a histogram showing number of unchanged, increased, and decreased spots between the two acquired images.
2. Use the exclude filter, or manually remove any obvious nonprotein background spots or streaks. Adjust histogram threshold parameters as required to display

spots, which are over- or underexpressed by the amount specified, such as greater than twofold difference.

3. Manually inspect histogram and images, to determine which spots indicated as being significantly changed in expression are present in sufficient quantity to enable identification by mass spectrometric techniques. In most analyses, many of the protein features that are indicated as being significantly differentially expressed are present in such low amounts that the quantitation is not reliable and subsequent identification is not feasible. The spots of greatest interest are usually those which are displayed as being of both moderate to high abundance with a significant degree of differential expression.
4. For more accurate quantitation of differential expression, especially in the case of subtle differences in expression level, the entire process should be repeated in triplicate and the results analyzed for statistical significance (*see* **Note 8**).

3.9. Post-Staining for Further Processing

Following image analysis, the spacer side plate of the glass cassette is removed. Gels are then stained with silver ***(29,30)*** while still immobilized on the remaining glass plate. This provides a visible stain that can then be used to manually cut out protein spots for identification by mass spectrometric techniques ***(6)***. It also ensures that the majority of protein present is actually stained, in contrast to CyDye labeling, which only labels a small proportion of the protein present ***(8)***. The majority of spots visible with CyDye staining are usually visible with silver staining, although we have noticed that this does vary between different samples. Care must be taken to ensure that the right spot is cut out, by matching the silver-stained image to the CyDye labeled image as closely as possible. This is not usually a problem, but at lower molecular weights the mass of the CyDye label can cause significant shifts, and also some discrepancies are caused by the fact that some proteins simply stain better with one technique than another (*see* **Note 9**).

3.10. Results Example: Examining Protein Expression Differences Caused by Salt Stress in the Roots of Tomato Plants

3.10.1. Salt Stress Experimental Design

1. Tomato plants (*Lycopersicon esculentum*, Betterboy variety, purchased from Mesquite Valley growers, Tucson, Arizona) were grown from 4-inch seedlings in a temperature-controlled greenhouse.
2. Twenty seedlings were grown in Hoglan's compost with the addition of Miracle-Gro fertilizer at the initial planting.
3. All 20 plants were watered daily with normal greenhouse water for 20 d, and then 10 plants were labeled as salt-stress plants and given 25 m*M* NaCl for 1 wk, which was ramped up to a 100 m*M* NaCl watering solution over the course of 1 mo.
4. The ten control plants were watered with water throughout the course of the experiment.

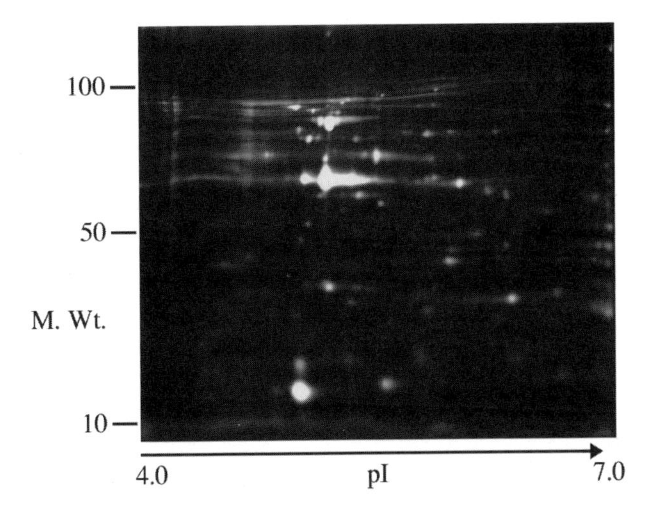

Fig. 1. Initial scanned gel image output from the DeCyder program for the analysis of samples prepared in the example shown in **Subheading 3.10.** The spots appear as yellow when corresponding to those proteins present at approximately equal amounts in both samples, green for those only present in the sample labeled with Cy3, and red for those only present in the sample labeled with Cy5. The *x*-axis shows *pI* values, and the *y*-axis shows apparent molecular weight in kilodaltons.

3.10.2. Harvesting Leaf and Root Tissue

1. At the 7-wk stage, after two consecutive weeks of 100 m*M* NaCl watering, 3 to 5 g of healthy leaf tissue were clipped from both control and salt-stress plants two times a week until leaves were necrotic.
2. As soon as leaves had become necrotic, each plant was uprooted, and roots were washed clean of soil and divided into proximal (nearest stem) and distal (furthest from stem) sections of half the total root length; they were immediately frozen in liquid nitrogen.

3.10.3. Extracting Protein from Proximal Roots

1. Harvested proximal root sections were ground to a fine powder using a mortar and pestle.
2. Protein precipitates were prepared using TCA/acetone as described in **Subheading 3.1.**
3. Since root tissue contains less extractable protein than leaf, for the gel shown below, it was necessary to prepare three aliquots of protein extract, as described in **Subheading 3.2.**, then perform an additional TCA/acetone precipitation step and combine the pellets from all three into a single tube prior to CyDye labeling.
4. The samples were combined, the gel was run and analyzed using the protocols described in **Subheadings 3.3.** to **3.6.**, and the resulting images are shown below in **Figs. 1** to **4**.

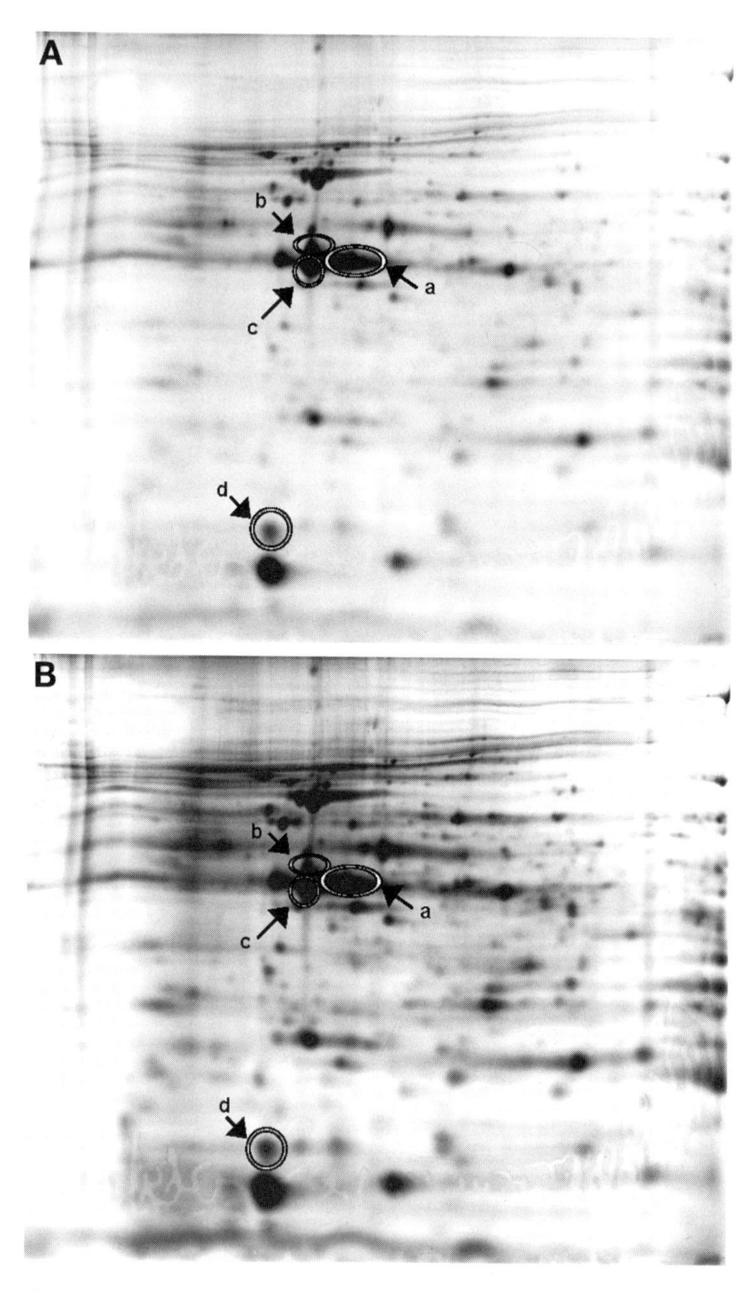

Fig. 2. **(A)** Single-channel fluorescence scanned image of the same gel as in **Fig. 1**, showing only those spots fluorescent at the Cy3 wavelength. The circled spots labeled a to d correspond to those labeled in the histogram in **Fig. 3** and represent relatively abundant spots that are overexpressed in the Cy3-labeled sample by greater than 1.5-fold. **(B)** Single-channel fluorescence-scanned image of the same gel as in (A), showing only those spots fluorescent at the Cy5 wavelength, with the same spots circled.

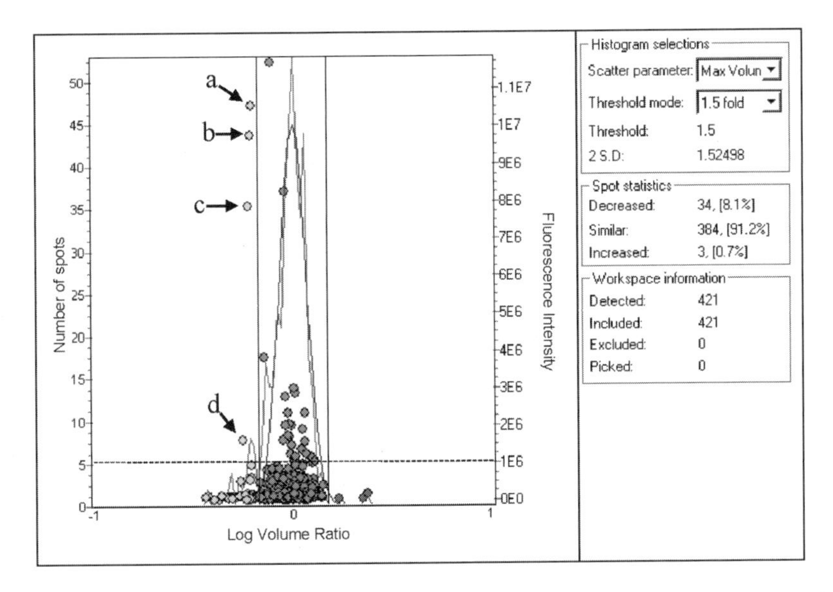

Fig. 3. Histogram of protein expression values for spots shown on gel in **Figs. 1** and **2**. The *x*-axis indicates log volume ratio of spot fluorescence between the two acquired images, the left *y*-axis indicates spot frequency on the histogram curves, and the right *y*-axis indicates absolute fluorescent intensity for the individual spots. The blue curve indicates a model spot frequency histogram, which is based on the actual histogram data shown in the red curve. The four spots labeled a to d, which are outside the normal distribution curve on the *x*-axis and above the background threshold level on the *y*-axis indicated by the dashed line, represent relatively abundant spots that are overexpressed in the Cy3-labeled sample by greater than 1.5-fold. These correspond to the same spots labeled and circled in **Fig. 2**.

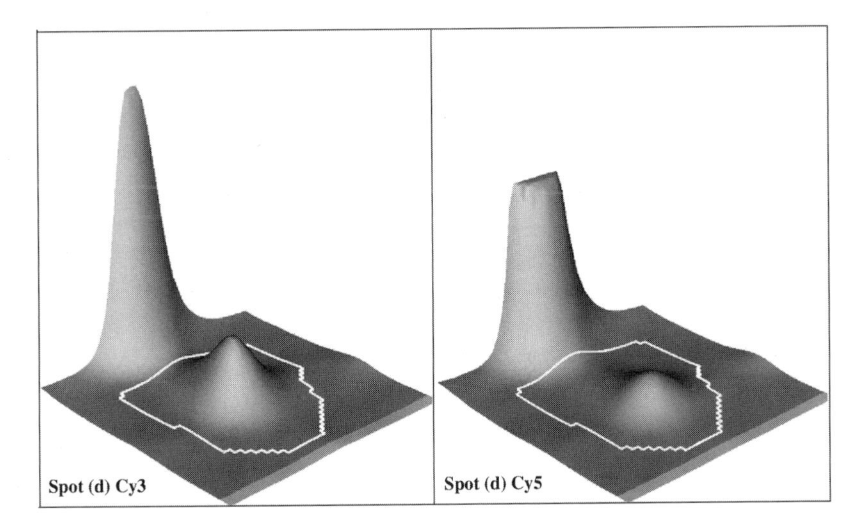

Fig. 4. Three-dimensional representation of fluorescent imaging of spot d indicated in **Figs. 2** and **3**, showing quantitation of differential expression.

3.10.4. Image Analysis of DIGE Gel Comparing Normal and Salt-Stressed Proximal Root Sections

The data produced in the gel from **Subheading 4.3.** were analyzed using the DeCyder Differential In-Gel Analysis (DIA) module as outlined in **Subheading 3.8.** The initial output from the DeCyder program is a false-colored representation, as shown in **Fig. 1**. Proteins present solely in the Cy5-labeled control material appear as red spots; those present solely in the Cy3-labeled material from salt-stressed plants appear as green spots. Proteins present in both samples at similar levels appear as yellow. This provides an easily readable initial scan for obvious protein expression differences, as clearly red and green spots can be readily detected by visual inspection. In the example shown here, the majority of spots are yellow, indicating that there are few gross differences in protein expression between the two samples.

The separated gel images of the two fluorescent wavelengths scanned are presented in **Fig. 2**. These images can be used for in-depth analysis of protein expression differences. The two images shown here look similar to each other, but there appears to be more protein present in **Fig. 2B**. There is some difference apparent in overall protein loading (or labeling efficiency), which serves to highlight one of the advantages of this approach. The software can be used to normalize the quantitative comparison of the two gel images, so that all the spots are considered, and only those with an expression ratio that is significantly more or less than the mean value are indicated as being differentially expressed.

Quantitative analysis of the two images produced the histogram shown in **Fig. 3**. As shown in the spot detection window, after normalization and processing, the software detects 421 protein spots, of which 384 are unchanged, 34 are of higher intensity in the Cy3-labeled sample, and 3 are of higher intensity in the Cy5-labeled material. The x-axis is in log volume ratio of spot fluorescence between the two acquired images, the left y-axis indicates spot frequency on the histogram curves, and the right y-axis indicates absolute fluorescent intensity for the individual spots, indicative of protein amount. The blue curve indicates a model spot frequency histogram, which is based on the actual histogram data shown in the red curve.

The two vertical lines indicate the x-axis points at which a 1.5-fold protein expression difference is reached in each direction, with spots between the two lines considered to be unchanged in expression between the two samples. The color scheme is the same as in **Fig. 1**: yellow indicates unchanged proteins, green indicates higher expression in the Cy3-labeled material, and red indicates higher expression in the Cy5-labeled material.

In this type of analysis, many spots are often detected that are significantly altered in expression but are present at such low levels that it is not feasible to identify them. In the analysis shown in **Fig. 3**, we imposed an absolute fluorescence threshold value of 10^6 fluorescence units, as indicated by a dashed line, to exclude these. There are four spots, labeled a to d, that are outside the normal distribution curve on the *x*-axis and above the specified 10^6 fluorescence unit background level on the *y*-axis. These spots therefore represent proteins that are overexpressed in the Cy3-labeled sample by greater than 1.5-fold and present at a level compatible with mass spectrometric identification. There are numerous other spots on the histogram that represent proteins with significantly different expression levels but that are below our specified abundance threshold.

The four highlighted protein features (a–d) thus represent proteins that are upregulated in expression by greater than 1.5-fold in response to salt stress. The first three of the indicated spots, labeled A, b, and c, highlight one of the disadvantages of this approach, as they all appear to be part of a single poorly resolved protein feature, which has been artificially divided into spots by the analysis software. One spot, however, labeled d in **Fig. 2**, is a well-resolved feature with a fluorescence volume ratio of 1.79, shown in **Fig. 4**. This gel spot is a good candidate for excision and identification by mass spectrometric techniques, in order to identify the corresponding protein.

4. Notes

1. The EDTA wash solution contains both mercaptoethanol and EDTA to help minimize proteolytic breakdown occurring during the protein extraction. The mercaptoethanol also helps prevent formation of disulfide bonds.
2. This washing step is essential to ensure that the pH of the extracted protein mixture is not too low owing to residual TCA. A certain indicator that residual acid is present is to extract another aliquot of the TCA/acetone plant tissue powder using standard SDS-PAGE sample buffer; if the Bromophenol Blue turns bright yellow, the pH is acidic.
3. The CyDye labeling reagents are commercial products of Amersham Biosciences, known as Cy2, Cy3, and Cy5. They are all cyanine derivatives, which are closely matched, but not identical, in terms of charge and molecular weight, but they have different fluorescent profiles.
4. The approach described here relies on minimal protein labeling *(8)*, hence the use of a 1:20 ratio off dye/protein. The majority of the protein remains unlabeled. One acknowledged problem with this approach is that the slight differences in size of the CyDyes can cause a difference in migration of the labeled and unlabeled protein, especially for low-molecular-weight proteins. This problem is addressed by the use of poststaining techniques to stain the remainder of the unlabeled protein and excising spots for further processing based on poststaining locations.
5. The protein concentration in each sample should be quantified as accurately as possible using standard quantitative protein assay techniques on a separate protein extract aliquot, and volumes should be adjusted so that equal amounts of protein

are labeled with each fluorescent dye. In the event that the extracted protein concentration is too low, multiple aliquots can be extracted, precipitated with TCA/acetone, and combined prior to labeling.

6. For a simple pairwise comparison experiment, two samples can be labeled with Cy3 and Cy5 and their relative fluorescent intensities compared directly. For comparing relative expression differences across replicates or across different members of a sample set, it is also necessary to include an aliquot of a pooled internal standard labeled with Cy2 to use in relative quantitation statistics ***(15)***.
7. When the aim of the experiment is to identify differentially expressed proteins by mass spectrometric techniques, it is usually necessary to spike the fluorescently labeled protein mixture with an equal volume of each of the unlabeled samples, for a final amount of 500 μg of total protein loaded in the gel. When doing this, care must be taken to ensure that the concentration of IPG buffer in the final mixture is between 0.5 and 1.0%.
8. For pairwise comparison gels, the image analysis is performed using the DeCyder DIA (differential in-gel analysis) module. If the experiment has been performed in at least triplicate and includes a Cy2-labeled pooled internal standard (*see* **Note 3**), the images can be analyzed in the DeCyder BVA (biological variation analysis) module. This analysis allows for the statistical evaluation, such as a Student's *t*-test, to determine the significance of any observed protein expression changes.
9. Fluorescent stains such as Sypro Ruby ***(31)*** or Deep Purple ***(32)*** can also be used for post staining, but this requires the use of a fluorescent-compatible spot-cutting robotic excision system.

References

1. O'Farrell, P. H. (1975) High resolution two-dimensional electrophoresis of proteins. *J. Biol. Chem.* **250,** 4007–4021.
2. Gorg, A., Obermaier, C., Boguth, G., et al. (2000) The current state of two-dimensional electrophoresis with immobilized pH gradients. *Electrophoresis* **21,** 1037–1053.
3. Haynes, P. A., Fripp, N., and Aebersold, R. (1998) Identification of gel separated proteins by liquid chromatography-electrospray tandem mass spectrometry: comparison of methods and their limitations. *Electrophoresis* **19,** 939–945.
4. Haynes, P., Miller, I., Aebersold, R., et al. (1998) Proteins of rat serum I: establishing a reference 2-DE map by immunodetection and microbore high performance liquid chromatography- electrospray mass spectrometry. *Electrophoresis* **19,** 1484–1492.
5. Cooper, B., Eckert, D., Andon, N. L., Yates, J. R., and Haynes, P. A. (2003) Investigative proteomics: identification of an unknown plant virus from infected plants using mass spectrometry. *J. Am. Soc. Mass Spectrom.* **14,** 736–741.
6. Andon, N. L., Hollingworth, S., Koller, A., Greenland, A. J., Yates, J. R. III, and Haynes, P. A. (2002) Proteomic characterization of wheat amyloplasts using identification of proteins by tandem mass spectrometry. *Proteomics* **2,** 1156–1168.

7. Unlu, M., Morgan, M. E., and Minden, J. S. (1997) Difference gel electrophoresis: a single gel method for detecting changes in protein extracts. *Electrophoresis* **18,** 2071–2077.
8. Tonge, R., Shaw, J., Middleton, B., et al. (2001) Validation and development of fluorescence two-dimensional differential gel electrophoresis proteomics technology. *Proteomics* **1,** 377–396.
9. Von Eggeling, F., Gawriljuk, A., Fiedler, W., et al. (2001) Fluorescent dual colour 2D-protein gel electrophoresis for rapid detection of differences in protein pattern with standard image analysis software. *Int. J. Mol. Med.* **8,** 373–377.
10. Zhou, G., Li, H., DeCamp, D., et al. (2002) 2D differential in-gel electrophoresis for the identification of esophageal scans cell cancer-specific protein markers. *Mol. Cell. Proteomics* **1,** 117–124.
11. Shaw, J., Rowlinson, R., Nickson, J., et al. (2003) Evaluation of saturation labelling two-dimensional difference gel electrophoresis fluorescent dyes. *Proteomics* **3,** 1181–1195.
12. Hu, Y., Wang, G., Chen, G. Y., Fu, X., and Yao, S. Q. (2003) Proteome analysis of *Saccharomyces cerevisiae* under metal stress by two-dimensional differential gel electrophoresis. *Electrophoresis* **24,** 1458–1470.
13. Kreil, D. P., Karp, N. A., and Lilley, K. S. (2004) DNA microarray normalization methods can remove bias from differential protein expression analysis of 2D difference gel electrophoresis results. *Bioinformatics* **20,** 2026–2034.
14. Karp, N. A., Kreil, D. P., and Lilley, K. S. (2004) Determining a significant change in protein expression with DeCyder during a pair-wise comparison using two-dimensional difference gel electrophoresis. *Proteomics* **4,** 1421–1432.
15. Alban, A., David, S. O., Bjorkesten, L., et al. (2003) A novel experimental design for comparative two-dimensional gel analysis: two-dimensional difference gel electrophoresis incorporating a pooled internal standard. *Proteomics* **3,** 36–44.
16. Lee, J. R., Baxter, T. M., Yamaguchi, H., Wang, T. C., Goldenring, J. R., and Anderson, M. G. (2003) Differential protein analysis of spasomolytic polypeptide expressing metaplasia using laser capture microdissection and two-dimensional difference gel electrophoresis. *Appl. Immunohistochem. Mol. Morphol.* **11,** 188–193.
17. Friedman, D. B., Hill, S., Keller, J. W., et al. (2004) Proteome analysis of human colon cancer by two-dimensional difference gel electrophoresis and mass spectrometry. *Proteomics* **4,** 793–811.
18. Somiari, R. I., Sullivan, A., Russell, S., et al. (2003) High-throughput proteomic analysis of human infiltrating ductal carcinoma of the breast. *Proteomics* **3,** 1863–1873.
19. Wang, D., Jensen, R., Gendeh, G., Williams, K., and Pallavicini, M. G. (2004) Proteome and transcriptome analysis of retinoic acid-induced differentiation of human acute promyelocytic leukemia cells, NB4. *J. Proteome Res.* **3,** 627–635.
20. Ruepp, S. U., Tonge, R. P., Shaw, J., Wallis, N., and Pognan, F. (2002) Genomics and proteomics analysis of acetaminophen toxicity in mouse liver. *Toxicol. Sci.* **65,** 135–150.

21. Kleno, T. G., Leonardsen, L. R., Kjeldal, H. O., Laursen, S. M., Jensen, O. N., and Baunsgaard, D. (2004) Mechanisms of hydrazine toxicity in rat liver investigated by proteomics and multivariate data analysis. *Proteomics* **4,** 868–880.
22. Swatton, J. E., Prabakaran, S., Karp, N. A., Lilley, K. S., and Bahn, S. (2004) Protein profiling of human postmortem brain using 2-dimensional fluorescence difference gel electrophoresis (2-D DIGE). *Mol. Psychiatry* **9,** 128–143.
23. Van den Bergh, G., Clerens, S., Vandesande, F., and Arckens, L. (2003) Reversed-phase high-performance liquid chromatography prefractionation prior to two-dimensional difference gel electrophoresis and mass spectrometry identifies new differentially expressed proteins between striate cortex of kitten and adult cat. *Electrophoresis* **24,** 1471–1481.
24. Yan, J. X., Devenish, A. T., Wait, R., Stone, T., Lewis, S., and Fowler, S. (2002) Fluorescence two-dimensional difference gel electrophoresis and mass spectrometry based proteomic analysis of *Escherichia coli*. *Proteomics* **2,** 1682–1698.
25. Damerval, C., de Vienne, D., Zivy, M., and Thiellement, H. (1986) Technical improvements in two-dimensional electrophoresis increase the level of genetic variation detected in wheat-seedling proteins. *Electrophoresis* **7,** 52–54.
26. Koller, A., Washburn, M. P., Lange, B. M., et al. (2002) Proteomic survey of metabolic pathways in rice. *Proc. Natl. Acad. Sci. U. S. A.* **99,** 11969–11974.
27. Thiellement, H., Zivy, M., Colas des Francs, C., Bahrman, N., and Granier, F. (1987) Two-dimensional gel electrophoresis of proteins as a tool in wheat genetics. *Biochimie* **69,** 781–787.
28. Tsugita, A., Kawakami, T., Uchiyama, Y., Kamo, M., Miyatake, N., and Nozu, Y. (1994) Separation and characterization of rice proteins. *Electrophoresis* **15,** 708–720.
29. Wilm, M., Shevchenko, A., Houthaeve, T., et al. (1996) Femtomole sequencing of proteins from polyacrylamide gels by nano-electrospray mass spectrometry. *Nature* **379,** 466–469.
30. Andon, N. L., Eckert, D., Yates, J. R. III, and Haynes, P. A. (2003) High-throughput functional affinity purification of mannose binding proteins from *Oryza sativa*. *Proteomics* **3,** 1270–1278.
31. Berggren, K., Chernokalskaya, E., Steinberg, T. H., et al. (2000) Background-free, high sensitivity staining of proteins in one- and two-dimensional sodium dodecyl sulfate-polyacrylamide gels using a luminescent ruthenium complex. *Electrophoresis* **21,** 2509–2521.
32. Mackintosh, J. A., Choi, H. Y., Bae, S. H., et al. (2003) A fluorescent natural product for ultra sensitive detection of proteins in one-dimensional and two-dimensional gel electrophoresis. *Proteomics* **3,** 2273–2288.

16

Quantitative Analysis of 2D Gels

Michel Zivy

Summary

The use of 2D electrophoresis for comparative proteomics allows the revelation of variations of protein relative amounts according to various physiological or genetic criteria. The statistical significance of these results is related to different factors, from the experimental design to the statistical tests. In this chapter we describe different parameters that should be taken into account in the experimental design and during image acquisition, and we present different programs for the normalization of quantitative data, the selection of reproducible spots according to user-defined criteria, and the selection of spots showing qualitative and significant quantitative variations.

Key Words: Proteomics; statistics; quantitative variations; two-dimensional gel electrophoresis.

1. Introduction

Comparative proteomics consists of the revelation of variations in protein amounts as a function of different physiological, developmental, or genetic criteria. Most studies are based on a comparison between Coomassie Blue– or silver-stained 2D electrophoresis gels. A large part of the expected protein variation is quantitative, and, although the human eye is very powerful in detecting spots on a gel, its assessment of quantitative changes is relatively crude. It is not able to take into account more than two or three intensity classes, and even simple experimental designs are impossible to handle without the help of dedicated software to reveal statistically significant variations among numerous gels. Several softwares have been developed for the detection, quantification, and matching of the revealed spots. The aim of this chapter is not to compare the different software products available, but to focus attention on different points related to image acquisition and experimental design that are important for obtaining good-quality data and to propose different procedures

From: *Methods in Molecular Biology, vol. 335: Plant Proteomics: Methods and Protocols*
Edited by: H. Thiellement, M. Zivy, C. Damerval, and V. Méchin © Humana Press Inc., Totowa, NJ

of first-step data analysis. Most 2D gel-dedicated softwares propose statistical tools, but they are in general more limited than general statistic packages, and they are not always explained in detail. In fact, once spots have been correctly quantified and matched, they can be analyzed like any other quantitative variable, and there is no reason to do without the multiple possibilities offered by statistics packages like SAS or R.

2. Materials

Examples of programs for statistical analysis of spot variations are given in SAS language (SAS Institute, Asheville, NC). SAS is a package available on different operating systems. Although the language is different, all the procedures shown can probably be programmed with the free open-source software R (http://www.r-project.org/). Spot data were extracted from csv files produced by exporting data from Progenesis (Nonlinear Dynamics, UK), but only small modifications should adapt them to files extracted from other packages like ImageMaster 2D Platinum (Amersham Biosciences) or PDQuest (Bio-Rad).

The free open-source software ImageJ (http://rsb.info.nih.gov/ij/) was used for examination of image dynamics.

Cluster and Treeview are programs allowing hierarchic classification that can be downloaded at http://rana.lbl.gov/EisenSoftware.htm.

3. Methods

3.1. Experimental Design

3.1.1. Equilibrated Design

Part of the quantitative variation of spots between gels is not controlled. The biological variation between replicated samples or any step from protein sample preparation to 2D gel staining can be the cause of uncontrolled variations. It has been shown that the batch effect of 2D electrophoresis (i.e., variations between different series of simultaneously run and stained 2D gels) is large. It is thus important to take this effect into account in the design of experiments. Batches must group the gels in a manner distinct from the analyzed factor. For example, to compare 12 treatments (three replicates per treatment) in a tank in which 12 gels can be run simultaneously, three 2D series should be run successively, and each treatment should be represented by one gel in each series. Thus, in case of uncontrolled variations affecting one of the three series (for example a silver staining a little darker than in the others), they will affect all treatments the same way. On the contrary, if the three replicates of four treatments are present in this tank, there will be no means to decipher between a batch effect and a real biological effect of these four treatments. In real life it is not always possible to build completely equilibrated designs, because of failed 2D gels that must be run a second time. Nevertheless, the supplementary batches should be as equilibrated as possible.

It can be noticed that the magnitude of the batch effect can be visualized *a posteriori* by running a principal component analysis (with the gels as observations and the spots as variables).

3.1.2. Technical and Biological Replicates

Replicates are essential for the detection of quantitative variations. They allow the measurement of uncontrolled variation (the residual variation), which is essential for evaluation of the effect of the studied factor.

When biological replicates are used (samples from different plants), the variation between replicates is caused by the biological and the technical variation. Of course when technical replicates are used (e.g., different gels from the same protein sample), only technical variations are taken into account. Thus technical replicates necessarily induce a lower residual variation than biological replicates, which leads to higher levels of significance in statistical tests. However, the conclusions drawn from the experiment can be applied only to the studied samples, because the source of the variation is not well determined: it can be caused by individual plant variations (e.g., environmental and developmental variations), or by technical variations during the preparation of the samples, as well as by the tested factor. In contrast, when biological replicates are used, the studied factor is the only cause of a possible significant result, because it is the only one that is different between the compared groups and identical within the compared groups. Thus biological replicates must be run, at least when the studied treatment is a qualitative variable (e.g., drought vs control).

When the studied factor is a continuous variable (e.g., several doses of the same molecule, or different time lengths of submission to the same treatment), it is not absolutely necessary to use replicates for all tested values (doses or time points): the statistical test (e.g., a linear regression) will compute the residual variation according to the difference between the observed values and the predicted values. However, replicates can be omitted only if (1) there are a sufficient number of data points (a regression to 3 points is not very informative) and if (2) one accepts *a priori* that the relationship between the protein amount and dose or time is linear (linear regression), or if the shape of the expected response is already known (nonlinear regression).

In some instances it is possible to use the homogeneity between contiguous data (e.g., between 3 time points) to interpret the results of an experiment without biological replicates: the coherence or continuity of the response at successive time points supports the hypothesis that the variation is related to the treatment. In contrast, a value that breaks the continuity of the response will be considered as possibly caused by individual variations. Consecutive values are thus virtually used as biological replicates. However, it is a fact that doing this leads to a loss of resolution in the analysis of the studied treatment.

The best way to take biological variations into account while minimizing individual plant variations is to mix several plants in each biological replicate.

3.1.3. Reference Gel

Even if it is not used in the quantitative analysis, a reference gel is in general useful. In most 2D packages, spot matching is based on the building of a reference virtual gel that contains all the matched spots. Building such a reference gel is a limiting step, and it is preferable to begin with a real gel that already contains almost all of them. It can be obtained by running a coelectrophoresis in which the different treatments are represented by equal amounts of proteins.

3.2. Image Acquisition

3.2.1. Digitalization

Image digitalization is the first step of the quantitative analysis. Gels can be scanned with a laser densitometer, a flat bed scanner (e.g., Pharmacia), or a CCD camera (e.g., ProXpress).

Whatever the system, transmission values must be acquired, i.e., the detector receives light emitted from the other side of the gel. The transmission value is the ratio between the intensity of the signal received by the detector in the presence of the gel and the intensity received in the absence of the gel (I/I_0). Of course, when a flat bed scanner is used, none of the functions of contrast enhancement (e.g., gamma correction) should be activated, because they would distort the real transmission value. Transmission values (range from 0 to 1) are in general encoded with 16 bits, i.e., the I/I_0 ratio is translated to a value ranging from 0 to 65,735. Thus the produced image is a matrix of values between 0 and 65,735 (the pixels). The gray-scale TIF format is generally used. Lossy compression (e.g., JPEG format) must not be used.

3.2.2. Image Resolution

The higher the resolution (pixels per length unit), the better small spots will be detected and quantified. Resolution is also a limiting factor for detection in groups of overlapping spots.

Most 2D software packages do not detect multiple spots in a group if there is no "valley" between the peaks of intensity. Thus, the accuracy of spot detection depends on the ability to detect these "valleys," which in turn depends on the number of pixels allowing the representation of intensity variations between spots. For gels of about 24 × 20 cm, the generally used resolution is 100 µm per pixel, which is not far from 300 ppi (84.7 µm per pixel). This value is actually a compromise, because two factors limit image resolution: (1) speed of image acquisition: the time spent for scanning one gel can become limiting when numerous gels must be scanned the same day and (2) image file size: files of 10 to 14 Mbytes are produced for gels of about 24 × 20 cm scanned with a resolution of 100 µm/pixel and digitalized on 16 bits. The larger the image file, the longer the time needed by the 2D software packages for detecting, quantifying, and matching spots (*see* **Note 1**).

3.2.3. Image Dynamics

During image acquisition, the 65,736 gray levels should be used as fully as possible. Depending on the type of scanner, it is possible to adjust exposure time or the aperture of the objective or to use filters. The image of the gel should not contain white surfaces (100% transmission): if the background is not detected, it is likely that small spots just above the background level will not be detected either.

The image should not contain black regions (0% transmission) because all spots actually darker than this threshold will be coded by the same value.

The freeware ImageJ can be used to look for these possible regions of saturated values. It also allows the inspection of image dynamics, i.e., the difference between the minimum and maximum values in the image. Image dynamics should be maximized: of course the precision of quantification depends on the number of gray levels used during image acquisition. It is a pity to use only a few hundreds of gray levels when more than 65,000 are available.

When a camera is used, optics can generate vignetting: peripheric regions of the image are darker than the central region. Vignetting can be taken into account by the system of image acquisition itself (e.g., ProXpress, Perkin). Otherwise it can be eliminated by background subtraction.

3.2.4. Conversion of Transmission Data into Optical Density

Transmission data must be converted into optical density (of course this conversion must not be done for fluorescent stains). In most 2D packages this is not required. However, protein concentration is linearly related to optical density, not to transmission. The relation between OD and transmission is as follows: $OD = -\log(I/I_0)$.

As this relation is not linear, a given increase of transmission can correspond to different OD increases, depending on the original value of transmission. Only the translation to optical density makes linear the relation between spot volumes and protein amounts. The result is that OD is additive: if the amount of a given protein is respectively A and B in two different samples with B = A + X, then $OD_B = OD_{(A+X)} = OD_A + OD_x$, which is not true for transmission. The conversion must be done before background subtraction (**Fig. 1**).

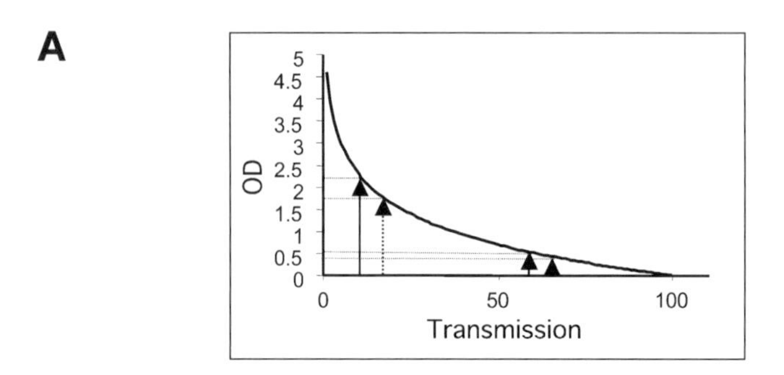

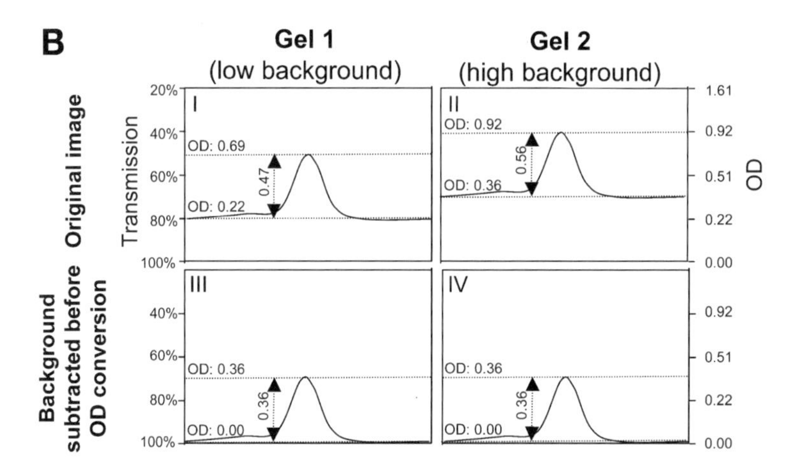

Fig. 1. Conversion of image data from transmission to optical density. **(A)** Because of the logarithmic relation between transmission percentage and optical density, a given difference of transmission (black and dotted arrows) corresponds to a higher variation of OD for low values of transmission than for high values of transmission. **(B)** Simulated 1D profile of a spot. According to transmission data, the height of the spot (maximum value minus background) is the same (30%) in images 1 and 2. As the background level is different in the two images, 30% transmission corresponds, respectively, to 0.47 and 0.56 OD units in images 1 and 2 (I, II). Thus protein content is actually different in the two gels, and the actual values were found by subtracting the background after conversion to OD. If backgrounds were subtracted before conversion to OD, i.e., if subtraction was performed in transmission units (III, IV), the same and false value of 0.36 OD units would be found in both gels.

In general the conversion to OD is done by scanning a Kodak strip. 2D software packages propose a tool for recording transmission data corresponding to known values of OD and for computing the adjusted curve. Note that the conversion must take into account the logarithmic nature of the relation between OD and transmission: a linear regression is of no use.

3.3. Spot Volume Normalization

As already discussed in **Subheading 3.1.1.**, a large part of the variation of spots is related to a gel effect: uncontrolled variations in the protein load, possible protein precipitation during 2D electrophoresis, and variations in the process of staining are responsible for a variation of gel global intensity. This variation more or less similarly affects all the spots of a given gel. The aim of normalization is to correct this general variation. It must be done after conversion to OD and background subtraction (*see* **Note 2**).

3.3.1. Definition of a Region for Spot Normalization

The region of interest is a surface determined by the user, in which spot detection will take place. In general, the same region is used for spot normalization. In this case it is very important to make sure that the same region is defined in all gels, because a widely used method of normalization is based on the total volume of all spots included in this region. Because of gel-to-gel variations (broken gels, regions of poor definition for various reasons), it is not always possible to define the same region for spot detection in all gels. It is then preferable to define another region for spot normalization. Actually, even if the same region of interest was delimited for spot detection, it can be useful to define a smaller region for the computation of normalization, to discard the most variable regions of the gels. The normalization of spot volumes according to the total volume of spots contained by a user-defined region can be done simply by programming outside of 2D packages. Data that should be exported from the 2D package are x, y, nonnormalized volume, and optionally the match number (i.e., the reference number) of every spot detected in each gel. These data are in general easily exported from 2D packages (e.g., by exporting data from the measurements window of Progenesis).

Let "firstgel.csv" be a text file in which the first line contains column titles and the following lines contain the spot number, the match number, x, y, and the nonnormalized volume for all spots detected in the first gel. **Figure 2** shows a program written in SAS language that normalizes spot volumes according to the sum of all spot volumes in a user-defined area and produces a single permanent table in which each spot is a variable (a column) and each gel is a line (an observation). Although not presented in **Fig. 2**, this method can easily be combined with another criterion. For example, the computation of the sum of spot volumes can be restricted to the spots present in all gels (*see* **Subheading 3.5.** for their selection). The computation can also be restricted to a specific list of spots. However, the number of spots that will finally be selected for normalization should not be too small: the smaller the number is, the more instable normalization will be.

```
/* normalization according to the sum of spot volumes in a user-defined rectangle */
libname dir2d '~zivy/demo2d'; /* (see Note 7) */
/* loading data into the table named "firstgel" (see Note 8) */
data firstgel;
infile 'firstgel.csv' firstobs=2;
/* (see Note 9) */
input spot match$ x y vol;
/* (see Note 10) */
/* Step 1: selecting spots according to their position */
data selspot; set firstgel;
xmin=500; xmax=1700;      /* change user-defined coordinates here */
ymin=300; ymax=1500;
/* selection of all spots in the region delimited by xmin, xmax, ymin and ymax: */
if x>xmin and x<xmax and y>ymin and y<ymax; /* (see Note 11) */

/* Step 2: compute the sum of the volumes of selected spots.
result in volsum in table selsum: */
proc means noprint data=selspot;
var vol;
output out=selsum sum=volsum;
/* Drop useless variables */
data selsum; set selsum;
a=1;
drop _type_ _freq_;
title 'sum of selected spot volumes ';
proc print data=selsum;
run;
data firstgel; set firstgel;
a=1;

/* Step 3: normalize. (see Notes 12 and 13)
Notice that all firstgel's matched spots are kept and normalized, while the scaling factor was computed only on the basis of a subset of its spots */
data firstgel;
merge firstgel selsum;
by a;
constant=10000;
vol=(vol/volsum)*constant;
/* unmatched spots were used for normalization but they are useless for gel to gel comparisons: */
if match="." then delete;
```

Fig. 2. Program for normalization according to the sum of spot volumes in a user-defined rectangle.

The programs shown in **Fig. 2** and the following figures might look complicated, because they contain many comments. They are actually relatively short and simple. As different methods can be used for normalization as well as for the analysis of qualitative and quantitative variations, it is worthwhile to program them with generalist statistical packages, which provide more flexibility and development means than the necessarily limited statistical tools provided by 2D packages.

```
/* step 4 : transpose the table, to put the volumes of all spots of the gel in one line */
data firstgel ;
set firstgel ;
drop spot x y constant xmin xmax ymin ymax volsum a;
/* (see Note 14) */
if length(match)=1 then match='s000'||match;
if length(match)=2 then match='s00'||match;
if length(match)=3 then match='s0'||match;
if length(match)=4 then match='s'||match;
proc transpose out=gel1;
id match;
data gel1;
set gel1;
gelname='firstgelname'; /* replace with the real gel name */
proc print data=gel1;

/* End of the program for one gel.
The same program will be run for all gels of the experiment, with adequate coordinates for delimiting selected
spots and gel names, producing tables gel1, gel2...geln. These tables should be concatenated, to produce a table
containing one line per gel: */
data gels;
set gel1 gel2 gel3 geln;
/* additional information relative to the gels and the samples can now be merged with this single table: let
samples be a table containing the variables gelname, batch, genotype, treatment */
proc sort data=samples; by gelname;
proc sort data=gels; by gelname;
data dir2d.alldata;
merge gels sample;
by gelname;
```

Fig. 2. (*continued*)

3.3.2. Other Methods of Spot Normalization

Another method of spot normalization is based on the ratio of spot volumes in a reference gel to the volumes of the same spots in the studied gel: vol_{ref}/vol_{gel} is computed for all spots that were matched in the gel and in the gel used as a reference. Normalization consists of multiplying the volume of each spot of the gel by the average (or the median value) of these ratios. As no total spot volume is computed, the accuracy of the method does not depend on a precise delimitation of the region of normalization. Thus the classical region of interest can be used, even if it is not very reproducibly defined. As it is based on spots matched in two gels (i.e., in the gel to be normalized and in the reference gel), the number of spots taken into account for normalization is higher than when it is based on the spots that are present in all the gels, because the latter decreases markedly when the number of gels in the experiment increases.

This method is also theoretically better than normalization relative to all spots in a region, because it is not biased by spots that are specific to a treatment. It can be improved by computing only the ratio of spots whose volume is within a given range. In fact, very faint spots should not be taken into account

```
/* Program for spot normalization according to the match ratio */

/* Each line of the file "quantif.dat" contains the matchname and the volume values of the spot in the different
gels of the experiment (there are 24 gels in this example). There are as many lines as spots.
Volume values will be given to variables x1 to x24 in table temp. */
data temp;
a=1;
infile 'quantif.dat' firstobs=2; /* change firstobs according to the number of lines to skip at the beginning of the
file */
input spot$ x1-x24;

/* Let us compute the ratio for all selected spots. Only not too small and not too large spots are selected. */
data select;
set temp;
large=300; /* change large and small according to your data */
small=50;
ref=x1; /* change ref according to the chosen reference gel */
array sp{*} x1--x24;
do i=1 to dim(sp);
  if sp{i}>small and sp{i}<large then sp{i}=ref/sp{i};
    else sp{i}=.;
end;
```

Fig. 3. Program for spot normalization according to the match ratio.

because small volume variations can induce large variations of the ratio, and very large spots should also be excluded because of the lack of linearity near saturated values of the stain.

As unmatched spots are useless in this method, only one table containing the match number and the raw volumes of all matched spots in all gels should be exported in one shot. In the program of **Fig. 3**, data are extracted from a csv file exported from the "comparison window" of Progenesis. The first column contains the match names, and the following columns contain the nonnormalized volume values of the spots in the different gels. Note that this way of exporting quantitative data is also the most convenient one if one is satisfied with the method of normalization proposed by the 2D software. The same exported file can also be used for the selection of reproducible spots and of qualitatively variable spots (*see* **Subheading 3.5.**).

Another method of spot normalization has been developed by Burstin et al. *(1)*. It is based on a principal component analysis and can be used when the variation owing to the studied parameter is small relatively to the residual variation, or if the variation involves a small number of spots. It will not be developed here.

3.4. Linearity of the Relation between Relative Intensity and Relative Amount

It is interesting to analyze the relation between protein amounts and measured spot volumes. A way of doing this is to compare gels loaded with a range of protein amounts of the same sample. However, this does not give a correct

```
/* Let us have a look to the distribution of the ratio in each gel: */
proc gchart;
vbar x2/midpoints=(0.1 to 10 by 0.2);
vbar x3/midpoints=(0.1 to 10 by 0.2);
vbar x4/midpoints=(0.1 to 10 by 0.2);
/*...*/

proc means  data=select;
var x1-x24;
output out=mmm median=d1-d24;

proc print data=mmm;
run;

data mmm;
set mmm;
a=1;

/* Normalization */
data tot;
merge temp mmm;
by a;
array sp{*} x1—x24;
array fact{*} d1—d24;
do i=1 to dim(sp);
  sp{i}=sp{i}*fact{i};
end;
drop a i mq _type_ _freq_ d1--d21;
run;

/* spotnames are changed as in Figure 2 */
data tot; set tot;
if length(spot)=1 then spot='s000'||spot;
if length(spot)=2 then spot='s00'||spot;
if length(spot)=3 then spot='s0'||spot;
if length(spot)=4 then spot='s'||spot;

/* the table is transposed to get a table with spots as columns and gels as lines */
proc transpose out=tt;
id spot;

/* table tt is now ready to be merged with a table containing the gels and information related to the samples (gels
in the 2 tables must be in the same order): */
data dir2d.alldata;
merge gels tt;
```

Fig. 3. (*continued*)

estimation of the quantification in real comparisons, in which spot volumes are normalized. In effect, one cannot normalize spot volumes from gels loaded with different amounts of the same sample, since normalization will precisely eliminate the global gel effect caused by differences in protein loading.

A better method is to use two very different samples that contain specific spots (e.g., a sample of interest and another from a different species or organ) and to prepare different mixes with different proportions, e.g., from 1:9 to 9:1, but with a constant total protein amount. 2D gels obtained from these different mixes and from the two pure samples can be normalized as usual, and the regression of the spots specific to the sample of interest can be computed as a function of the known ratio of the sample in the mix. This method, used in David et al. *(2)*, allows the study of the linearity of spot response to protein concentration in the same conditions as in normal experiments. If the response is linear, this also allows the determination of the minimum significant difference in protein concentration.

3.5. Qualitative Variations

Qualitative variations, i.e., the presence/absence variation of spots, are a priori more easily identified than quantitative variations. However, it can be relatively tricky, at least when dealing with large experiments, because one has to tolerate a certain amount of missing data.

A reproducible spot should not be present in all gels because by definition it would make impossible the detection of reproducible qualitatively varying spots. Thus it is preferable to deal with the idea of consistency, which takes into account the fact that a spot can be consistently present or absent. The most severe criterion of consistency is to consider that a spot must be present in all the replicates of a given group (treatment, genotype, and so on) to be declared "present" and absent in all replicates of another group to be declared "absent." However, when the experiment contains numerous gels, this criterion is too severe, because of possible accidents in the experiment (e.g., gels less stained then others, broken gels, and so on).

Figure 4 shows a program that takes into account various user-defined criteria to determine consistent spots and qualitatively variable spots.

3.6. Quantitative Variations

The goals of quantitative proteomic analyses can be very different, from global analyses, in which the interest is in the identification of the principal causes of protein variations, to the identification of the few proteins that respond to a particular treatment.

Quantitative variations can be used to analyze relationships between proteins, e.g., to determine groups of coregulated proteins. In general, a hierarchic classification is used in this case, which allows the visualization of clusters of proteins according to their accumulation in different tested conditions. This can be done by using the "cluster" program.

Principal component analysis (PCA), with the spots as variables and the samples as observations, allows visualization of the distribution of the different samples according to variables that represent the principal variation of protein spots (*see* Chapter 17 in this book). PCA can also be used to detect automatically abnormal gels that will stay outside the scatter of points representing the other gels (*see* **Note 3**).

```
/* determination of reproducible spots and identification of qualitative variations */

/* read the spot names and their value in the 24 gels, put these values in variables x1 to x24 in table temp */
data temp;
infile 'quantif.dat' firstobs=3;
input spot$ x1-x24;

/* transpose the matrix of data like in step 4 of figure 2 */
data temp; set temp;
if length(spot)=1 then spot='s000'||spot;
if length(spot)=2 then spot='s00'||spot;
if length(spot)=3 then spot='s0'||spot;
if length(spot)=4 then spot='s'||spot;
proc transpose out=tt;
id spot;

/* Merge the table of spot volumes "tt" with a table containing the characteristics of the samples ("gels"). In this
example, gels contains two variables: gelname and geno (for genotype). Gels in the table "gels" must be in the
same order than in the table "tt". */
data total;
merge gels tt;
drop _name_;

/* counting the frequency of a spot in each genotype */
proc sort data=total;
by geno;
proc means noprint data=total;
by geno;
output out=pa n=;
*proc print data=mmm;
/* Now the table mmm contains as many lines as there is different genotypes and the value at the ncol column
and the ngeno line is the number of times the ncolth spot was present in the ngenoth genotype. A reproducible
spot should not necessarily be present in all gels because this would preclude the identification of spots showing
qualitative variation.
Thus it is preferable to deal with the idea of consistency, which takes into account the fact that a spot can be
consistently present or absent in the different groups or genotypes. The most severe criterion of consistency is
that a spot must be present in all the replicates of a genotype to be considered as present, and absent in all
replicates to be considered as absent. However when the experiment contains numerous gels this criterion is too
severe, because of all possible accidents in the experiment (e.g. a gel less stained then others, a problem of
detection on another one, etc...).
*/
```

Fig. 4. Program for the determination of reproducible spots and identification of qualitative variations.

Another approach toward quantitative variations is to look for proteins that show a significant variation as a function of controlled factors of the experiment (e.g., treatment, genotype) or other factors that are measured during the experiment (e.g., hormone dose). Significant spots are not particularly expected to be detected in global analyses such as PCA, because they are not necessarily numerous and their variation can be highly specific relative to the variation of most spots. In general, analysis of variance is the method of choice for detecting spots that are significantly variable according to one or several factors. When more than two treatments have to be compared, it is preferable to use

```
/* selecting consistent spots */
data reprod;
set pa;
okmin=0; /* replace with your criterion for absence: to be considered as reproducibly absent a spot must be
detected at most okmin times in the same genotype */
okmax=0; /* replace with your criterion for presence: to be considered as reproducibly present, a spot must be
missed at most okmax times in the same genotype */
array sp{*} s0001--s3254; /* replace with the names of the first and the last spot */
/* a consistent spot is either consistently present (value=1) or consistently absent (value=0). */
do i=1 to dim(sp);
  if sp{i}<=okmin then sp{i}=0;
    else if sp{i}>=_freq_-okmax then sp{i}=1;
      else sp{i}=.;
end;
*proc print;

/* compute the number of genotypes where the spots where considered as consistent */
proc means noprint data=reprod;
var s0001--s3254; /* replace with the names of the first and the last spot */
output out=mmm2 n=;
/* compute the number of genotypes where the spots where considered as present */
proc means noprint data=reprod;
var s0001--s3254; /* replace with the names of the first and the last spot */
output out=mmm3 sum=;

/* let us transpose both results and merge them */
proc transpose data=mmm2 out=mt2;
proc transpose data=mmm3 out=mt3;

data mt4;
merge mt2 (rename= (col1=nbcons)) mt3 (rename=(col1=nbpres));

/* A spot should be considered as globally reproducible when it is consistently present or absent in all genotypes.
However, when there are numerous genotypes, it would be too bad to consider as non reproducible a spot that is
consistent in almost all genotypes except one or two. */

data cons nocons;
okcons=6; /* replace with the number of genotypes in which a spot must be consistent to be declared as
reproducible */
set mt4;
if nbcons>=okcons and nbpres>0 then output cons;
          else output nocons;

proc print data=cons;
title "consistent spots";
run;
```

Fig. 4. (*continued*)

analysis of variance rather than running Student tests directly, because it allows better computation of the residual variation (*see* **Note 4**).

After analysis of variance, different methods of comparisons of means can be used according to the biological problem. For instance, a Dunnet test can be used when several treatments must be compared with the same control, and a Duncan or a Student-Newman-Keuls test can be run to compare all treatments

```
/* let us create a permanent table (rep.reprod) that contains only consistent spots */
data nocons;
set nocons;
if _name_='_TYPE_' then delete;
data _NULL_;
set nocons;
file 'noreprod.out';
put _name_;
data rep.reprod;
set totprov;
drop
%include 'noreprod.out';
;

*proc print data=rep.reprod;
title "reduct";
run;

/* the spots showing qualitative variation are those for which the frequency of presence is smaller than the
frequency of consistent: */
data quali;
set cons;
if nbpres<nbcons;

title « qualitative variations »;
proc print data=quali;
run;

/* Let us have a look on bar graphs of qualitatively variable spots
Of course the list of qualitatively variable spots can also be exported and used in Excel for making other graphs
or in a 2D software for display */
data _null_;
set quali;
file 'quali.out';
put "%look(" _name_ ");";
run;

%macro look(spot);
proc gchart data=rep.reprod;
hbar gt / type=mean sumvar=&spot freqlabel='Effectif'
      meanlabel='Volume moyen' clm=50 noframe
      midpoints='P1/SS' 'P1/WW' 'P2/SS' 'P2/WW' '151/SS' '151/WW' '245/SS' '245/WW' 'mix/ss'; /* replace
with the values of genotype */
%mend look;

%include 'quali.out';

run;
```

Fig. 4. (*continued*)

with each other (*see* **Note 5**). Linear regression can be used to compute the relationship between spots and a continuous variable (e.g., a dose of hormone). **Figure 5** shows a SAS program for the selection of all spots that show a significant variation in an analysis of variance with two factors and their interaction (*see* **Note 6**).

```
/* analysis of variance */

/* take the table of reproducible spots, that was created by the program of figure 4. */
data bid;
set rep.reprod;

/* analysis of variance. In this example there are 2 factors (genotype and treatment), that are discrete variable.
They are declared in the "class" command. The interaction between the 2 factors is also declared in the model.
Statistical results are stored in the table ggg. The analysis is carried out on all spots contained in the table "bid".
*/
proc glm noprint outstat=ggg;
class geno treat;
model s0002--s3254=geno treat geno*treat/ss3; /* s0002 and s3254 are the first and last spots in rep.reprod */

/* let us arrange the results to get a single table containing the level of significance of the various factors */
data ggeno;
set ggg;
if _source_='geno';
proc sort; by _name_;

data gtreat;
set ggg;
if _source_='treat';
proc sort; by _name_;

data ginter;
set ggg;
if _source_='geno*treat';
proc sort; by _name_;
```

Fig. 5. Program for analysis of variance.

In general, 0.05 or 0.01 significance levels are used in statistical tests. This means that a variation can be considered significant when there is a 5 or 1% probability that observed data are actually random. In other words, the significance level is a probability of false-positive detection. Thus, if a 0.01 significance level is used for 1000 spots, it is certain that about 10 of them will be false positives. One way of taking this into account is to divide the significance level by the number of comparisons (Bonferroni correction). In the present case, this would lead to a significance level of 10^{-5}. This way, the overall probability of detecting a false positive among the 1000 spots is kept at 0.01. This is a conservative method, but it lowers the sensitivity, because variations must be very large to be significant at the 10^{-5} level. It is likely that many true positive spots are lost by using this method.

Benjamini and and Hochberg *(3)* proposed the false discovery rate (FDR) method to take into account multiple comparisons. The principle of this method is to accept that few percent (e.g., 5 or 1%) of the detected variations are false positives. Whereas the Bonferroni correction consists in keeping at 1% the risk of discovering one false positive among the 1000 spots, the FDR methods

```
/* in the permanent table resanov, variables cgeno, ctreat and cinter will contain "***", " **", "  *", "   ", or "  ."
according to the level of significance of respectively the genotype, the treatment and the interaction. */
data rep.resanov;
merge ggeno (rename=(prob=pgeno) keep=_name_ prob)
    gtreat (rename=(prob=ptreat) keep=_name_ prob)
    ginter (rename=(prob=pinter) keep=_name_ prob);
by _name_;
array proba{3} pgeno ptreat pinter;
array cproba{3} $ cgeno ctreat cinter;
do i=1 to 3;
  if proba[i}=. then cproba{i}='  .';
    else if proba{i}<0.001 then cproba{i}='***';
      else if proba{i}<0.01 then cproba{i}=' **';
        else if proba{i}<0.05 then cproba{i}='  *';
          else cproba{i}='   ';
end;
proc print;
var _name_ cgeno ctreat cinter;

/* let us select significant spots and draw bar graphs */
data signif;
set rep.lanov;
if (substr(cgeno,2,2)='**'
   or substr(ctreat,2,2)='**' or substr(cinter,2,2)='**');
proc print;
var _name_ cgeno ctreat cinter;

data _null_;
set signif;
file 'signif.out';
put "%look(" _name_ ");";

data bid;
set rep.reprod;
gt=trim(geno)||'/'||treat;

%macro look(spot);
proc gchart;
hbar gt / type=mean sumvar=&spot freqlabel='Effectif'
       meanlabel='Volume moyen' clm=50 noframe
       midpoints='P1/SS' 'P1/WW' 'P2/SS' 'P2/WW' '151/SS' '151/WW'
              '245/SS' '245/WW' 'mix/ss'; /* replace with the values of genotype/treatment combinations */

%mend look;
%include 'signif.out';
```

Fig. 5. (*continued*)

accepts that 1% of the detected positives are actually wrong. This method is less conservative but is more sensitive than the Bonferroni correction. It is an intermediate solution between the total absence of correction (all spots tested at 1%) and an "excessive" degree of prudence (Bonferroni correction). **Figure 6** shows a SAS program for the selection of significant spots according to the FDR method.

```
/* FDR method */
/* Assume that an analysis of variance was run as in figure 5.
We will use the FDR method for the genotype effect. We just need the name of the spots and the p-values
associated to the genotype effect. */
data pvalues;
set rep.resanov;
keep _name_ pgeno;
/* get rid of missing data */
if pgeno=. then delete;
a=1;

/* number of spots */
proc means noprint;
output out=mmm;
data mmm;set mmm;
if _stat_='N';
a=1;
keep _freq_ a;
data pvalues;
merge pvalues mmm; by a;

proc sort data=pvalues;
by pvalue;

data prov;
set pvalues;
alpha=0.05; /* change the threshold here */
j=_n_;
b=j*alpha/_freq_;
c=pvalue-b;
if c<0;
run;

proc sort; by descending j ;
title 'Benjamini threshold p-value';
data prov;
set prov;
if _n_=1;
a=1;

proc print;
var pvalue;
run;

title 'Significant spots';
data pvalues;
merge pvalues prov(rename=(pvalue=threshold)); by a;
if pvalue<=threshold;
proc print;
var spot pvalue;
```

Fig. 6. Program for the selection of significant spots with the FDR method.

4. Notes

1. When minigels are used, the resolution must be increased because spots are in general smaller than in large gels.
2. During the interactive step of spot editing, it is in general possible to delete spots. It should be noticed that some normalization methods are based on the total volume of spots. Thus one should avoid deleting too many spots and be sure to treat all gels the same way.
3. Clustering and PCA are methods that do not accept missing data. Missing data must be replaced; otherwise the entire observation that contains one missing data will be deleted from the dataset. One way of replacing missing data is to replace them with zeroes. However, this is not necessarily the most conservative approach. For example, if there are many zeroes (i.e., many qualitative variations), they will represent a large part of the variation and the risk is that these qualitative variations eventually mask the quantitative variation of other spots. Other statistical tools, like correspondence analysis (e.g., in **ref. *4***), should be used to analyze qualitative variations globally. In PCA it can be more prudent to replace missing data with the mean value of the spot. This operation can be done very easily in SAS programs by using the "standard" procedure.
4. It must be pointed out that analyses of variance will not take into account the treatments when a spot is absent. For example, if a spot is absent in treatment A and present in treatments B and C, the analysis of variance will not declare that this spot is significantly variable if it is not variable between B and C. The qualitative variation must be detected independently with a qualitative analysis. It is not a good idea to replace missing values with zeroes, because it will alter the computation of the residual variation and make the comparison unreliable.
5. It is tempting to select the spots according to a rate of increase or decrease relatively to a control. This is not a good idea for several reasons: (1) some of the spots selected this way will not be actually statistically significant, because this method of selection does not take into account the magnitude of the residual variation; it is also likely that some spots that are actually significant will not be selected for the same reason; (2) the choice of the control is sometimes arbitrary, and some spots will be selected or not depending on the chosen control; and (3) this method is biased in favor of decreasing spots, because the difference between the control and a spot decreased by a factor of 0.5 is two times smaller than the difference between the control and a spot increased by a factor of 2. However, once the statistically significant spots have been selected according to appropriate tests, one can select among them the spots that show the maximum of variation.
6. There is often a relationship between the average of the spot and its variance: the standard deviation is smaller when the spot is small than when it is large. This is not suitable for analysis of variance. This problem can be overcome by using the log of the spot's volume instead of the volume itself (e.g., **ref. *1***).

For the following notes, *see* **Fig. 2**.

7. Defines as "dir2d" the directory in which permanent SAS tables will be stored.

8. "/*" and "*/" mark the beginning and the end of a comment, respectively. Thus it is possible to copy the entire code, including comments.
9. A dollar sign must follow alphanumeric variables. Match is a number, but in this case it is easier to declare it as a string, because it will be used as a variable name later in the program.
10. Missing data, i.e., undetected spots, must be represented as a period in the csv file. They must not be represented as zeroes. Zeroes in place of periods would make the computations of quantitative and qualitative variations wrong.
11. If <condition>: keep only the observations in which the condition is true.
12. The table firstgel contains one line per spot, whereas the table sumfstgel contains only one line. We need to merge the two tables, because the spot's volumes are in firstgel and sumgel is in sumfstgl. By merging by the variable "a", which was set to the same value in all lines of table firstgel and in the single line of sumfstgel, the single line of sumfstgl is duplicated for every line of firstgel.
13. The user can change the value of the constant. This has no effect on statistical tests. (Of course the same constant must be used for all gels.)
14. Match numbers are converted into strings beginning with an alphabetic character in a way that still allows numeric sorting.

References

1. Burstin J., Zivy, M., de Vienne, D., and Damerval, C. (1993) Analysis of scaling methods to minimize the experimental variations in two-dimensional electrophoresis quantitative data. Application to the comparison of maize inbred lines. *Electrophoresis* **14,** 1067–1073.
2. David J. L., Zivy, M., Cardin, M. L., and Brabant, P. (1997) Protein evolution in dynamically managed populations of wheat: adaptative responses to macro-environmental conditions. *Theor. Appl. Genet.* **95,** 932–941.
3. Benjamini, Y. and Hochberg, Y. (1995) Controlling the false discovery rate: a practical and powerful approach to multiple testing. *J. R. Statist. Soc.* B, **57,** 289–300.
4. Marquès, K., Sarazin, B., Chané-Favre, L., Zivy, M., and Thiellement, H. (2001) Comparative proteomics to establish genetic relationships in the Brassicaceae family. *Proteomics* **1,** 1457–1462.

17

Multivariate Data Analysis of Proteome Data

Kåre Engkilde, Susanne Jacobsen, and Ib Søndergaard

Summary

We present the background for multivariate data analysis on proteomics data with a hands-on section on how to transfer data between different software packages. The techniques can also be used for other biological and biochemical problems in which structures have to be found in a large amount of data. Digitalization of the 2D gels, analysis using image processing software, transfer of data, multivariate data analysis, interpretation of the results, and finally we return to biology.

Key Words: Multivariate data analysis; proteomics; 2D-electrophoresis; 2D-gels; PCA; PLSR; image processing.

1. Introduction

It is obvious that a technique like 2D electrophoresis, which gives a wealth of information in each experiment, will not lend itself to ordinary statistics. If we were to apply ordinary statistical methods, we would face the problem of not having enough degrees of freedom in the normal situation, in which we have a few 2D gels with a large number of spots. Ordinary statistics would only apply when we wanted to look at the up- and downregulation of a few proteins and when we had a sufficient amount of samples. In the normal situation (limited number of 2D gels and a large number of spots) in which you want to make an intelligent selection of interesting spots to focus on, we prefer the multivariate approach ***(1–4)***. Thus we propose a new workflow involving a multivariate approach that is hypothesis generating instead of hypothesis driven. This will give us the freedom to explore the data without prejudice and then use available biochemical knowledge to finally set up relevant hypotheses.

Hypothesis-generating analysis is a natural consequence of the entire concept behind multivariate analysis. In traditional statistical terms, a hypothesis is set up first and then experiments are carried out to approve or disprove the

From: *Methods in Molecular Biology, vol. 335: Plant Proteomics: Methods and Protocols*
Edited by: H. Thiellement, M. Zivy, C. Damerval, and V. Méchin © Humana Press Inc., Totowa, NJ

hypothesis. This is known as deductive analysis. In contrast to traditional statistical methods, multivariate analysis is an inductive analysis, whereby hypotheses can be set up after one has carried out the computational experiments.

Multivariate analysis builds on the application of statistical and mathematical methods and includes the analysis of data with many observed variables, as well as the study of systems with many important types of variation *(5)*.

The methods we introduce here are principal component analysis (PCA) and partial least squares regression (PLSR). PCA is used to get an overview of the data and to see whether any connections exist that could be worthwhile looking into. PCA analysis can be used to find hidden structures in a dataset in order to describe these structures. PCA provides a low-dimensional plot of the data, e.g., it projects many dimensions onto a few dimensions. In this process it is possible to identify outlying observations, clusters of similar observations, and other data structures.

The technique is based on principal components, a mathematical technique for an orthogonal orientation to principal axes. A principal component is also referred to as a latent variable. This variable cannot be measured directly but must be expressed as a linear combination of a set of input variables *(5)*. The data matrix X is decomposed into a structural part and an error part. The structural part consists of a score matrix, T, and a transposed loading matrix, P^T; the error part is termed E *(6)*. The following equation shows the mathematics behind the principal component method:

$$X = T \cdot P^T + E$$

PCA is capable of transforming a large number of possible correlated variables to a smaller number of uncorrelated variables or principal components. The original axes are being replaced by principal components axes, each one of which is a linear combination of the original variables.

Data are structured so that the rows are the samples and the columns are the variables. In the present context, this means that the rows are the gels and the columns are the spots (spot intensities). The relationship of the principal components to the samples is called scores, and that to the variables is called loadings. The first principal component covers as much as possible of the variation in the dataset, and each subsequent component covers as much as possible of the remaining variation.

The PLS is used to relate a y-matrix, which is usually the property to be calibrated for (the response data), with the x-matrix (the descriptor data), which is defined as the output of the instrument *(5)*. In the present context, Y contains the properties of the samples, and X contain the spots.

Calibration involves relating the two sets of data by regression modeling:

$$Y = X \cdot B$$

B is a matrix containing b-regression vectors expressing the link between variation in the predictors and variation in the response.

The major procedures involved in the evaluation of 2D gels using multivariate data analysis are the following:

1. Create 2D gels of the proteins under investigation.
2. Digitalize the gels using a scanner capable of scanning in transparency mode.
3. Analyze the digitalized 2D gels using an image analyzer software.
4. Generate a spot list.
5. Import the spot list into a multivariate data analysis software.
6. Conduct a PCA on the spot list.
7. Interpret scores and loadings plots.
8. Return to the biological interpretation.
9. Analyze response variables connected to the protein samples using PLSR.

2. Materials

1. A scanner capable of scanning in transparency.
2. Image analysis software.
3. A version of the Excel program.
4. Software able to perform multivariate data analysis.

2.1. Multivariate Data Analysis Software

Independent software solutions with user-friendly interface and with easily interpretative graphic plots of the results.

1. The Unscrambler from Camo (to download a trial version) (http://www.camo.com).
2. SIMCA-P from Umetrics (to download a demo version) (http://www.umetrics.com).
   Software working within the MATLAB computational environment.
3. PLS-Toolbox from Eigenvector (http://eigenvector.com).
   Software able to run PCA.
4. MVSP from Provalis Research (to download a demo version) (http://www.simstat.com).
5. XLSTAT from Addinsoft (add-in to Excel) (http://www.xlstat.com).
6. MINITAB from MINITAB Inc. (http://www.minitab.com).
7. Pirouette from Infometrix (http://www.infometrix.com).
8. Chemometrics Toolbox from Applied Chemometrics (working within the MATLAB environment) (http://www.chemometrics.com).
9. Evince from UmBio (http://www.umbio.com).
10. Extract from Extract Information (http://www.extractinformation.com).

2.2. 2D Electrophoresis Image Analysis Software

1. ImageMaster from Amersham Biosciences (http://www.amershambiosciences.com).
2. PDQuest from Bio-Rad Laboratories (http://www.bio-rad.com).
3. Z3 from Compugen (http://www.2dgels.com).
4. Phoretix 2D and Progenesis from Nonlinear Dynamics (http://www.nonlinear.com).

3. Methods

Multivariate data analysis of 2D gels using Progenesis, Excel, and The Unscrambler.

3.1. Create 2D Gels of the Proteins Under Investigation

This is outside the scope of this chapter. However, be sure to use a staining procedure that satisfies your need for quantitative evaluation of the gels (*see* **Note 1**).

3.2. Digitalize the Gels Using a Scanner Capable of Scanning in Transparency Mode

This is outside the scope of this chapter. However, be sure that you scan the images with high color depth and a high resolution (*see* **Note 2**) and that you use an appropriate file format for the image processing software (*see* **Note 3**).

3.3. Analyze the Digitalized 2D Gels Using Image Analysis Software

After production and digitalization of the 2D gel of interest (**Fig. 1**), the image analysis program Progenesis detects the spots and match these spots to a reference gel. Either the reference gel can be automatically selected, or the operator can select a specific 2D gel to be the reference gel. Unmatched spots can be added to the reference gel.

3.4. Generate a Spot List

After spot detection and matching a spot list can be generated, which should contain the values of interest. Most often this is the volume data. In Progenesis the spot list can be found in the Comparison Window. This spot list (**Table 1**) can be exported to Excel by selecting Copy to Excel in the Edit menu. A binary spot list, i.e., a spot with 1 for spot present and zero for spot absent can be very useful in certain cases (*see* **Note 4**). Another point is that it is important to synchronize the spot designation (*see* **Note 5**). Problems with the importing of data into Excel can occur for different reasons (*see* **Note 6**).

3.5. Import the Spot List into Multivariate Data Analysis Software

1. Prior to importation into The Unscrambler, it is useful to replace missing values with zeroes (*see* **Note 7**), because The Unscrambler can have problems with the missing values.
2. In The Unscrambler the spot list is imported using the Import submenu in the File menu. In the Import window select the Excel spreadsheet with the spot list, and press the Import button (*see* **Note 8**).
3. In the following window, the Import Worksheet window, select the correct worksheet, and fill in the correct areas. Sample Names covers the spot designations and Variable Names cover the gel names.

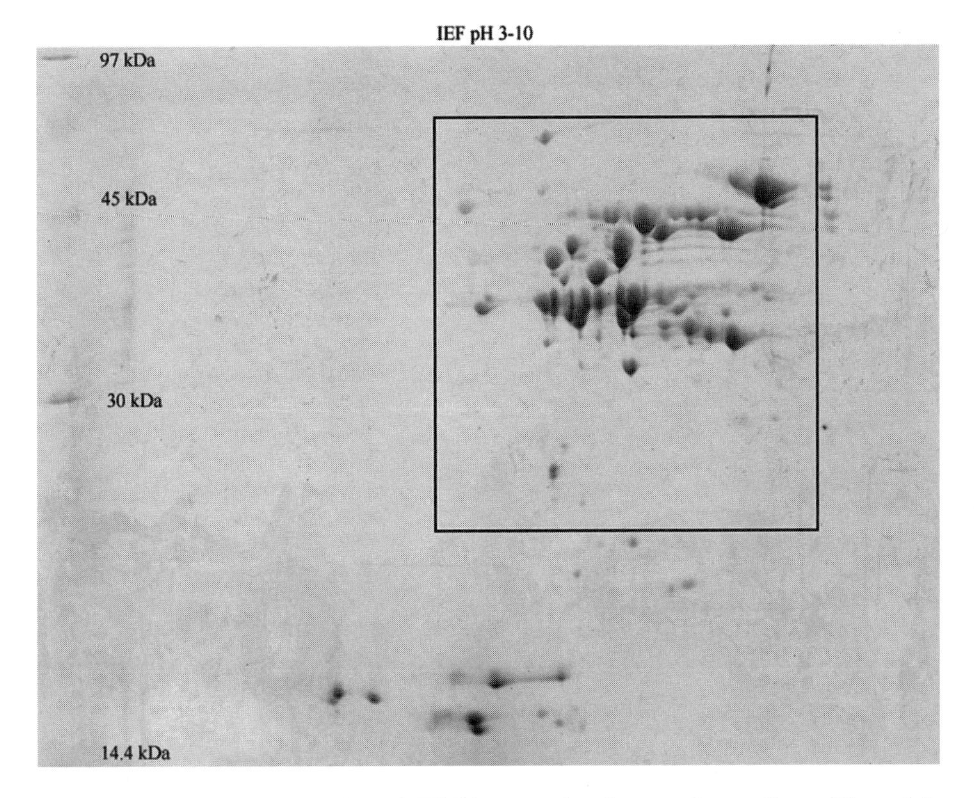

Fig. 1. A 2D gel of an ethanol (70%) extraction from a durum flour. The gel is stained with CBB. The proteins in the area within the square and proteins from two other durum flours in the corresponding area will in the following be analyzed. There are two gels for each flour, giving a total of six gels.

4. After the import, the data matrix needs to be transposed to ensure that the sample names correspond to the objects and that the spot designations correspond to the variables. The transpose function is found in the Transform submenu under the Modify menu.

3.6. Conduct a PCA on the Spot List

You are now ready to perform a PCA of your data. In The Unscrambler a PCA is run by pressing PCA in the Task menu; subsequently the PCA window pops up, and different settings can be chosen. Often a PCA is run with autoscaling of the variables and centering of the data. Autoscaling is equivalent to Weights with 1/SDev. By dividing by the standard deviation (SDev), the covariance matrix is changed to the correlation matrix, so that now we are conducting our PCA on the correlation matrix instead of on the covariance matrix. The covariance measure depends on the units of the variables, which

Table 1
Normalized Volume of a Section of the Spot List for the Six Gels

	Flour 1		Flour 2		Flour 3	
Ref. spot[a]	Sample 1	Sample 2	Sample 1	Sample 2	Sample 1	Sample 2
8	0.512	0.425	0.325	0.338	0.000	0.000
19	8.427	7.403	7.539	8.491	6.451	7.107
24	2.643	3.165	2.192	2.583	1.960	1.562
27	0.570	0.593	0.391	0.492	0.000	0.000
28	4.768	4.894	5.549	6.361	4.438	4.753
34	2.248	3.261	0.000	0.788	1.172	0.953
42	7.490	8.922	6.772	8.116	5.529	6.162
45	1.316	1.242	1.690	1.716	1.177	0.972
46	0.606	0.244	0.910	0.724	0.058	0.170
47	0.459	0.000	0.371	0.377	0.335	0.299
48	0.596	0.815	0.000	0.000	3.164	0.000
49	1.805	1.756	1.196	1.438	2.035	2.135
55	0.799	0.746	1.273	1.244	1.338	1.226
57	1.404	1.529	0.981	1.116	1.694	1.714
64	1.361	1.015	0.457	0.308	0.247	0.493
65	0.370	0.401	0.267	0.320	0.106	0.232
66	0.273	0.347	0.206	0.209	0.114	0.093
67	0.480	0.443	0.350	0.458	0.415	0.264
68	0.351	0.267	0.152	0.178	0.000	0.000
69	0.052	0.038	0.000	0.000	0.115	0.097
71	1.351	1.143	1.146	1.123	0.919	0.843
72	3.949	2.799	2.573	3.122	3.439	3.297
73	3.398	4.326	4.042	3.559	2.186	2.016
77	2.325	2.070	1.600	1.957	1.277	1.945
86	0.315	0.311	0.753	0.988	0.644	0.709
88	1.105	0.000	0.986	1.013	0.000	0.000
97	0.507	0.529	1.554	0.453	0.467	1.691

[a] Ref. spot is the spot designation for the various spots given by the 2D analysis software. Normalized volume is calculated by dividing the spot volume by the total volume of the gel and multiplying by 100. Other kinds of volume data could also be used.

makes it unsuitable for a PCA on mixed variables, whereas by dividing each column in the *X*-matrix by the SDev, we ensure that each variable gets the same variance, 1.0. Standardization with 1/SDev permits small variations to participate with the same weight as larger variations in the analysis. Therefore we generally advise beginners to use autoscaling (centering and weighing with 1/SDev).

3.6.1. Validation Method

The next step is the choice of validation method, which is highly dependent on the number of samples and the possibility of making another dataset. If the dataset contains many gels, the preferred validation method is the test set; otherwise cross-validation is used.

1. Test set validation is a based on two different datasets, one to be used during calibration of the PCA (the calibration set) and creation of the model and the other to be used to test the computed model of the PCA calibration (the test set/the validation set).
   The test set has several requirements. First, all samples should be chosen from the same population as the calibration set, and sampling conditions should be similar to those of the calibration set. Furthermore, the two datasets should be representative of future use. It is not sufficient to simply divide a large dataset into two, because the two datasets might then be too similar and the only difference between them would be the sampling variance, i.e., the variance owing to the independent samplings from the target population *(6)*. The calibration set must be large enough to calibrate a model satisfactorily, and the test set must be large enough to test the model. Often we do not have enough samples to conduct a test set validation which brings us to the other two types of validation, leverage and cross-validation.
2. Leverage validation should only be used if you have a very small sample set in which all samples are equally important. Leverage validation frequently leads to optimistic results, because the concept is to use the whole dataset for calibration and later use the leveraged corrected dataset for testing. We recommend that you do not use leverage validation.
3. Cross-validation can be used on medium to large datasets. The dataset is segmented, and each segment is left out in turn while the submodel (the dataset without a segment) is used to calibrate and the segment is used to test the model. This is done for each segment. The segment size and construction (random, systematic, or manual) can vary depending on the type of dataset. Each segment may consist of 25% of a large dataset, meaning that four submodels are calculated and tested. On smaller datasets, segments of only one sample are frequently used. This is called full cross-validation and means that each sample in turn is left out during calibration and the left-out sample is used for validation. Full cross-validation constructs as many submodels as there are samples. Because each time a sample is left out, only because only one sample (the left-out sample) is used each time to test the model, full cross-validation tends to lead to overoptimistic validation result when it is run on a balanced dataset *(6)*.
4. Variable selection is also an option in The Unscrambler (*see* **Note 9**).

An example of a PCA analysis is shown in **Fig. 2**.

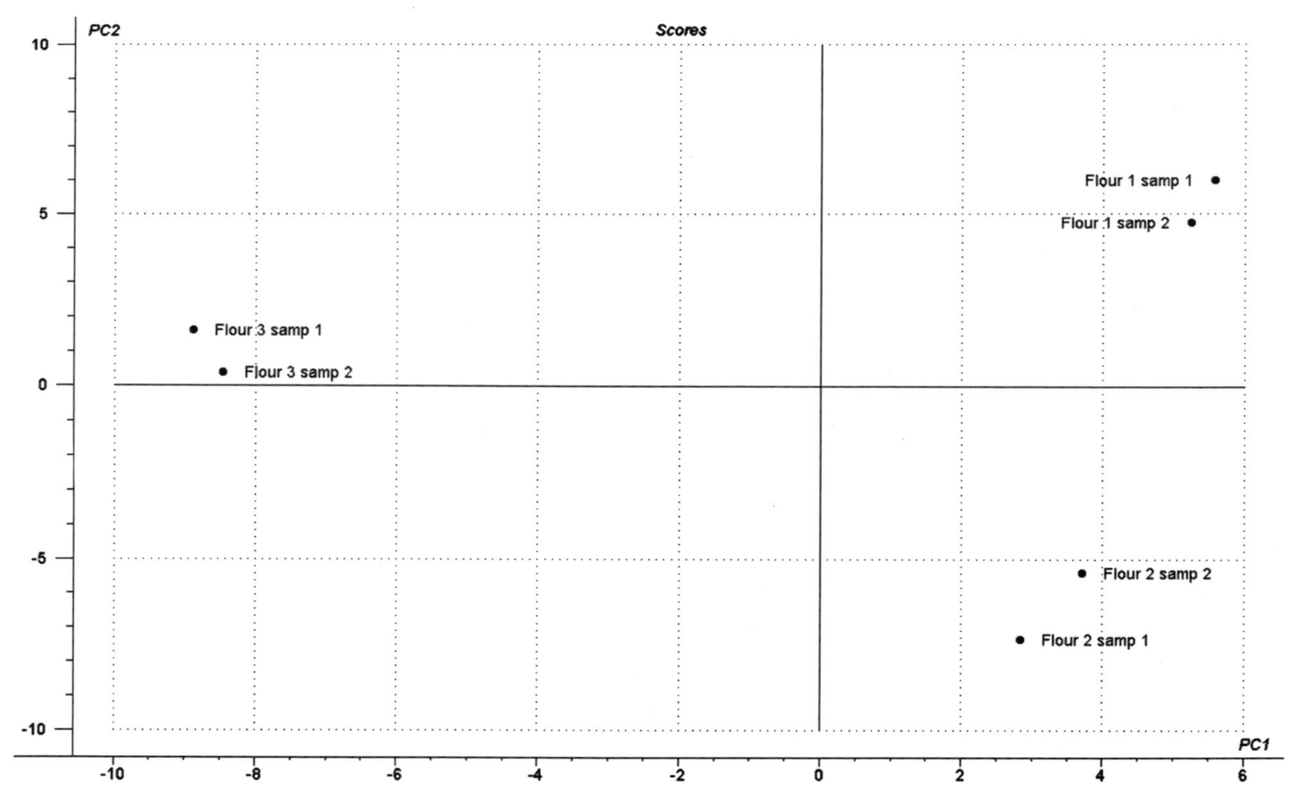

Fig. 2. The score plot, after a Principal Component Analysis (PCA) for Principal Components (PC) one and two. It is apparent that PC1 separates flour 3 from the two other flours and that PC2 to some extent separates flour 1 and flour 2. Sometimes it can be necessary to investigate higher orders of PCs to find interesting separations.

3.7. Interpret Scores and Loading Plots

1. The principal components (PCs) are linear functions of the original variables and contain the structured information in the data. Most of the information is contained in the first PC; progressively less information is in the higher numbered PCs. PCs are also called latent variables or score vectors.
2. A score plot, then, is a plot of the sample position along two or three PCs; therefore samples that are similar (with respect to the actual PCs) will be closer together in the score plot. In the first interpretation of a score plot one looks for clusters, i.e., samples that have features in common, and for distribution of samples over the entire plot. This gives information about the samples and also about the variables that distinguish the samples. Furthermore, it is possible to spot outliers, i.e., samples that for some reason are very different from the rest. Outliers should be investigated further because they could be very interesting samples or they could be indicators of errors in the analysis or errors in the data collection (meaning they would have to be removed).
3. The score plot should be interpreted together with the information contained in the loading plot for the same PCs, as this can help to determine the variables responsible for the differences between the samples that are seen in the score plot. The loading plot (**Fig. 3**) describes the data from the variable point of view: each variable has a loading on each PC that reflects both how much the variable contributes to that PC and how well that PC takes into account the variation of that variable over the data points.
4. Interpretation of the loading plot begins by looking at the variables with high loading, as this will help in interpretation of the meaning of a particular PC (**Fig. 4**). Also it is important to note that two variables with high loading are highly correlated. Loadings can have values in the interval [–1:+1] because they are actually the cosine to the angle between the variable and the PC. Variables with high loading and the same sign are positively correlated; those with opposite signs are negatively correlated. To help with the interpretation, it is also possible to make a biplot, in which both the scores and the loadings are plotted in a scatter plot (**Fig. 5**).

3.8. Return to the Biological Interpretation

Once sample distribution has been interpreted in the score plot and the variables that are responsible for that distribution have been determined in the loading plot, it is possible to return to the biology or biochemistry of your samples. In 2D electrophoresis the spots are the variables, which means that it is possible with this kind of analysis to point out which of the spots are responsible for a certain distribution of the 2D gels (**Fig. 6**). Then it is up to the investigator to propose a hypothesis that can account for the observed distribution. This is called exploratory data analysis When used with sufficient caution and knowledge of the biochemistry behind the samples, it is one the most efficient tools available for analysis of the proteome.

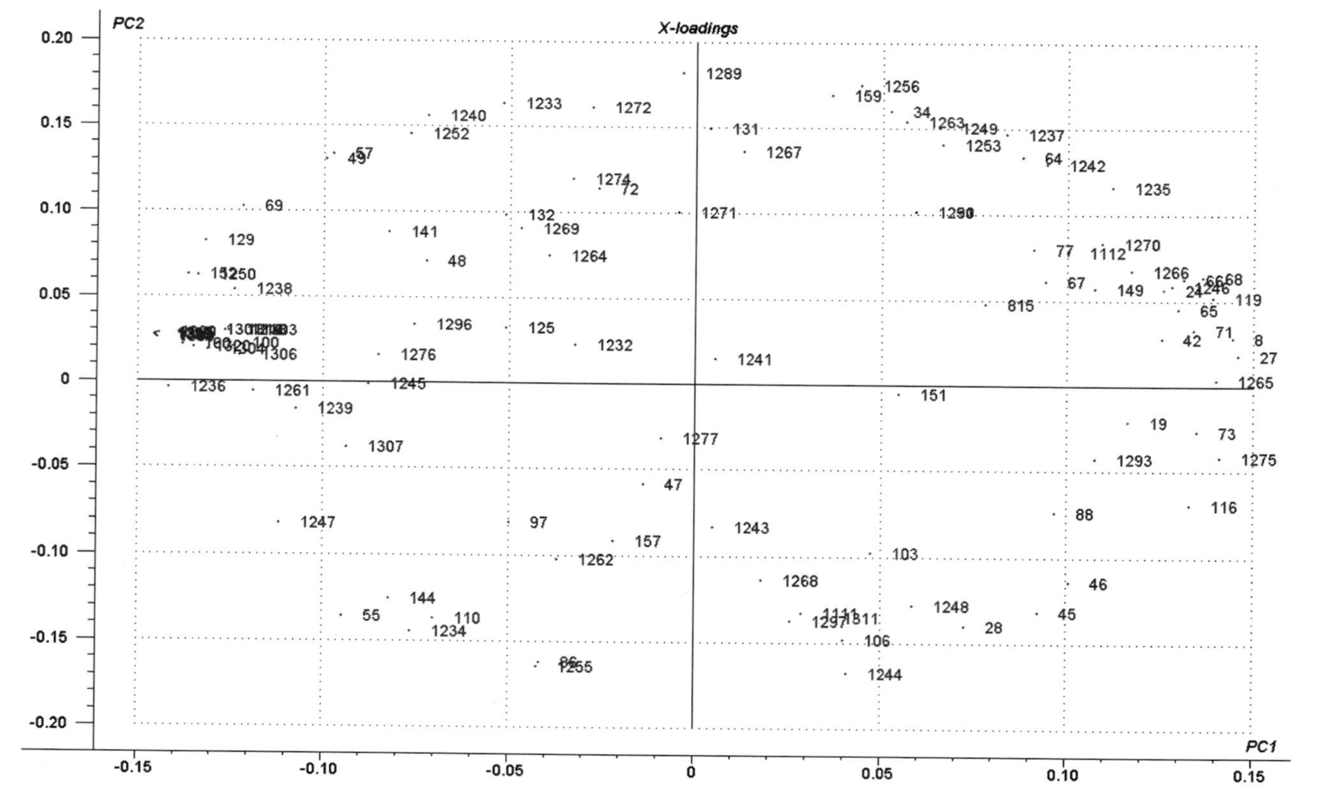

Fig. 3. The loading plot for PC1 and PC2. The loading plot is just as important as the score plot because it shows which variables, in this case spots, are responsible for the separation seen in the score plot. In other words, the loadings show the relationship between the multidimensional space stretched out by the objects (gels in our case) and the reduced space made up of the PCs.

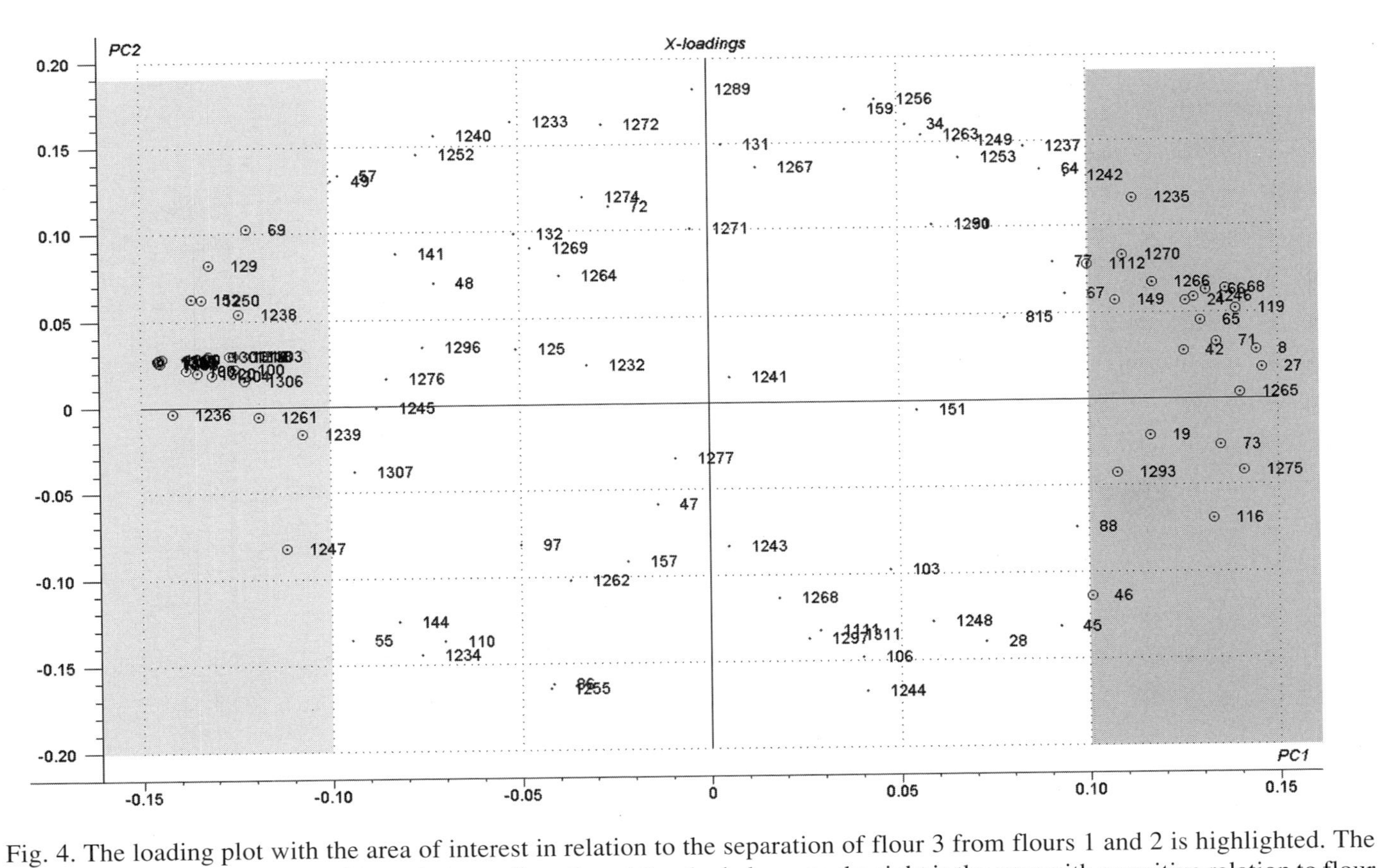

Fig. 4. The loading plot with the area of interest in relation to the separation of flour 3 from flours 1 and 2 is highlighted. The shaded area to the left is the area with a relation to flour 3, and the shaded area to the right is the area with a positive relation to flour 1 and 2.

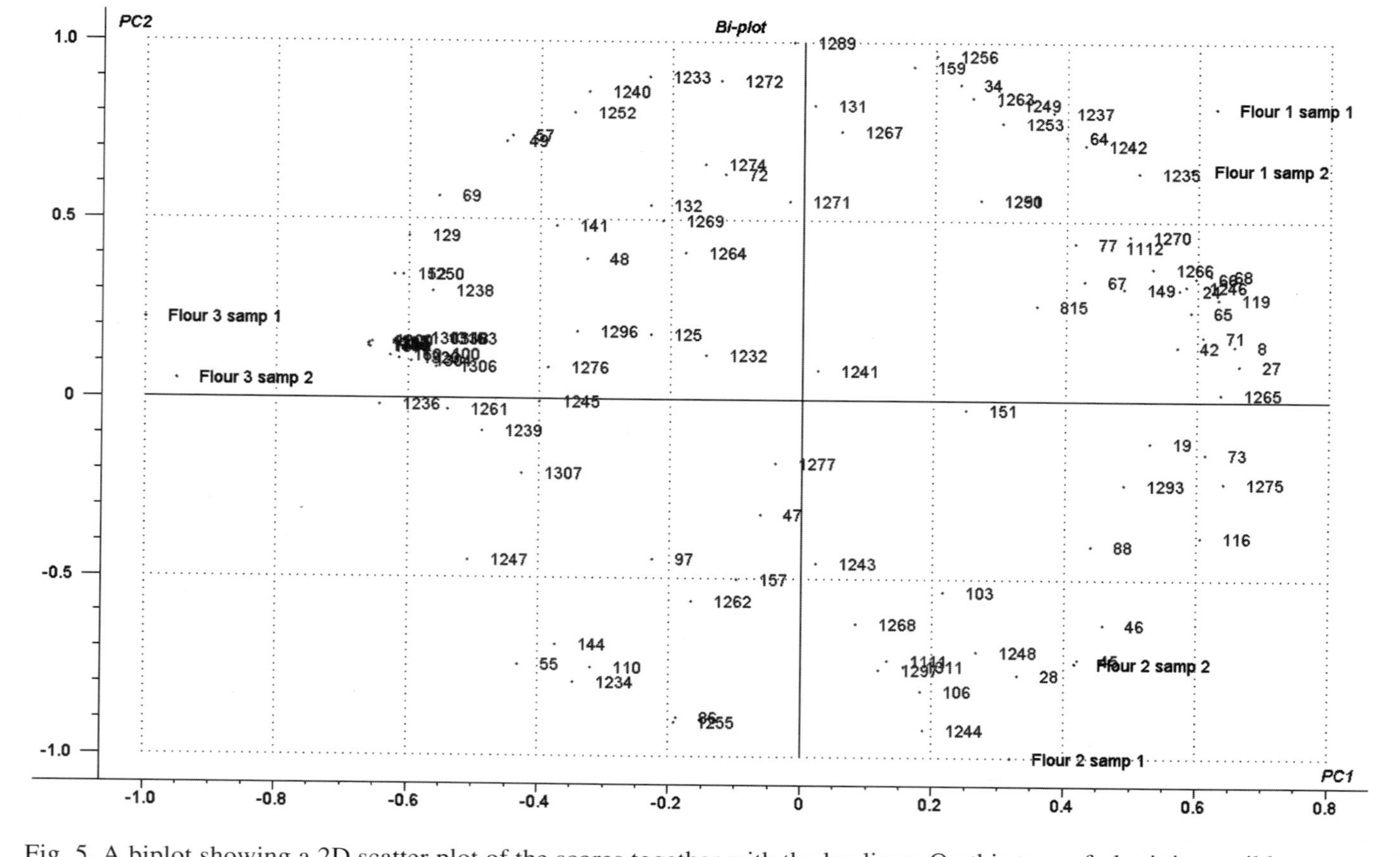

Fig. 5. A biplot showing a 2D scatter plot of the scores together with the loadings. On this type of plot it is possible to see and relate the samples to the variables. In this case it is possible to see which spots relate to which variety.

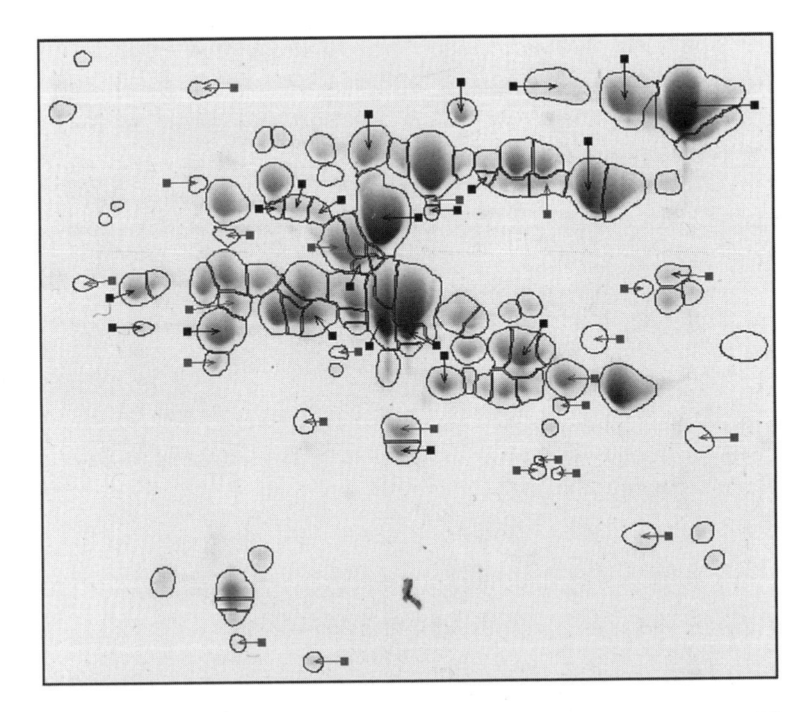

Fig. 6. Going back to the reference gel, we can use the information from **Figs. 2**, **3**, and **4** to interpret the biological relevance of the information from the PCA analysis. The reference gel is an artificial gel image produced by the 2D image analysis software. In this case all unmatched spots have been added to the gel, which means that all spots from the various gels are present on this gel. This makes it suitable for an overview of the result of the experiments. Arrows mark the spots that seem to be of interest based on the loading plot. Light gray arrows point to the spots that should be specific for flour 3 based on the loading plot, and black arrows point to the spots that should be specific for flours 1 and 2 based on the loading plot.

3.9. Analyze Response Variables Connected to the Protein Samples Using Partial Least Square Regression

A partial least square regression (PLS-R) relates two sets of data matrices (X and Y) by regression and is a supervised method. The principle in PLS-R is the same as that in PCA: to find the best straight line explaining most of the variation through the data points in a multidimensional space *(6)*. In PLS-R, the purpose is to build a linear model from a data table in order to predict a desired characteristic in another data table. Thus, whereas PCA is used for extracting hidden information from one data table, the X-matrix, PLS-R is used for examination of the relationships between two data tables, the X- and Y-matrices *(6)*. The X-matrix has dimension ($N \times K$), and the Y-matrix has dimension

$(N \times J)$, where N are the samples and K and J the X- and Y-variables, respectively *(5)*.

PLS-R works by performing a PCA for the X and Y data matrices, respectively, but these are not performed independently.

The **X**-variables are modeled by X-scores **T**, in terms of X-loadings **P** and X-residuals **E**, just as in PCA *(5)*:

$$\mathbf{X} - \bar{X} = \mathbf{T} \times \mathbf{P'} + \mathbf{E}$$

Likewise, the **Y**-variables are modeled by X-scores **T**, in terms of Y-loadings **Q** and Y-residuals **F** *(5)*:

$$\mathbf{Y} - \bar{Y} = \mathbf{T} \times \mathbf{Q'} + \mathbf{F}$$

The Y-variables can then be modeled directly from the X-variables using the regression coefficient matrix **B** *(5)*:

$$\mathbf{Y} = \mathbf{B}_0 + \mathbf{X} \times \mathbf{B} + \mathbf{F}$$

A prediction model then becomes the result as the equations are combined, whereby Y can be predicted from a set of X-variables.

3.9.1. Validation of PLS-R

In PLS-R the significance of validation is given by two parameters: the residual validation Y-variance (RVYV) gives the residual between the measured (Y_{ref}) and the predicted (Y_{pred}) that is the modeling error. It gives the optimal dimensionality with the lowest number *(6)*. A comparison of different modelings is given by the square root of RVYV, called the RMSE *(6)*.

$$RVYV = \frac{\sum_{i=1}^{n} \left(Y_{pred} - Y_{ref}\right)^2}{n} \qquad RMSE = \sqrt{\frac{\sum_{i=1}^{n} \left(Y_{pred} - Y_{ref}\right)^2}{n}}$$

The X-data are used to make a model. The X-values are then inserted into the model in order to predict Y. The modeling error is obtained by the subtraction of Y-data from Y-predicted *(6)*:

1. X_{cal} Y_{cal} $\rightarrow$ model

2. $X_{\text{cal}} + \text{model} \rightarrow \hat{Y}_{\text{cal}}$

3. $\text{Modeling error} = \hat{Y}_{\text{cal}} - Y_{\text{cal}}$

The deviation is the expression of how similar the prediction samples are to the calibration samples used when making the model. The deviation is small when the predicted samples are similar to the calibration samples. When the deviation is high, the predicted *Y*-values cannot be trusted ***(6)***.

The last important parameter is the correlation coefficient (r), defined as the correlation between *X* and *Y* as shown in the following equation:

$$r = \frac{\text{covariance between } x \text{ and } y}{S_{dev}(\mathrm{x}) \times S_{dev}(\mathrm{y})}$$

$$r = \frac{\dfrac{\sum_{i=1}^{n}\left(X_i - \bar{X}\right)\left(Y_i - \bar{Y}\right)}{n-1}}{\left(\sqrt{\dfrac{\sum_{i=1}^{n}\left(X_i - \bar{X}\right)^2}{n-1}}\right)\left(\sqrt{\dfrac{\sum_{i=1}^{n}\left(Y_i - \bar{Y}\right)^2}{n-1}}\right)}$$

The correlation is a measure of the linear relationship between two variables. The value 1 means a linear relationship exists between the variables, and the value 0 means the variables are not correlated.

4. Notes

1. If the gels are going to be analyzed based on spot volumes, the preferred procedures are staining with Coomassie brilliant blue (CBB) or fluorescent dyes, to ensure that the spot intensities correlate linearly with the protein amount.
2. After production of the 2D gels, the gels need to be digitalized. This should be done with as high a color depth as possible, to enable the image analysis software to split adjacent spots with higher precision. Therefore a color depth of 16-bit gray scale and a resolution of 250 to 300 are preferable.
3. Make sure that the file format for the digitalized 2D gels is compatible with the analysis software.
4. The quantitative difference can be analyzed by generating a binary spot list consisting of ones and zeroes corresponding to the presence or absence of a spot. The binary spot list can be generated on a normal volume spot list using the IF procedure in Excel.
5. It is important to synchronize the spot designation prior to generation of the spot list, to ensure that matched spots have the same designation. In Progenesis this is done in the Edit menu by selecting Synchronize Spot Numbers.

6. It is important to ensure that Excel interprets the spot list values as Numbers; otherwise the multivariate data analysis software may have problems importing the spot list correctly. It may also be necessary to ensure that the decimal separator is a period.
7. Replacement of the missing values with zeroes can be done in Excel with the Replace option in the Edit menu.
8. It is our experience that the import of the data matrix from Excel into The Unscrambler can be accomplished most easily by naming the cell range for the data, the spot designation, and the gel names, because these names reappear in The Unscrambler during the import. Naming a range of cells in Excel is done by selecting the cells to be named and typing the name in the Name box to the left of the formula bar. Naming a range of cells can also be done in the Define Name dialog, which is found by selecting the Insert menu, pointing to Name, and clicking Define.
9. In The Unscrambler there is a variable selection tool named Uncertainty Test (jack-knife) that can be used to select interesting variables.

References

1. Gottlieb, D. M., Schultz, J., Bruun, S. W., Jacobsen, S., and Søndergaard, I. (2004) Multivariate approaches in plant science. *Phytochemistry* **65,** 1531–1548.
2. Jessen, F., Lametsch, R., Bendixen, E., Kjærsgård, I. V. H., and Jørgensen, B. M. (2002) Extracting information from two-dimensional electrophoresis gels by partial least squares regression. *Proteomics* **2,** 32–35.
3. Radzikowski, L., Nešić, L., Hansen, H. B., Jacobsen, S., and Søndergaard, I. (2002) Comparison of ethanol-soluble proteins from different rye (*Secale cereale*) varieties by two-dimensional electrophoresis. *Electrophoresis* **23,** 4157–4166.
4. Schultz, J., Gottlieb, D. M., Petersen, M., Nešić, L., Jacobsen, S., and Søndergaard, I. (2004) Explorative data analysis of 2-D electrophoresis gels. *Electrophoresis* **25,** 502–511.
5. Martens, H. and Martens, M. (2001) *Multivariate Analysis of Quality. An Introduction.* John Wiley & Sons, New York.
6. Esbensen K. H. (2000) *Multivariate Data Analysis—In Practice.* Camo, Oslo, Norway.

18

Edman Sequencing of Proteins from 2D Gels

Setsuko Komatsu

Summary

The Western blotting/sequencing technique using polyvinylidene difluoride (PVDF) membrane is one of the most popular technique for Edman sequencing. A protein sample is transferred from a 2D polyacrylamide gel electrophoresis (2D-PAGE) gel onto a PVDF membrane by electroblotting. The membrane carrying the protein is directly subjected to protein sequencing. If sequencing fails after a few cycles, the PVDF membrane is removed from the sequencer and treated with deblocking solution. If this attempt at sequencing fails, alternative methods such as the Cleveland method are required. Because the resolution of 2D-PAGE is high, the combined use of 2D-PAGE, Western blotting, and Edman sequencing often allow effective sequence determination of proteins that could not be easily purified by conventional column chromatography.

Key Words: Edman sequencing; 2D-PAGE; blocked protein; deblocking; Cleveland method.

1. Introduction

Edman sequencing of proteins separated on 2D gels became possible with the introduction of protein electroblotting methods that allow efficient transfer of sample from the gel matrix onto supports suitable for gas-phase sequencing or related techniques *(1)*. Picomole amounts of protein are first separated by 2D polyacrylamide gel electrophoresis (2D-PAGE) *(2)* and then electroblotted from 2D-PAGE gels onto polyvinylidene difluoride (PVDF) membrane. The amino acid sequences of the electroblotted protein are determined by Edman sequencing. Direct N-terminal sequencing is the most sensitive method (1–5 μg of protein), but, when gaps or ambiguous assignments are seen, verification of the sequence by other means often demands much more material.

Proteins are often posttranslationally modified, and N-terminal blockage is one of the more common posttranslational modifications. Proteins can become

From: *Methods in Molecular Biology, vol. 335: Plant Proteomics: Methods and Protocols*
Edited by: H. Thiellement, M. Zivy, C. Damerval, and V. Méchin © Humana Press Inc., Totowa, NJ

N-terminally blocked not only in vivo but also in vitro. However, it is possible to prevent in vitro blocking, which is generated during protein extraction, 2D-PAGE, and blotting. The use of very pure reagents during these procedures, the addition of thioglycolic acid as a free radical scavenger to the extraction buffer, electrophoresis and electroblotting buffers, and preelectrophoresis to remove the free radicals from the gel might all be effective in preventing in vitro blocking *(3)*. However, if proteins are blocked in vivo, a chemical or enzymatic deblocking procedure or peptide map procedure is required to determine the N-terminal or internal sequence.

2. Materials

1. Sodium dodecyl sulfate (SDS) sample buffer: 0.06 *M* Tris-HCl (pH 6.8), 2% SDS, 10% glycerol, and 5% mercaptoethanol *(4)*.
2. Acrylamide for separating gel (acrylamide/BIS = 30:0.135): 30.00 g acrylamide, 0.135 g BIS. Make volume to 100 mL with MQ water. Keep in the dark (brown bottle).
3. Separating gel buffer (pH 8.8): 12.11 g Tris-HCl to a 1 *M* final concentration, 0.27 g SDS, to a 0.27% final concentration. Dissolve in 80 mL MQ water, adjust pH to 8.8, and make the volume to 100 mL.
4. Acrylamide for stacking gel (acrylamide/BIS = 29.2:0.8): 29.2 g acrylamide, 0.8 g BIS. Make volume to 100 mL with MQ water. Keep in the dark (brown bottle).
5. Stacking gel buffer (pH 6.8): 3.03 g Tris-HCl, to a final concentration of 0.25 *M*, 0.20 g SDS, to a final concentration of 0.2%. Dissolve in 80 mL MQ water, adjust pH to 6.8, and make the volume to 100 mL.
6. SDS-PAGE running buffer: 9 g Tris-HCl, 43.2 g glycine, and 3 g SDS. Dissolve in 3 L MQ water.
7. Bromophenol blue (BPB) solution: dissolve 0.1 g BPB in 100 mL 10% glycerol.
8. Blotting solution A: 36.33 g Tris-HCl, final concentration 0.3 *M*, 200 mL methanol, final concentration 20%, and 0.20 g SDS, final concentration 0.02%. Make volume to 1000 mL with MQ water, and keep at 4°C.
9. Blotting solution B: 3.03 g Tris-HCl, final concentration 25 m*M*, 200 mL methanol, final concentration 20%, and 0.20 g SDS, final concentration 0.02%. Make volume to 1000 mL with MQ water, and keep at 4°C.
10. Blotting solution C: 3.03 g Tris-HCl, final concentration 25 m*M*, 5.20 g β-amino-n-caproic acid, final concentration 40 m*M*, 200 mL methanol, final concentration 20%, and 0.20 g SDS, final concentration 0.02%. Make volume to 1000 mL with MQ water, and keep at 4°C.
11. Separating gel solution (amounts for one gel [18%]): 10 mL acrylamide for separating gel, 6.3 mL separating gel buffer (pH 8.8), 120 µL 10% APS, 20 µL TEMED.
12. Stacking gel solution (amounts for one gel [5%]): 1 mL acrylamide for stacking gel, 3 mL stacking gel buffer (pH 6.8), 2 mL MQ water, 30 µL 10% APS, 20 µL TEMED.

3. Methods

3.1. Cleveland Peptide Mapping (5)

1. Prepared samples are separated by 2D-PAGE. Then the gels are stained with Coomassie brilliant blue (CBB), and gel pieces containing protein spots are removed.
2. The protein is electroeluted from the gel pieces using an electrophoretic concentrator run at 2 W constant power for 2 h. After electroelution, the protein solution is dialyzed against MQ water for 2 d and lyophilized.
   a. Cut out stained protein spots from 2D gels and soak in MQ water.
   b. Fill 750 µL of electroelution buffer in the 2-mL Eppendorf tube containing protein spots (5–20 gel pieces). Shake for 30 min.
   c. Cut 12- to 15-cm-long seamless cellulose tubing (small size, no. 24, Wako, Osaka, Japan) as space is needed for clipping. Fill a 300-mL beaker with 250 mL MQ water, boil it for 5 min, and keep the tubing membrane in it. Wet the small pieces of cellophane film in a small beaker with MQ water.
   d. Close the bottom of the small part of cup with a cellophane film, open the bottom of the large part of cup, connect and twist the tubing membrane, and close the distal end by clipping (**Fig. 1A**).
   e. Fix the cup in the electrophoretic concentrator (**Fig. 1B**; Nippon Eido, Tokyo, Japan). Deposit gel pieces containing proteins on the cellophane film (the small part of the cup), and add 750 µL of the electroelution buffer from the Eppendorf tube. Fill the small part of the cup with electroelution buffer, and then fill the large part of the cup with electroelution buffer in such a way that a layer of buffer joins both the cup parts, allowing movement of protein from the small part of the cup to the tubing membrane. Fill the apparatus with electroelution buffer. The small part of cup containing protein spots should be toward the positive side.
   f. Run at 2 W constant power for 2 h.
   g. Remove the tubing membrane and clip to close the end. Dialyze in a cold room (4°C). Change the deionized water three times the first day. The next day, change the MQ water two times.
   h. Transfer the protein solution to two to six 2-mL Eppendorf tubes. Freeze-dry overnight.
   i. Dissolve the protein in 30 µL of SDS sample buffer.
3. The protein is dissolved in 20 µL of SDS sample buffer (pH 6.8) and applied to a sample well of an SDS-PAGE gel. The sample solution is overlaid with 20 µL of a solution containing 10 µL of *Staphylococcus aureus* V8 protease (Pierce, Rockford, IL) (0.1 µg/µL) in MQ water and 10 µL of SDS sample buffer (pH 6.8). Electrophoresis is performed until the sample and protease are stacked in the stacking gel. The power is switched off for 30 min to allow digestion of the protein, and then electrophoresis is continued.
   a. Fix glasses (100 × 140 × 1 mm) with clip, keeping a 1-mm space between the plates.

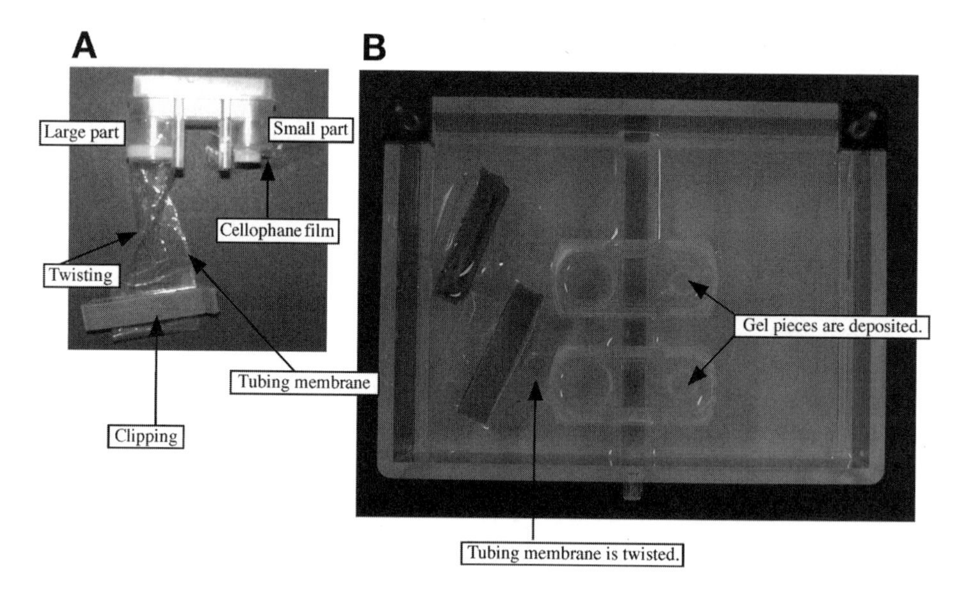

Fig. 1. Electrophoretic concentrator. **(A)** Close-up of the cup with long seamless cellulose tubing. **(B)** The cups fixed to the electrophoretic concentrator.

b. Prepare separating gel solution in a 100-mL beaker. Mix the solutions, and fill the plates about 3 cm from top. (**Caution:** Pour the solutions into the plates immediately after adding 10% APS and TEMED.)
c. Overlay the separating gel solution with 1 mL MQ water.
d. Leave the gel for 40 to 60 min at room temperature for polymerization.
e. Remove the overlaid water and pour the following stacking gel solution.
f. Prepare the stacking gel solution in a 100-mL beaker. Mix well, pour on the separating gel, and insert comb.
g. Leave the gel for 20 min at room temperature for polymerization.
h. Take out the comb, clips, and silicon tubes.
i. Clean the wells with a syringe.
j. Fix the gel plates with the apparatus. Pour SDS-PAGE running buffer.
k. Dissolve the protein in 20 μL of SDS sample buffer (pH 6.8), and apply to a sample well in SDS-PAGE. Overlay the sample with 20 μL of a solution containing 10 μL of *Staphylococcus aureus* V8 protease with 1 μg/μL in MQ water and 10 μL of SDS sample buffer (pH 6.8). Add 30 μL BPB solution.
l. Electrophoresis was performed until the sample and protease were stacked in the upper gel and interrupted for 30 min to digest the protein.
m. Run the gel at 35 mA until the BPB line reaches about 5 mm near the bottom.
n. Disconnect the electricity, and take out the plates.
o. Separate two plates with a spatula.
p. Separate the stacking gel, and take out the separating gel.

3.2. N-Terminal and Internal Amino Acid Sequence Analysis and Homology Search of Amino Acid Sequence

1. Following separation by 2D-PAGE or by the Cleveland method, the proteins are electroblotted onto a PVDF membrane (Pall, Port Washington, NY), using a semidry transfer blotter (Nippon-Eido, Tokyo Japan), and detected by CBB staining.
   a. Cut the PVDF membrane equal to the size of the gel.
   b. Cut the Whatman 3MM filter paper equal to the size of the gel.
   c. Wash the PVDF membrane in methanol for a few seconds, and transfer the membrane to 100 mL blotting solution C; shake for 5 min.
   d. Wet two blotting papers (3 MM) in A, B, and C blotting solutions.
   e. Place the separating gel in 100 mL blotting solution C, and shake for 5 min.
   f. Wet the semidry blotting apparatus with MQ water. Place blotting paper A on the blotting plate and then blotting paper B. Remove air bubbles, if any. Place the PVDF membrane on the plate and then the gel and blotting paper C.
   g. Connect the power supply. Run the blotting at 1 mA/cm^2 for 90 min.
   h. Wash the membrane in 100 mL MQ water.
   i. Stain the gel for 2 to 3 min in CBB stain.
   j. Destain the gel in 60% methanol for 3 min twice.
   k. Wash with MQ water and air-dry at room temperature.
2. The stained protein spots or bands are excised from the PVDF membrane and applied to the upper glass block of the reaction chamber in a gas-phase protein sequencer, Procise 494 or cLC (Applied Biosystems, Foster City, CA). Edman degradation is performed according to the standard program supplied by Applied Biosystems. The released phenylthiohydantoin (PTH) amino acids are separated by an online high-performance liquid chromatography (HPLC) system and identified by retention time.
3. The amino acid sequences obtained are compared with those of known proteins in the Swiss-Prot, PIR, Genpept, and PDB databases with the web-accessible search program FastA.

3.3. Deblocking of Blotted Proteins (6) (see Note 1)

3.3.1. Acetylserine and Acetylthreonine

Proteins with acetylserine and acetylthreonine at their N-termini separated by 2D-PAGE are electroblotted onto a PVDF membrane. The region of the PVDF membrane carrying the protein spot is excised and treated with trifluoroacetic acid (TFA) at 60°C for 30 min. They are then directly subjected to protein sequencing (*see* **Note 2**).

3.3.2. Formyl Group

N-formylated protein separated by 2D-PAGE is electroblotted onto a PVDF membrane. The region of the PVDF membrane carrying the protein spot is excised and treated with 300 µL of 0.6 *M* HCl at 25°C for 24 h. The membrane is washed with MQ water, dried, and applied to the protein sequencer.

3.3.3. Pyroglutamic Acid

1. Proteins with pyroglutamic acid at their N-termini separated by 2D-PAGE are electroblotted onto a PVDF membrane.
2. The region of the PVDF membrane carrying the protein spot is excised and treated with 200 µL of 0.5% (w/v) polyvinylpyrrolidone (PVP)-40 in 100 m*M* acetic acid 37°C for 30 min (*see* **Note 3**).
3. The PVDF membrane is washed at least 10 times with 1 mL of MQ water.
4. The PVDF membrane is soaked in 100 µL of 0.1 *M* phosphate buffer (pH 8.0) containing 5 m*M* dithiothreitol and 10 m*M* EDTA acid.
5. Pyroglutamyl peptidase (5 µg) is added, and the reaction solution is incubated at 30°C for 24 h.
6. The PVDF membrane is washed with MQ water, dried, and applied to the protein sequencer ***(6)***.

4. Notes

1. These deblocking techniques may be combined to allow the sequential deblocking and sequencing of unknown proteins that have been immobilized onto PVDF membranes. A protein on the PVDF membrane is directly subjected to gas-phase sequencing. If sequencing fails at step, the PVDF membrane is removed from the sequencer, and performed to remove acetyl group, formyl group and then pyroglutamic group.
2. The advantage of this method is that the debloking can be easily and rapidly done, although overall sequencing yields obtained by this procedure were low compared with acylamino acid-releasing enzyme (AARE) digestion. N-acetylated protein: N-acetylated proteins are enzymatically deblocked with AARE after on-membrane digestion with trypsin to generate the N-terminal peptide fragment. This tryptic digestion required since AARE can only remove the acetylamino acid from a short peptide ***(6)***.
3. PVP-40 is used to unbind pyroglutamic acid from the PVDF membrane, while the rest of the protein stays bound to it.

References

1. Eckerskorn, C., Mewes, W., Goretzki, H., and Lottspeich, F. (1988) A new siliconized-gas fiber as support for protein-chemical analysis of electroblotted proteins. *Eur. J. Biochem.* **176,** 509–512.
2. O'Farrell, P. F. (1975) High resolution two-dimensional electrophoresis of proteins. *J. Biol. Chem.* **250**, 4007–4021.
3. Hirano, H., Komatsu, S., Kajiwara, H., Takagi, Y., and Tsunasawa, S. (1993) Microsequence analysis of the N-terminally blocked proteins immobilized on polyvinylidene difluoride membrane by Western blotting. *Electrophoresi* **14,** 839–846.

4. Laemmli, U. K. (1970) Cleavage of structural proteins during the assembly of the head of bacteriophage T4. *Nature* **227,** 680–685.
5. Cleveland, D. W., Fischer, S. G., Krischer, M. W., amd Laemmli, U. K. (1977) Peptide mapping by limited proteolysis in sodium dodecyl sulphate and analysis by gel electrophoresis, *J. Biol. Chem.* **252,** 1102–1106.
6. Hirano, H., Komatsu, S., Nakamura, A., et al. (1991) Structural homology between semidwarfism-related proteins and glutelin seed protein in rice (*Oryza sativa* L.). *Theor. Appl. Genet.* **83,** 153–158.

19

Peptide Mass Fingerprinting

Identification of Proteins by MALDI-TOF

Nicolas Sommerer, Delphine Centeno, and Michel Rossignol

Summary

MALDI-TOF peptide mass fingerprinting (PMF) is the fastest and cheapest method of protein identification; the studied genome is sequenced and annotated, and the protein is amenable to separation and detection in 2D gel electrophoresis. In plant proteomics there are two main difficulties: few plant genomes are sequenced, and major contaminants are non-plant specific. This chapter describes the classical "bottom-up" method (i.e., from peptide to protein identification) of gel cutting, in-gel digestion, peptide recovery and purification, MALDI-TOF mass spectrometry, and critical survey of protein database queries.

Peptide mass fingerprint (PMF); MALDI-TOF; TOF/TOF; trypsin; in-gel tryptic digestion; protein identification; database search.

1. Introduction

It is noteworthy that protein identification by mass spectrometry (MS) was made possible by the development of "soft" ionization techniques developed in the late 1980's in Europe by Michael Karas and Franz Hillenkamp for matrix-assisted laser desorption ionization (MALDI) and in the United States by John Fenn for electrospray ionization (ESI). Half of the 2002 Chemistry Nobel Prize for chemistry was Awarded to Fenn and Tanaka "for their development of soft desorption ionisation methods for mass spectrometric analyses of biological macromolecules."

Sufficient separation of the protein extract by 2D gels and compatible staining (*see* Chapters 16 and 17), make them amenable to protein identification by MS. However, MS of the entire protein does not lead to the direct identifica-

From: *Methods in Molecular Biology, vol. 335: Plant Proteomics: Methods and Protocols*
Edited by: H. Thiellement, M. Zivy, C. Damerval, and V. Méchin © Humana Press Inc., Totowa, NJ

tion of the protein. To obtain a good sensitivity, enough mass accuracy, and access to the most of the data (i.e., to cover the sequence), the protein must be digested (cut) by an endoprotease to generate multiple protein-specific fragments. If the endoprotease has sufficient specificity and if the protein is available in databases, a computed comparison of the experimental peptide masses (generated by the endoproteic digestion and detected by MALDI-time of flight [TOF] mass spectrometry) with the peptide masses of the in-silico digestion of all the proteins present in the database will allow the identification of protein candidates.

Concerning plant proteomics, the techniques used are similar to those of other proteome studies. However, particular attention must be paid to the specificity of the biological origin of the sample. Indeed, partly owing to their very large genomes (and often to polyploidy) compared with animal genomes, very few plant species genomes have been sequenced and annotated. *Arabidopsis thaliana* genome was the first to be completed, in December 2000. For nonsequenced plants, partial protein information may be available, but for the vast majority of plants (mainly trees and cereals, apart from rice), very few genomic information is available. Thus, for nonsequenced species, protein identification may be achieved by homology. For unidentified proteins, MS/MS peptide sequencing will be the only possibility of identification (either by MALDI-TOF/TOF or by nano-liquid chromatography [LC]-ESI-MS/MS). Moreover, plant-specific databases cannot be used first, as most protein contaminants have a human or mammalian origin.

The protocol steps are as follows:

1. Spot excision from the gel and in-gel digestion with trypsin (*see***Subheading 3.1.**).
2. Digest deposition on the MALDI target and MALDI-TOF spectrum acquisition and annotation (*see* **Subheading 3.2.**).
3. Database search and careful survey of search engine hits (*see* **Subheading 3.3.**).

2. Materials

2.1. Equipment

1. Clean vacuum centrifuge.
2. Clean oven (up to 56°C)
3. Modern (i.e., end of the 1990s as the limit) MALDI-TOF mass spectrometer equipped with delayed or pulsed ion extraction technology and an electrostatic mirror (reflectron [MALDI-reTOF]).

2.2. Reagents

2.2.1. In-Gel Digestion

1. Water (HPLC grade, or Milli-Q grade).
2. High-purity ammonium bicarbonate.

3. Acetonitrile (high-performance liquid chromatography [HPLC] grade).
4. Trifluoroacetic acid (TFA; HPLC grade).
5. *n*-Octyl-glycopyranoside (*n*-OGP).
6. Sequencing grade porcine trypsin. Although other good-quality sequencing-grade porcine trypsin sources are available, for homogeneity and convenience, the protocols and data presented below are optimized with Promega Porcine trypsin (sequencing grade, modified).
7. Ammonium bicarbonate buffer: 25 m*M* ammonium bicarbonate buffer, pH 7.8.
8. Acetonitrile/ammonium bicarbonate buffer: 50/50 (v/v) acetonitrile/25 m*M* ammonium bicarbonate buffer (pH 7.8).
9. Digestion buffer (prepare in ice): 0.0125 µg/µL sequencing grade trypsin in 25 m*M* ammonium bicarbonate buffer containing 5 m*M* *n*-OGP.

2.2.2. MALDI-TOF Mass Spectrometry

1. α-Cyano-4-hydroxycinnamic acid powder, re-crystallized.
2. Water (HPLC grade, or milli-Q grade).
3. Acetonitrile (HPLC grade).
4. Acetone (HPLC grade).
5. Ethanol (HPLC grade).
6. TFA solution: 0.1% TFA (HPLC grade) solution in water.
7. Acetonitrile/TFA solution: 3:2 (v/v) acetonitrile/water acidified by 0.1% TFA.
8. Acetonitrile/TFA solution: 1:1 (v/v) acetonitrile/water acidified by 0.1% TFA.
9. Ethanol/acetone/TFA solution: 6:3:1 (v/v/v) ethanol/acetone/0.1% TFA in water.

3. Methods

Many gel staining protocols are suitable for subsequent MALDI-TOF peptide mass fingerprinting (PMF) analysis, although some give better results than others. Comparative performance and compatibility are discussed in Chapter 14.

3.1. In-Gel Digestion

The in-gel digestion protocol is adapted from Jensen et al..

For gel handling and digestion steps, careful sample handling is compulsory, to minimize sample contamination with exogenous keratins (hair, skin, wool clothes, draughts, air-cooling, and so on) (*see* **Note 1**).

3.1.1. Excision of Protein Spots from 2D Polyacrylamide Gels

1. Cut with a razor blade a 1 mL pipet tip 5 mm above the tip to make a punch (stamp-out diameter of 1–2 mm) (*see* **Note 2**).
2. Stamp out (excise) the gel spot, and transfer it to a microcentrifuge tube.

3.1.2. Washing of Gel Pieces

1. Wash the spot with 100 µL of ammonium bicarbonate buffer (15 min vortex), and then discard the supernatant.

2. Wash the spot twice with 100 µL of acetonitrile/ammonium bicarbonate buffer (15 min vortex), and then discard supernatant.
3. Wash the spot with 100 µL HPLC grade acetonitrile to shrink the gel piece (15 min vortex), and then discard supernatant.
4. Dry gel fragments under vacuum on a centrifugal evaporator.

If needed, after these washing steps, the gel fragment can be stored at –20°C for few weeks before in-gel digestion.

For 2D gels, reduction and alkylation of cystein residues prior to in-gel digestion are not necessary (*see* **Note 3**).

3.1.3. *In-Gel Digestion* (see ***Notes 4*** *and* ***5***)

1. On ice (*see* **Note 6**), add 8 µL of cooled digestion buffer.
2. After 20 min, adjust with the minimum ammonium bicarbonate buffer volume to cover the gel.
3. Incubate at 37°C for 4 h (or overnight if more convenient; *see* **Note 7**).

3.1.4. *Peptide Extraction, Concentration, and Desalting*

3.1.4.1. Peptide Extraction

1. Extract the resulting tryptic peptide fragments by adding 20 µL of the TFA solution (sonicate for 15 min). Preserve the supernatant in a 500-µL microcentrifuge tube.
2. Extract with 20 µL of 3:2 acetonitrile/TFA solution in an ultrasonic bath for 15 min. Pool the supernatants.

3.1.4.2. Peptide Concentration and Desalting

1. Concentrate the pooled supernatant to a final volume of approx 10 µL in a vacuum centrifuge. This step removes acetonitrile (which otherwise will not allow peptide fixation on the hydrophobic chromatographic media) and reduces the volume to an acceptable one for next chromatographic step.
2. Firmly fix the ZipTip to a 10-µL adjustable pipet (back pressure may be high).
3. Wash the ZipTip five times with 10 µL of 3:2 acetonitrile/TFA solution. Discard the dispensed liquid.
4. Equilibrate the ZipTip five times with 10 µL of TFA solution. Discard the dispensed liquid.
5. Fix the peptides by slowly aspirating and dispensing 10 µL of the concentrated supernatant ten times (*see*) without lifting the pipet tip.
6. Wash (desalt) the fixed peptides four times with 10 µL of TFA solution. Discard the dispensed liquid.
7. Elute the peptides in a 500-µL microcentrifuge vial with 2 µL of 1:1 acetonitrile/TFA solution.

3.2. MALDI-TOF Peptide Mass Fingerprinting

1. In the MALDI processes, the sample is cocrystallized with an organic aromatic, usually an acidic compound (the matrix), whose main property is to absorb at the wavelength of the UV laser (generally 337 nm for nitrogen lasers or 355 nm for the tripled frequency of Nd:YAG lasers). During the desorption step, when the UV laser is firing, the aromatic moiety of the matrix (present in a very large excess of approx 10,000:1) absorbs the UV energy and the matrix sublimates, protecting the sample from decomposition or dissociation and driving it to a gaseous phase.
2. The ionization step is a proton exchange in the dense but dilating and cooling gas phase (the plume) between initially charged matrix ions or clusters and neutral peptides (ion-molecule reactions).
3. Once desorbed and ionized, the sample is accelerated in the TOF tube toward the detector, its speed (i.e., TOF in the fixed-length flight tube) being directly proportional to the root square of its m/z ratio.
4. The reflectron (an electrostatic mirror) corrects the initial kinetic energy dispersion, improving mass resolution and thus mass accuracy. (Isobaric ions with a larger kinetic energy will have a larger path in the reflectron and will be focalized with lower kinetic energy ions at the detector focal plane.)
5. Delayed or pulsed ion extraction corrects the initial spatial dispersion, thus improving again resolution and mass accuracy. (Isobaric ions spatially dispersed in the desorption plume will be differentially accelerated according to their position at the beginning of the ion extraction from the source to the analyzer.)

For more details concerning MALDI-TOF and the basics of MS, many good references are available.

MALDI-TOF mass spectrometry is the technique of choice for PMF, thanks to several features:

1. Soft ionization technique (i.e., the peptide remains intact after the MALDI process).
2. Ionization method very tolerant to mixtures of different peptides and relatively tolerant to contaminants (buffer, salts, plasticizers, and so on).
3. Highly sensitive method (owing to both the discrete MALDI process and TOF mass analyzer).
4. Good resolution, currently exceeding 15,000 (full width, half mass [FWHM]), allowing detection of monoisotopic peaks in the mass range of interest.
5. Excellent mass accuracy of the TOF analyzer (regularly in the range of 10–30 ppm).
6. Fast data accumulation.

3.2.1. On-Target Sample Deposition

3.2.1.1. Classical Dried Droplet Method

1. Prepare α-cyano-4-hydroxycinnamic acid matrix at half-saturation (approx 5 mg/mL): dissolve (sonicate for 10 min) α-cyano-4-hydroxycinnamic acid (a small

spoon tip) in 300 μL of 1:1 acetonitrile/TFA solution. Centrifuge for a few seconds to obtain a clear supernatant. Put 200 μL of the clear supernatant in a new microcentrifuge vial, and add the same volume (200 μL) of 1:1 acetonitrile/TFA solution to obtain half-saturation.

2. Rapidly (to avoid crystallization in the pipet tip) mix in a microcentrifuge vial 0.8 μL of digest solution with 0.8 μL of matrix and deposit rapidly on the MALDI target 0.8 μL of the mix. The remaining 0.8 μL can be deposited at another position on the MALDI target. Do not touch the target with the pipet tip. The droplet should be preformed at the pipet tip and deposited on the target by capillarity.
3. Allow the mix to dry and cocrystallize (*see* **Note 8**).

3.2.1.2. Dried Droplet Method on Prestructured Hydrophobic Target (*see* **Note 9**)

1. Prepare α-cyano-4-hydroxycinnamic acid matrix: weigh 10 mg of α-cyano-4-hydroxycinnamic acid, and dissolve (sonicate for 10 min) with 1 mL of 1:1 acetonitrile/TFA solution. In a new microcentrifuge vial, put 56 μL of that matrix solution and add 944 μL of 1:1 acetonitrile/TFA solution.
2. Take 0.8 μL of the supernatant obtained after the in-gel digestion step (*see* **Subheading 3.1.3.**), and mix it rapidly with 0.8 μL of matrix solution. Deposit on the MALDI target 0.8 μL of the mix. The remaining 0.8 μL can be deposited at another position on the MALDI target (*see* **step 2** of **Subheading 3.2.1.1.**).
3. Allow the mix to dry and cocrystallize.
4. Wash on-target the crystallized spot with 4 μL of TFA solution, let the solution stay in contact for 30 s, and remove the liquid by aspirating (without touching the crystals with the pipet tip).
5. Recrystallize the sample with 0.8 μL of ethanol/acetone/TFA solution.
6. Allow the mix to dry and cocrystallize.

3.2.2. Spectrum Acquisition

1. Insert the target in the mass spectrometer and wait for vacuum stabilization (below the 10^{-6} Torr range).
2. Turn the high voltage on (and ideally wait 20 min for voltage and temperature [Joule effect] stabilization).
3. Tune laser power (attenuation) and target position to obtain a good signal-to-noise ratio and monoisotopic resolution over the mass range 700 to 4000 Th.
4. Fire 10 laser shots and discard them, as they are usually noisy in the low mass range (owing to matrix and/or salts clusters).
5. Acquire and sum 80 to 200 laser shots.
6. Calibrate the mass spectrum internally with autoproteolytic trypsin ions at 842.5099 and 2211.1046 Th
7. Save spectrum (*see* **Note 10**).

3.2.3. Spectrum Annotation

Automatic monoisotopic mass assignment is possible, but we recommend a careful survey of the annotation (*see* **Notes 11** and **12**).

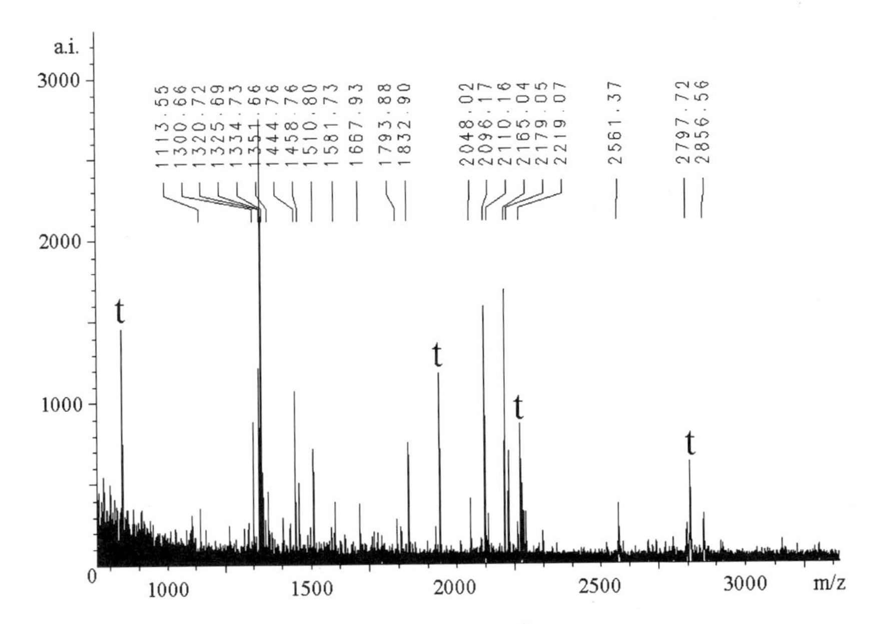

Fig. 1. MALDI-TOF peptide mass fingerprint of an in-gel digest of a 2D spot from a *Pisum sativum* protein extract.

Without spectrum smoothing, annotate (by centroiding at 70% height, or, better, by using advanced algorithms, e.g., SNAP from Bruker Daltonics) all first monoisotpic peaks apart from the main trypsin known autoproteolytic peaks (**Fig. 1**; *see* **Notes 13** and **14** for details).

3.3. Database Search Using MASCOT and Critical Review of the Search Result

The steps described below are for the MASCOT PMF search engine (Matrix Science, London, UK) that can be licensed in-house (*see* **Note 15**) or run on the distant London server (http://www.matrixscience.com/cgi/search_form.pl?SEARCH=PMF).

However, the proposed search strategy is directly adaptable to other search engines (*see* **Note 16**).

3.3.1. First Search Step

The objective of this first step (**Fig. 2.**, left raw) is to gain an overview of the data quality, to remove possible contaminants, and possibly to identify the plant protein directly.

1. Specify a global database (e.g., MSDB, NCBInr, or the cleaner Swiss-Prot), not a species-specific one, to have non-plant protein contaminants in the database.

2. Do not specify a species in the taxonomy field (i.e., "all entries"), to allow the identification of non-plant protein contaminants.
3. Allow one missed cleavage.
4. Set a large mass tolerance of 100 ppm, to evaluate the mass spectrum calibration. (The two trypsin autoproteolytic peptides used for calibrating the mass spectrum may have too low abundance and ion statistics to allow good calibration.)
5. Fixed modifications: carboxymethyl (C) or carbamidomethyl (C). For 2D-gel samples, cysteine residues are reduced and alkylated usually either with iodoacetic acid or iodoacetamide, giving carboxymethyl-cysteine or carboxyamidomethyl-cysteine (contracted in carbamidomethyl-cysteine), respectively.
6. Variable modification: none.

For good-quality data (signal-to-noise ratio, mass accuracy, low level of contaminants) and for protein referenced in the database searched, a candidate protein may appear (**Fig. 2B**).

Criteria to evaluate the candidate protein are as follows:

1. Score superior to the significant threshold value.
2. Large difference between the first-ranked protein and the following nonrelated one.
3. Protein from the studied species (or a close species with high homology) at the first rank.
4. Candidate protein MW and p*I* compatible with 2D-gel data.
5. Good mass accuracy (ideally) or bad mass accuracy (owing to calibration failure or miscalibration problem) but associated with a low and linear distribution of mass dispersion over the mass range.
6. A maximum of one miscleaved peptide for three matching peptides.
7. A minimum of five matching peptides.
8. Homogenous localization of the matching peptides in the protein sequence and sequence coverage (*see* **Note 17** and **Fig. 2C**).

3.3.2. Second and Following Search Steps

One or more steps may be required if the previous search failed to give a hit, or if too many peaks (e.g., more than 10) remain unmatched (which may indicate the presence of two or more proteins in the excised spot) (*see* **Note 18** and right raw of **Fig. 2**).

Fig. 2. Database search of **Fig. 1** peptide mass fingerprint. Left raw illustrates the first search step (described in **Subheading 3.3.1.**). Right raw illustrates subsequent searches (described in **Subheading 3.3.2.**) **(A)** First-step search parameters: large database, taxon not restrained, large mass tolerance, no variable modification. **(B)** Search engine answer. Significant hit (score above the significance threshold) for two proteins. **(C)** Incorrect top score hit details. Wrong taxon identified (but no contaminant found), too many miscleaved peptides, low mass accuracy with high mass

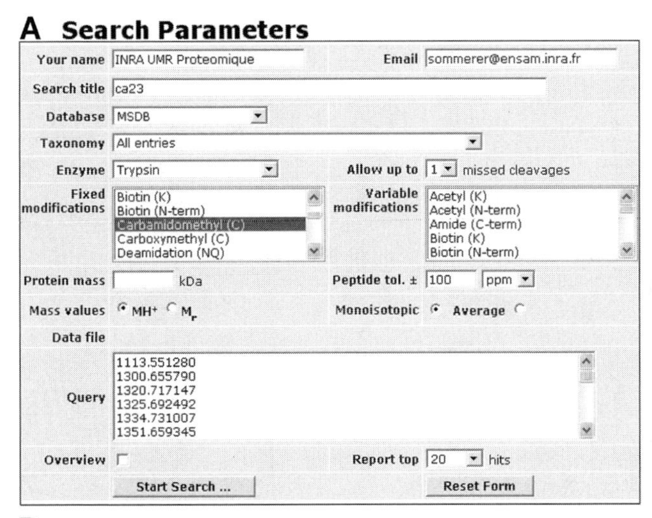

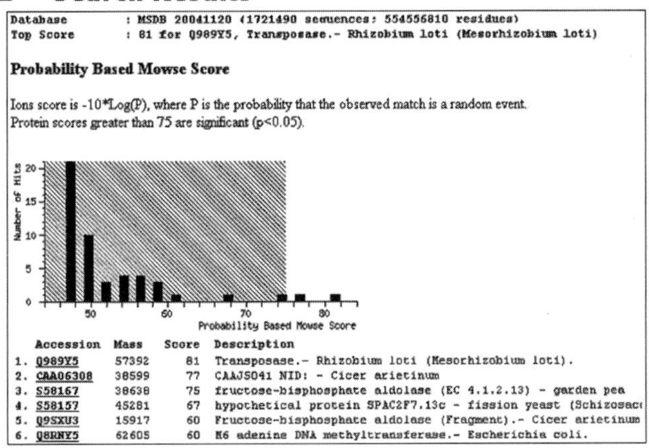

	Accession	Mass	Score	Description
1.	Q989Y5	57392	81	Transposase.- Rhizobium loti (Mesorhizobium loti).
2.	CAA06308	38599	77	CAAJ5041 NID: - Cicer arietinum
3.	S58167	38638	75	fructose-bisphosphate aldolase (EC 4.1.2.13) - garden pea
4.	S58157	45281	67	hypothetical protein SPAC2F7.13c - fission yeast (Schizosac
5.	Q9SXU3	15917	60	Fructose-bisphosphate aldolase (Fragment).- Cicer arietinum
6.	Q8RNY5	62605	60	M6 adenine DNA methyltransferase.- Escherichia coli.

C Top score hit is incorrect

Q989Y5 **Mass:** 57392 **Score:** 81 **Expect:** 0.013 **Peptides matched:** 8

Transposase.- Rhizobium loti (Mesorhizobium loti).

Observed	Mr(expt)	Mr(calc)	Delta	Start	End	Miss	Peptide
1300.66	1299.65	1299.66	-0.01	415 -	426	0	AWQALDAALDAR
1351.66	1350.65	1350.65	-0.00	374 -	384	1	HAERGHNIDFR
1444.76	1443.75	1443.66	0.09	219 -	229	1	DFTRNYEALCR
1510.80	1509.80	1509.83	-0.03	333 -	345	1	VTYSVPSRLIGYR
1581.73	1580.72	1580.82	-0.10	311 -	324	1	RTTDFTLVSADVTR
1667.93	1666.92	1666.87	0.05	378 -	391	1	GHNIDFRHFIGNLK
2179.05	2178.04	2178.10	-0.05	114 -	132	1	ALNGGEKEIFFPQHHEIGR
2219.07	2218.06	2218.21	-0.15	58 -	75	1	TRPDPLEQIWPRVEELLK

Fig. 2. (*continued from opposite page*) dispersion. **(D)** Second search step. Taxon restrained to Viridiplantae, two variable modifications selected according to sample knowledge, mass accuracy restrained to 40 ppm after calibration survey. **(E)** Search engine answer. Significant hit for two proteins, large gap between significant hits and nonsignificant ones. **(D)** Correct top score. Right taxon identified, protein MW consistent

D Search Parameters

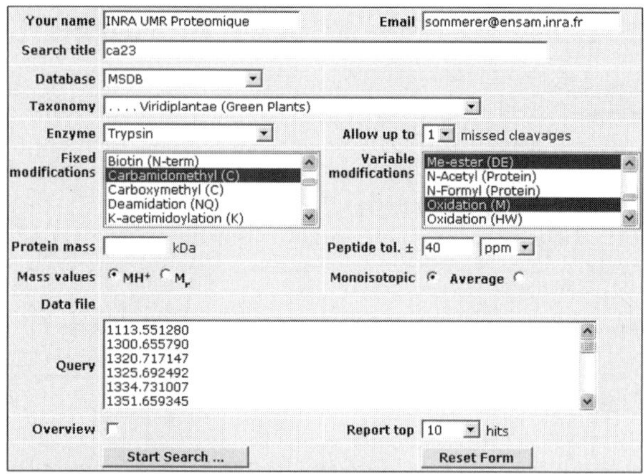

E Search Results

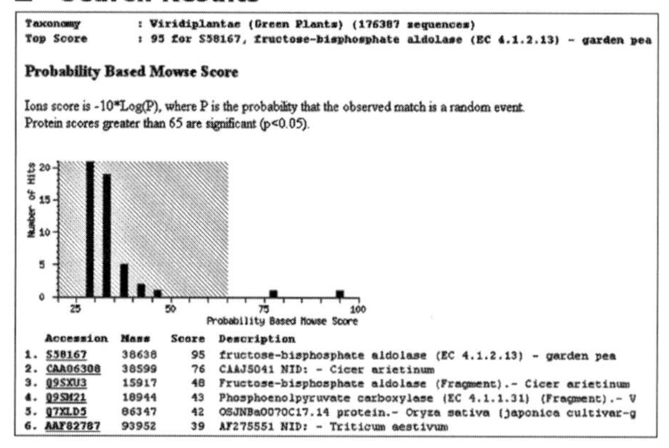

Taxonomy : Viridiplantae (Green Plants) (176387 sequences)
Top Score : 95 for S58167, fructose-bisphosphate aldolase (EC 4.1.2.13) - garden pea

Probability Based Mowse Score

Ions score is -10*Log(P), where P is the probability that the observed match is a random event.
Protein scores greater than 65 are significant (p<0.05).

	Accession	Mass	Score	Description
1.	S58167	38638	95	fructose-bisphosphate aldolase (EC 4.1.2.13) - garden pea
2.	CAA06308	38599	76	CAJ5041 NID: - Cicer arietinum
3.	Q9SXU3	15917	48	Fructose-bisphosphate aldolase (Fragment).- Cicer arietinum
4.	Q9SM21	18944	43	Phosphoenolpyruvate carboxylase (EC 4.1.1.31) (Fragment).- V
5.	Q7XLD5	86347	42	OSJNBa0070C17.14 protein.- Oryza sativa (japonica cultivar-g
6.	AAF82787	93952	39	AF275551 NID: - Triticum aestivum

F Top score hit is correct

S58167 Mass: 38638 Score: 95 Expect: 5.6e-005 Peptides matched: 13
fructose-bisphosphate aldolase (EC 4.1.2.13) - garden pea

Observed	Mr(expt)	Mr(calc)	Delta	Start	End	Miss	Peptide
1320.72	1319.71	1319.72	-0.01	239	250	0	VAPEVIAEHTVR
1325.69	1324.69	1324.69	-0.01	289	299	0	KPWTLSFSFGR
1334.73	1333.72	1333.74	-0.01	239	250	0	VAPEVIAEHTVR + Me-ester (DE)
1510.80	1509.80	1509.81	-0.01	287	299	1	GKKPWTLSFSFGR
1832.90	1831.90	1831.91	-0.01	8	24	0	YHDELIANAAYIGTPGK
2048.02	2047.01	2047.04	-0.02	6	24	1	SKYHDELIANAAYIGTPGK
2096.17	2095.17	2095.17	-0.00	84	104	0	TAAGKPFVDVLNEAGVLPGIK
2110.16	2109.15	2109.18	-0.03	84	104	0	TAAGKPFVDVLNEAGVLPGIK + Me-ester (DE)
2165.04	2164.03	2164.03	0.00	150	169	0	IGANEPSEHSIHENAYGLAR
2179.05	2178.04	2178.05	-0.00	150	169	0	IGANEPSEHSIHENAYGLAR + Me-ester (DE)
2219.07	2218.06	2218.07	-0.01	108	130	0	GTVELAGTDGETTTQGLDGLGAR
2561.37	2560.36	2560.26	0.10	105	130	1	VDKGTVELAGTDGETTTQGLDGLGAR
2856.56	2855.55	2855.49	0.07	212	238	0	ALSDHHVILEGTLLKPNMVTPGSDAPK + Oxidation (M)

Fig. 2. (*continued from previous page*) with in-gel data, good mass accuracy with low dispersion (out of the calibration range peptide masses are less accurate), correctly cleaved/miscleaved ratio below 3:1 ratio limit. Not shown data: protein p*I* consistent with in-gel data (6.7), sequence coverage of 38%, matching peptides localized all over the protein sequence. Note: the second significant hit corresponds to an homologous protein on a close species (chickpea), with fewer matching peptides.

One or more search criteria can be adapted (**Fig. 2D**):

1. Remove peptide digest peaks from identified non-plant protein contaminants ("search unmatched" option button in Mascot).
2. Set the mass tolerance according to the real observed mass accuracy of the mass spectrum (e.g., as evaluated on non-plant protein contaminants or verified on trypsin autoproteolytic fragments). Typically, a good-quality search needs mass accuracy to be below 30 ppm (the lower value the better).
3. Allow the search with nonstoichiometric or unpredicted chemical modifications (mainly methionine oxidation, aspartic and glutamic acid methylesterification, N-terminal pyroglutamylation) (*see* **Note 19**).
4. Restrain the taxon search range (e.g., to Viridiplantae).
5. Allow the search of a specific database.

The evaluation and validation criteria described in **Subheading 3.3.1.** remain necessary (**Fig. 2E** and **F**).

Identification of secondary (or minor) components in the mixture is important.

After successful identification of a protein by PMF, several peaks in the MALDI-TOF mass spectrum usually remain unidentified. These peaks can have different origins:

1. Peptide ion from other protein(s) comigrating in the electrophoresis gel.
2. Peptide ion from the same protein whose sequence differs slightly from the protein in the database (either by genome annotation mistakes or mutation).
3. Peptide ion from the same protein whose mass differs from the predicted peptide mass because of posttranslational modification (PTM).
4. Peptide ion from a contaminant protein (e.g., keratin or trypsin).
5. Nonpeptidic contaminant ion from a polymer residue or plasticizer.
6. Nonpeptidic contaminant ion with a chemical or biochemical origin.

3.4. MALDI-TOF/TOF Strategies

Proteins identified after PMF are only candidates, the scoring indicating the level of confidence. Thus, one should not forget that no sequencing step was performed to confirm this tentative identification.

Peptide sequencing on a MALDI mass spectrometer was difficult before the recently developed TOF/TOF tandem analyzers. In TOF/TOF analyzers, the first TOF allows the isolation of the peptide ion of interest; all other peptides from the path tube are discarded. A collision cell after this first TOF breaks (by collisionally induced dissociation) the peptide into sequence-specific fragments. The second TOF separates and measures the masses of these sequence-specific fragments.

3.4.1. Confirmation of Identification

To confirm an identification, one to three peptides identified from a single protein can be fragmented successively by TOF/TOF. The fragment ions must correlate with the predicted peptide sequence.

3.4.2. Identification of Unknown Proteins

In the case of an unsuccessful database search (*see* **Subheading 3.3.**), some peptides (from two to five usually) can be analyzed by TOF/TOF. De novo sequencing and/or database search of the TOF/TOF fragments of these peptide ions may allow identification, either in the studied taxon or by cross-species homology (*see* also Chapter 20 for LC-MS/MS approaches).

4. Notes

1. Automation of the gel cutting (spot excision), in-gel digestion, and on-target sample deposition is possible with specific liquid handling automates. To increase throughput, large proteomic facilities are equipped with such technologies. Automation also limits sample handling and contamination possibilities by human keratins.
2. Cutting spots that are too large would be detrimental to protein identification. Indeed, it would increase background chemical noise, lower trypsin ability to penetrate the core of the spot, and reduce the diffusion of proteolytic peptides out of the gel.
3. In 2D-PAGE, proteins are reduced (typically with DTT or mercaptoethanol) and alkylated (usually with iodoacetamide or iodoacetic acid) during equilibration of the isoelectrofocusing gel prior to SDS-PAGE. This is to break the disulfide bridges between two cystein residues and thus prevent the secondary and tertiary structure of the protein from interfering with the migration of the protein in the polyacrylamide mesh. For 1D-PAGE separated proteins, reduction and alkylation has to be done before in-gel digestion (*see* Chapter 22).
4. Certain features of trypsin make it an appropriate endoprotease choice for MS identification of proteins after in-gel digestion:
   a. Quite low molecular weight enzyme (approx 24 kDa), penetrating easily into the dehydrated gel with the rehydratation buffer.
   b. Good specificity (cuts at the C-terminus of K and R, does not cut if the next amino acid is P, cuts less efficiently if acidic moieties of the lateral chains of D/E amino acid are present in the neighborhood, cuts less efficiently if the bulky F/W/Y amino acids are present in the neighborhood because of a steric effect of the lateral chains). Commercial porcine trypsin is treated with *N*-tosyl-L-phenylalanine chloromethyl ketone (TPCK) to avoid nonspecific chymotrysic activity
   c. Occurrence of K and R generates most of the proteolytic peptides in the mass range 600 to 4000 Daltons, which is compatible with high mass accuracy MALDI-TOF-MS.

   d. Commercial porcin trypsin lysine side chains are treated with reductive methylation to give few autoproteolytic ions, but two intense ions at 842.5099 Th and 2211.1046 Th can be used for internal mass spectra calibration.
   e. Proteolytic peptides are easily protonated (and thus give good signal in mass spectra) because of their two basic sites: one at the N-terminus of the peptide and one on the lateral chain of the last C-terminus amino acid (K or R). Moreover, this property is very useful in ESI, giving mainly doubly charged species that generate informative b/y fragment ion series by MS/MS (fixed charge fragmentation; *see* Chapter 25).
5. For a specific purpose (e.g., trying to increase the sequence coverage of the protein), other endoproteases can be used with the same protocol:
   a. Chrymotrypsin cuts after F/W/Y, generating more and thus shorter peptides.
   b. Endo Lys-C cuts after K (not R).
   c. Protease V8 cuts after DE or E (depending on the digestion buffer pH).
6. On-ice fresh trypsin preparation and trypsin in-gel penetration are recommended to minimize autoproteolysis.
7. Minimum digestion time is 3 h at 37°C, and overnight digestion should be considered as a maximum. Overnight digestion may be the most convenient. For long digestion time (e.g., overnight), digestion can take place at room temperature instead of 37°C. This reduces the amount of trypsin autoproteolytic signal.
8. In the dried droplet method, the goal is to obtain homogeneous and small crystals. Long crystallization will give large crystals, whereas rapid crystallization will give smaller crystals. To hold the crystallization process without changing the solvent composition, the target can be dried under vacuum or heated (30–40°C) in an oven for few minutes.
9. Some MALDI-TOF mass spectrometer manufacturers have developed prestructured hydrophobic/lyophobic targets (e.g., AnchorChip™ technology from Bruker Daltonics). These targets allow precise positioning of the sample and concentration on the target. Thus the sensitivity is increased by a factor 5 to 20. Careful cleaning of the target is required to avoid contaminant (mainly plasticizers) concentration on the anchor. Peptide digest can be spotted on the target directly after in-gel digestion (*see* **Subheading 3.1.1.**), or after peptide extraction and concentration (*see* **Subheading 3.1.4.2., step 1**).
10. From the mass spectrometry point of view, two criteria are essential to allow reliable protein identification:
    a. Mass accuracy (ideally below 30 ppm, nowadays MALDI-TOF instruments being able to achieve 10–15 ppm with careful internal calibration of the spectra).
    b. Sensitivity to detect as many ions as possible and to increase sequence coverage.
11. Annotation of the spectrum is very often done automatically with good efficiency algorithms, but if the protein identification fails, we recommend surveying the annotation and limiting it to:
    a. The 40 most intense peaks as a maximum, assuming that the 2D-gel spot contains one or two identifiable proteins (i.e., with close expression level).
    b. A minimum signal-to-noise ratio of 3:1.

12. For intense mass spectra, annotation of low signal ions may be detrimental for protein identification. We recommend exclusion of signals whose intensity is below 2% of the most intense protein-specific peptide ion.
13. The main known autoproteolytic peaks from Promega modified sequencing grade porcine trypsin are: 842.51, 870.54, 1045.56, 1940.94, 2211.10, 2225.12, 2239.14, 2283.18, 2299.18, and 2807.31.
    A complete list of less intense ions can be found at: http://prospector.ucsf.edu/ucsfhtml4.0/misc/trypsin.htm.
14. We recommend annotating even known contaminants (i.e., non-plant keratin peaks), performing the first database search step with these contaminants, and filtering (remove) them only after precise database identification of these contaminants. Indeed, keratin proteolytic peaks (e.g., 1475.74 Th) may belong to different keratin subclasses (e.g., keratin 10 and keratin 2), the other keratin proteolytic detected peaks being specific to one or both of these subclasses.
15. In-house licensing of the search engine (e.g., Mascot):
    a. Allows the installation of specific or proprietary databases.
    b. Preserves the confidentiality needed for some studies.
    c. Makes the search time independent of the network speed and the distant server busyness.
    d. Allows batch searches.
    e. Allows the archiving of all searches without time limit.
16. Search engine URLs:
    a. Mascot: http://www.matrixscience.com/cgi/search_form.pl?FORMVER=2&SEARCH=PMF.
    b. MS-Fit (protein prospector): http://prospector.ucsf.edu/ucsfhtml4.0/msfit.htm
    c. ProFound: http://prowl.rockefeller.edu/profound_bin/WebProFound.exe.
    d. Peptide Search: http://www.mann.embl-heidelberg.de/GroupPages/PageLink/peptidesearch/Services/PeptideSearch/FR_PeptideSearchFormG4.html.
    e. PepMAPPER: http://wolf.bms.umist.ac.uk/mapper/.
    f. Aldente: http://www.expasy.org/tools/aldente.
17. The sequence coverage information alone has limited sense, partly depending on the candidate protein predicted MW.
    For a high MW protein candidate, a low sequence coverage may indicate that:
    a. Some parts of the protein are less accessible to trypsin.
    b. The protein contains little tryptic cleavage sites (e.g., PRP: proline-rich proteins), so tryptic peptide sizes are too big to be detected (out of the mass range).
    c. Peptide ionization competition does not allow the ionization of some peptides.
    d. The in-gel spot protein is only a fragment of this candidate protein.
    e. The database does not contain exactly the sequence of the in-gel protein (mutation, or PTM, but more typically sequence polymorphism owing to cross-species identification).

On the other hand, for a low MW protein candidate, a high sequence coverage may solely indicate that the protein described in the database is too short: the sequence coverage may be irrelevantly too high.

For these reasons, it is more appropriate to survey the localization of the matching peptides in the protein sequence than the sequence coverage: are the matching peptides localized all over the protein sequence (which may indicate a valid candidate)? Or, on the contrary, are they localized on a limited part of the protein (which may indicate the identification of a fragment protein if this correlates with in-gel MW and p*I* measurements)?

18. Multiple proteins are likely to be present in gel pieces excised from a monodimensional gel. This has to be taken into account to adapt database searches (*see* Chapter 22 for a more specific discussion).
19. Widely opening the variable modification search parameter will increase both search time and nonspecific protein hits dramatically. If more variable modification are needed for the search, we recommend constraining the search (e.g., lower the mass tolerance, describe the taxon more precisely, set the number of missed cleavage sites to null).

References

1. Karas M., Bachman D., Bahr U., and Hillenkamp F. (1987) Matrix-assisted ultraviolet laser desorption of non-volatile compounds. *Int. J. Mass Spectrom. Ion Proc.* 53–68.
2. Karas M. and Hillenkamp F. (1988) Laser desorption ionization of proteins with molecular masses exceeding 10,000 daltons. *Anal. Chem.* 2299–2301.
3. Karas M., Bahr U., Ingendoh A., and Hillenkamp F. (1989) Laser desorption/ionisation mass spectrometry of proteins of mass 100,000 to 250,000 Dalton. *Angewandte Chemie Int. Ed.* 760–761.
4. Fenn J. B., Mann M., Meng C. K., Wong S. F., and Whitehouse C. M. (1989) Electrospray ionization for mass spectrometry of large biomolecules. *Science* 64–71.
5. http://nobelprize.org/chemistry/laureates/2002/index.html.
6. Tanaka K., Waki H., Ido Y., Akita S., Yoshida Y., and Yoshida T. (1988) Protein and polymer analysis up to m/z 100,000 by laser ionisation time-of-flight mass spectrometry. *Rapid Comm. Mass Spectrom.* 151–153.
7. *Arabidopsis* Genome Initiative (2000) Analysis of the genome sequence of the flowering plant. *Nature* 796–815.
8. Jensen O., Wilm M., Schevschenko A., and Mann M. (1999) Sample preparation methods for mass spectrometric peptide mapping directly from 2-DE gels, in *Methods in Molecular Biology: 2-D Proteome Analysis Protocols*, vol. 112 (Link, A. J., ed.), Humana, Totowa, NJ, pp. 513–530.
9. De Hoffmann, E., Charette, J., and Stroobant, V. (1999) *Spectrométrie de Masse, Cours et Exercices Corrigés*, 2ème édition. Dunod, Paris.
10. De Hoffmann, E. and Stroobant, V. (2002) *Mass Spectrometry: Principles and Applications*, 2nd ed. John Wiley & Sons, Hoboken, NJ.

11. Throck Watson, J. (1997) *Introduction to Mass Spectrometry*, 3rd ed. Lippincott-Raven: New York.
12. James, P. (ed.) (2000) *Proteome Research: Mass Spectrometry (Principles and Practice)*. Springer Verlag, New York.
13. Cotter, R. J. (1997) *Time-of-Flight Mass Spectrometry: Instrumentation and Applications in Biological Research*. American Chemical Society, Washington, DC.
14. Vestal M. L., Juhasz P., Hines W., and Martin S. (1998) in *Proceedings of the 46th ASMS Conference on Mass Spectrometry and Allied Topics*, Orlando, FL.

20

Protein Identification Using Nano Liquid Chromatography–Tandem Mass Spectrometry

Luc Negroni

Summary

Tandem mass spectrometry is an efficient technique for the identification of peptides on the basis of their fragmentation pattern (MS/MS scan). It can generate individual spectra for each peptide, thereby creating a powerful tool for protein identification on the basis of peptide characterization. This important advance in automatic data acquisition has allowed an efficient association between liquid chromatography and tandem mass spectrometry, and the use of nanocolumns and nanoelectrospray ionization has dramatically increased the efficiency of this method. Now large sets of peptides can be identified at a fentomole level. At the end of the process, batch processing of the MS/MS spectra produces peptide lists that identify purified proteins or protein mixtures with high confidence.

Key Words: Tandem mass spectrometry; MS/MS; nanoLC; peptide analysis; protein identification; EST databases.

1. Introduction

Proteomic analysis is a multistep process that typically involves protein extraction, fractionation, separation, and mass spectrometry. The method of choice for protein separation is 2D electrophoresis. Then the in-gel-purified proteins are identified on the basis of the peptide analysis with MS or MS/MS (tandem mass spectrometry). MS spectrometry with matrix-assisted laser desorption ionization-time of flight (MALDI-TOF) is generally used first when the complete sequence of the organism is available. It is a high-throughput method for protein identification and does not require peptide purification. On the other hand, MS/MS analysis is more laborious but can identify individual peptides with or without databases (de novo sequencing). Theoretically, only one peptide identified with MS/MS is sufficient to find the corresponding pro-

From: *Methods in Molecular Biology, vol. 335: Plant Proteomics: Methods and Protocols*
Edited by: H. Thiellement, M. Zivy, C. Damerval, and V. Méchin © Humana Press Inc., Totowa, NJ

tein, whereas a minimum of four to five peptides are needed for identification with MS and peptide mass fingerprinting. This explains why MS/MS is more convenient with incomplete sequences databases (for example, expressed sequence tag [EST] databases) or when conserved sequences of other species are used for identification. In contrast to 2D electrophoresis, which separates proteins before their identification, another approach proposes a simple hydrolysis of the protein mixture without protein separation; then multidimensional liquid chromatography (LC) and MS/MS are used to separate the complex mixtures and to identify thousands of peptides ***(1–3)***. Between these two approaches, an intermediate method uses sodium dodecyl sulfate-polyacrylamide gel electrophoresis (SDS-PAGE) for partial protein separation and LC-MS/MS for protein identification ***(4–6)***.

MS/MS for proteomics mainly uses ion trap (IT) and quadrupole-time of flight (Q-TOF) instruments equipped with an electrospray source. Electrospray utilizes a potential difference between the needle and the inlet of the mass spectrometer. This induces charged droplets at the needle outlet. In the mass spectrometer, the droplets are evaporated and the ions (i.e., the charged peptides) are detected according to their mass-to-charge ratio (*m*/*z*). In the Q-TOF mass spectrometers, the first cell transfers the ions from the electrospray source to the quadripole. The quadripole is set to transmit only one ion (parent or precursor ion) to the collision cell. In the collision cell, the parent ion is fragmented and the fragments (daughter or fragment ions) are measured in a second mass analyzer (TOF tube). As the fragmentation occurs along the peptide backbone, the amino acid sequence can be determined from the fragment ion scan (MS/MS scan). In the IT mass spectrometers, ions selection, fragmentation, and fragment analysis are performed in the same cell, the IT. In practice, the MS/MS instruments switch between MS and MS/MS. The MS scans are obtained when the ion selection and the fragmentation are off. This MS scan allows detection of all the peptide ions and choice of the parent ion to fragment.

One limitation of electrospray is its sensitivity to salts. Therefore desalting procedures must be performed to clean up peptide samples. The most efficient way is to use reverse-phase high-performance liquid chromatography (RP-HPLC). RP-HPLC not only removes salts but also purifies and concentrates peptides. These three points explain why MS/MS coupled with liquid chromatography (LC-MS/MS) is the most frequent configuration. Since electrospray is a concentration-dependent technique, the decrease of column diameter is a major parameter to increase the sensitivity ***(7)***. Nanocolumns from 50 to 100 μm ID are now currently used to increase peptide concentration at the inlet of the mass spectrometer. In addition, reduction of the flow rate to 150 to 300 nL/min with these nanocolumns increases the effectiveness of spray ***(8)***.

During the last decade, use of the nanoLC-MS/MS technique has increased dramatically. It has become a widespread technique used in most of the mass spectrometry laboratories. Because of the electrospray ionization (ESI) process, analyzed peptides are often different from those detected with MALDI ionization. These differences explain why nanoLC-MS/MS is as an interesting complement to MALDI-TOF MS analysis ***(9,10)***.

2. Materials

2.1. Trypsin Digestion

1. Natural polypropylene polymerase chain reaction (PCR) 12-well tube strips (0.2 mL/well) and corresponding storage box holder (VWR International, Fontenay-sous-Bois, France).
2. Multichannel pipetor (VWR International).
3. Eppendorf Repeater® and Eppendorf Combitips (VWR International).
4. Thermomixer (Eppendorf, Le Pecq, France) and modular heating blocks for well tube strips (VWV International).
5. Modified trypsin (Promega, Madison, WI) stock solution at 200 ng/μL.
6. Destaining solution: 10% acetic acid, 40% methanol in dH_2O.
7. Washing solution: 50 m*M* NH_4HCO_3 in dH_2O (stock solution 1 *M*).
8. Digestion buffer: 25 m*M* NH_4HCO_3, 20% methanol in dH_2O.
9. Shrinking solution: 100% acetonitrile.
10. Extraction solution: 0.5% trifluoroacetic acid (TFA), 40% acetonitrile, 10% methanol in dH_2O.

2.2. HPLC (NanoLC)

1. NanoHPLC Ultimate system: Famos-Switchos-Ultimate (Dionex-LC Packings, France).
2. C18 reverse-phase column, 300 μm × 5 mm (trap column) and 75 μm × 10 cm (analytical column; PEPMAP C18, Dionex LC-Packings).
3. Fluidic fittings and tubing (Upchurch).
4. Fused silica tubes (Upchurch).
5. Sample solution: 0.1% TFA, 2% acetonitrile in dH_2O.
6. Loading solvent (for trap column washing at 7 μL/min; *see* **Fig. 1**): 0.1% formic acid, 0.02% TFA, 2% acetonitrile.
7. HPLC solvent A: 0.1% formic acid, 2% acetonitrile in dH_2O.
8. HPLC solvent B: 0.1% formic acid in 95% acetonitrile, 5% dH_2O.

2.3. Mass Spectrometry (Ion Trap)

1. Nanospray needle 360/20/10, uncoated (New Objective, Woburn).
2. Ion trap mass spectrometer, LCQ XP Plus (Thermo Electron, Courtaboeuf, France).
3. Acquisition software Xcalibur 1.3 (Thermo Electron).
4. Data processing, BIOWORKS 3.1 (Thermo Electron).

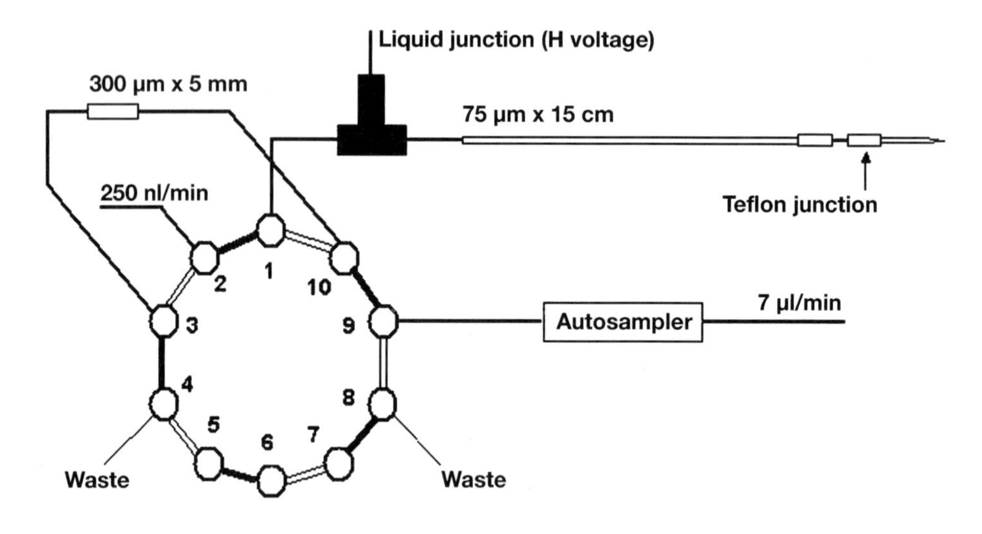

Fig. 1. Schematic of the columns on the switching valve. A robust system is obtained with the liquid junction between the preconcentration column and the 75-µm column. The column acts as a filter that prevents needle clogging. At this position, the liquid junction does not degrade chromatographic resolution. A short, 20-µm ID tube is set between the column and the needle to protect the column if the needle needs to be changed. Positions 1-2 and 10-1 correspond to the two positions of the valve for sample loading on a precolumn and back-flush elution, respectively.

3. Methods

3.1. In-Gel Protein Hydrolysis and Peptide Extraction (One-Day Protocol)

This protocol is efficient with in-gel proteins from SDS-PAGE, 2D-PAGE, and BN-PAGE ***(11)*** gels stained with Coomassie Blue-G250. It does not involve the use of a digestion robot. The use of a multichannel pipetor and a microplate format allows the processing of a plate of 96 samples per day. The standard method does not use reduction and alkylation. These two steps are laborious and have no effect on most of the identifications, even with SDS-PAGE and BN-PAGE sample. In our hands, these two steps appeared to be an important source of keratin contamination. However, in case of cysteine-rich proteins, the reduction and alkylation step may increase the protein coverage (*see* **Note 1**). A multichannel pipetor is used to remove the different liquids and an Eppendorf repeator is used to dispense the solutions. The different steps are performed at 37°C with agitation using a thermomixer.

1. Put the pieces of gel (1.5 mm diameter) in the tube strips. They can be stored without liquid during several weeks at 4°C and for several months at –20°C. Sealing is carried out with simple painter's tape.
2. Destaining: 2X 30 min with 50 µL destaining solution.
3. Washing: 15 min with 50 µL washing solution.
4. Shrinking: 15 min with 50 µL shrinking solution.
5. Washing: 15 min with 50 µL washing solution.
6. Shrinking: 15 min with 50 µL shrinking solution.
7. Shrinking: 15 min with 50 µL shrinking solution.
8. Put the strip holder on ice for 10 min, add 5 µL/well trypsin (20 ng/µL in digestion buffer), wait 10 min, and put the strip holder back in the thermomixer for 4 h. Add 5 µL of digestion buffer after 2 h of trypsin digestion to compensate for evaporation.
9. Extraction 1: add 15 µL of extraction solution, and wait 15 min. Transfer the extract.
10. Extraction 2: add 15 µL acetonitrile, and wait 15 min. Transfer the extract.
11. Drying: evaporate at room temperature overnight or use a Speed Vac for 30 min to dryness (*see* **Note 2**).

Optional steps: reduction and alkylation:

1. Reduction and alkylation steps are carried out with 10 m*M* (1.5 mg/mL) dithiothreithiol (DTT) and 55 m*M* (10 mg/mL) iodoacetamide (IAA), both in 50 m*M* NH_4 HCO_3.
2. Add 25 µL of 10 m*M* DTT after **step 4**. Allow the reaction to proceed at 56°C for 1 h.
3. Remove liquid and add 25 µL of 55 m*M* IAA. Allow reaction to proceed in the dark for 45 min at room temperature. Remove the liquid and go to **step 5**.

3.2. HPLC and ESI Settings

1. The major problems with nanoLC are misalignments or bad tube cutting. They generate time delay and liquid dilution. If a dead volume is present between the 75-µm column and the mass spectrometer, a tailing effect is observed. If a dead volume is present before the 75-µm column, the peptides are eluted with an increased retention time but without important chromatographic degradation.
2. The tubing ends should be checked with a magnifying glass before connection to be sure they are clean and perpendicularly cut.
3. **Figure 1** presents a schematic of nanoLC set up with an RP trap column and an RP nanocolumn, as described in the following steps. A liquid junction is used to apply voltage. This method extends the lifetime of the needle (*see* **Note 3**).
4. Use a 50-µm ID tube from the autosampler to the trap column, a 30-µm ID tube from the trap column to the switching valve, a 20-µm ID tube from the switching valve to the analytical column, and a 20-µm ID tube from the column to the nanospray needle.
5. The connection of capillary tubing to the valve port (standard port for 1/16-inch OD tubing) is carried out with Tefzel sleeves, Tefzel ferrules, and PEEK nuts.

Table 1
HPLC Gradient and Valve Position

Time (min)	Flow rate (nL/min)	%B	Valve position
Part	0	250	0 1–2
2.5	250	4	1–2
Part 2			
2.5	250	4	10–1
22	250	20	10–1
27	250	40	10–1
Part 3			
27	250	40	10–1
31	250	100	10–1
37	250	100	10–1
38	250	0	10–1
41	250	0	1–2
50	250	0	1–2

These fittings hold a lower pressure than stainless ferrules and nuts, but they limit the risk of breaking the capillary tubing during tightening.

6. The LC Packings nanoLC column is equipped with a 20-μm ID outlet tube. This tube is cut at 3 cm (i.e., 9 nL). A Teflon junction (Teflon tubing, 300 μm ID) is used to connect it to the nanospray needle, which is shortened to 3 cm. This connection does not stand pressure in case of clogging of the nanospray needle. This disadvantage may be of interest: leakage of the Teflon junction means that of the needle tip is clogging, which is generally associated with an irregular spray.
7. Use a liquid junction (microtee connector) to apply the voltage. In our hands, it is the more robust configuration. The microtee is set before the 75-μm column, and the electrode is a platinum wire (*see* **Note 4**). With this liquid junction, the needles for the spray are uncoated (360 OD, 20 μm ID, 10-μm tip ID).
8. If the pump device uses a split to deliver a nL/min flow rate without a flow detector (old versions of LC-Packings instruments), the flow rate must be checked with a 25-μL syringe connected to the 75-μm column. The LC Packings system is a nanoLC that uses a 1:1000 split; the pump flow rate should be 150 to 200 μL/min for 150 to 250 nL/min on a 75-μm column (*see* **Note 5**).

3.3 HPLC Separation

1. **Table 1** presents a typical gradient at 250 nL/min. Flow optimization is possible: for example, 300 nL/min at the beginning and during regeneration, to decrease delays and 200 nL/min during peptide detection. A development of this aspect was optimized with the peak parking system ***(12)***.

2. During part 1, from 0 to 3 min, the valve is in position 1-2 (*see* **Table 1** and **Fig. 1**). An autosampler injects 5 μL of sample, which is loaded and trapped on the reverse-phase precolumn at 7 μL/min. At the beginning of part 2, the valve switches to position 10-1 and connects the trap column to the 75-μm column. Then the sample is eluted in back-flush mode using the nanoHPLC gradient at 250 nL/min. Part 3 is a regeneration step for column cleaning.
3. The gradient depends on column and fittings. It should be adjusted with a control such as a bovine serum albumin (BSA) tryptic digest (100–300 fmols injection).
4. The digests are dissolved in 10 to 15 μL of sample solution with TFA. TFA is used as an ion pairing agent in the loading solvent because it allows better trapping of peptides on the precolumn in comparison with formic acid. On the other hand, formic acid is used as an additive for HPLC solvents A and B because ionization is strongly suppressed with TFA.
5. In these conditions a complete run is achieved in 45 to 50 min. **Figure 2** is an example of a chromatogram obtained with a complex mixture of peptides. The peptides are eluted from 11 to 35 min. The MS base peak trace is similar to the UV profile and allows one to check the quality of the separation. TIC of MS/MS scans shows a high signal-to-noise ratio in the range of peptide separation.

3.4. Mass Spectrometer Settings

1. With nanoLC, peptides are eluted in 10 to 40 s. Thus the instrument should perform one cycle in less than 10 s (i.e., one survey scan in MS mode and the associated MS/MS scans). Some of the steps described here are specific to an LCQ IT equipped with Xcalibur 1.3 software for instrument control, but the main principle is the same for all MS/MS instruments: MS/MS is data dependent, i.e., MS/MS acquisition is automatically triggered when a peak is detected in MS mode.
2. An efficient acquisition method of Xcalibur is the "Nth" method. This method is a succession of three scans: (1) full MS scan (m/z 350–1900), (2) ZoomScan (scan of the two major ions with a higher resolution), and (3) MS/MS of these two ions (*see* **Note 6**).
3. The ionization voltage with a liquid junction is set from 1.2 to 1.6 kV. The need to increase the voltage to more than 1.6 kV is generally the sign of problems with the needle or buffers. In this case, first change the nanospray needle and test it with 1.3 kV. If the problem persists, change the buffers (*see* **Note 7**).
4. The fragmentation parameters for an LCQ instrument are $Q_z = 0.22$, activation time = 50 ms, collision energy = 35 to 45%. A Q_z decrease from 0.25 to 0.22 allows detection of lower masses in an MS/MS scan (*see* **Note 8**).
5. Dynamic exclusion is used to prevent successive MS/MS analyses on the same ion (*see* **Note 9**).
6. An exclusion list is used to avoid noise analysis. Some polysiloxane ions are generally present at masses 371, 445, and 462 Th *(13)*.
7. With an LCQ IT, the "Nth" method, and a 50-min acquisition time, the size of the raw file is about 12 Mo.

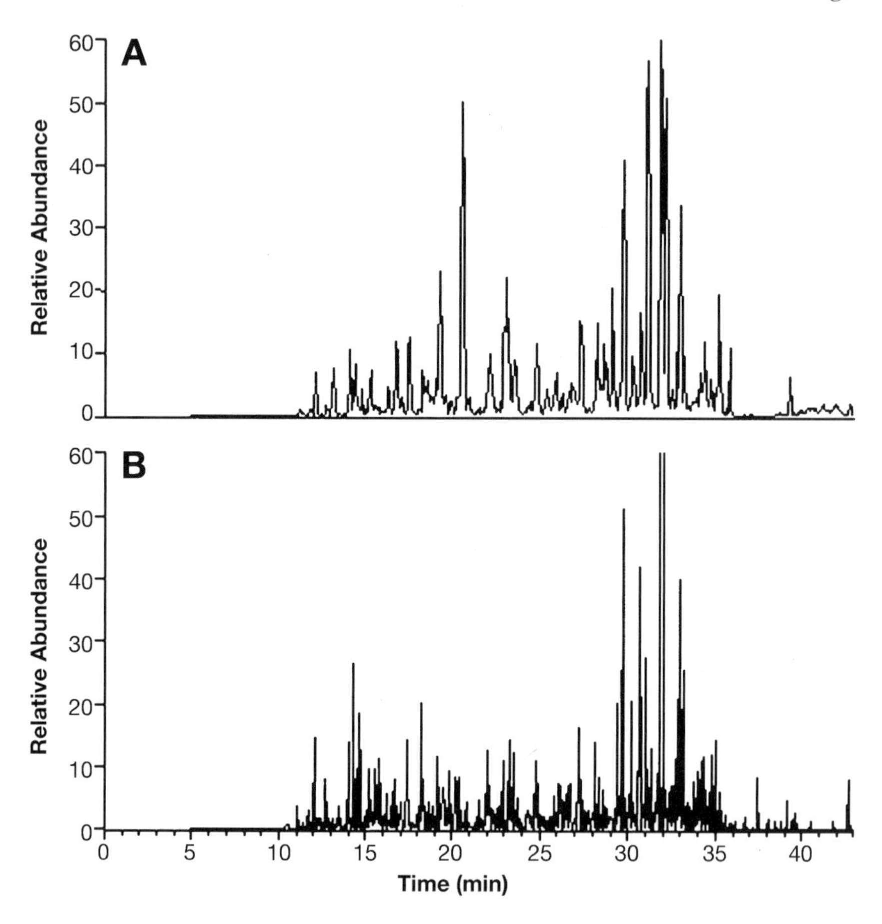

Fig. 2. Example of a nanoLC-MS/MS chromatogram. **(A)** MS base peak trace. This corresponds to the intensity of the most intense ion in the different MS scan. Each peak corresponds to the elution of one or several peptides. **(B)** Corresponds to these total ion current (TIC) of MS/MS scans. In this example, peptides are eluted between 11 and 35 min after sample injection. The last peaks correspond to column regeneration and valve switching artifacts.

3.5. Data Processing and Interpretation

1. Data analysis is a three-step procedure.
   a. In the first step, the MS/MS spectra are extracted from the raw acquisition files and filtered.
   b. In the second step, they are compared with the theoretical MS/MS spectra calculated from the chosen sequence database. For one experimental MS/MS spectrum, the mass of the parent ion is used to extract the "isobaric" peptides from the database. If a specific enzyme like trypsin is used, only the isobaric

peptides that can be produced according to its specificity are extracted from the database. Then the software converts the peptide sequences in MS/MS theoretical spectra using the fragmentation rule and annotation proposed by Roepstorff and Fohlman ***(14)***.

c. The last step is the computation of a correlation score between the experimental spectrum and the different theoretical spectra extracted from the sequence database. Here data are analyzed with BIOWORKS software (previously named Sequest). With this software, the correlation score is called XC.

2. In contrast to Edman sequencing, the identified sequences are not based on the actual detection of the successive amino acids. If the exact sequence is not present in the database used for the query, peptide identification is not possible. Moreover, the peptide identifications are based on correlations to preexisting sequences. BIOWORKS, like the other algorithms, uses a scoring system to propose the most probable peptide for each MS/MS spectrum ***(15,16)***. This means that a wrong peptide sequence can be presented as the most probable candidate, leading to false protein identification. To limit this problem, protein identification should be considered significant when at least two MS/MS spectra identify two peptides of the same protein.
3. Protein databases are generally used for the identification of proteins from model sequenced organisms. For other species, EST contig databases are suitable (*see* **Note 10**). The protein and EST databases are available on FTP sites (*see* **Note 11**).
4. The databases can be preprocessed using a database indexation. This function allows a 10-fold decrease of the query time. Unfortunately, the indexing fixes the cleavage specificity and the authorized mass modifications. If you choose to index a database, set methionine oxidation (+16) as a variable modification (i.e., Met will possibly be considered oxidated) (*see* **Note 12**).
5. The extraction of MS/MS spectra is based on TIC intensity and time range interval. If peptides are eluted between 20 to 40 min, use these values as time range parameters. The TIC intensity must be set according to the minimal TIC of correct MS/MS spectra (usually 1 to 5×10^5 with an LCQ). The MS/MS extraction produces 50 to 200 MS/MS spectra.
6. The sample processing is processed in batch mode with the sample queue used for acquisition. In the case of the Swiss-Prot database and BIOWORKS installed on a personal computer with a Pentium® 4 processor and 512 MB DDR SDRAM, each spectrum is compared with the indexed database in 1 to 2 s. The result for a nanoLC analysis is thus obtained in 1 to 2 min.
7. First use the Swiss-Prot database. This database allows checking for keratin contamination and trypsin peptides and contains well-annotated sequences.
8. Second, use the specific plant database. If an EST database is used, it will be translated to a protein database by the software. Set the frame of translation to the six possibilities (i.e., the three forward and the three reverse frames).
9. The results corresponding to the different samples are stored in directories in accordance with the sample name. Retrieve this result with the Multiconsensus result selection.

10. Filter the peptide identification with XC threshold values of 1.4 to 1.7, 2 to 2.5, and 2.5 to 3 for, respectively, mono-, di-, and tri-charged peptides. Some false peptide identifications remain in the list. They often correspond to noisy spectra. This explains the need to check the identification manually when only two MS/MS spectra are used for identification.
11. Remove all the proteins identified with only one MS/MS spectrum. If the proteins are identified with only two MS/MS spectra, visually check that the most intense peaks of the experimental spectra are used for the matching.
12. If "no enzyme" is selected for the query parameter and trypsin is used for digestion, check that candidate peptides are really tryptic peptides. This checking appears to be a relevant criterion that unambiguously validates identification.
13. **Table 2** is an extract of BIOWORKS summary output and corresponds to the nanoLC-MS/MS analysis of an in-gel trypsin digest from 2D gel spot. Among the hundreds of MS/MS spectra produced during the nanoLC-MS/MS run, less than 100 MS/MS spectra had a sufficient intensity to be extracted and used for the database query. The EST database used for the query was not indexed, no enzyme specificity was set, and methionine oxidation was considered variable. The peptide identification was filtered with an XC threshold (XC magnitude up to 1.5, 2, and 3 for, respectively, mono-, di-, and tri-charged peptides). The 19 MS/MS spectra that passed the filter identified trypsic peptides (*see* **Table 2**). As the search was performed without specifying the enzyme, lysine or arginine at the C terminus of the identified peptides increases the credibility of the match. The Charge value is the charge of the parent ions. Most of the trypsic peptides are doubly charged. However three mono-charged ions identify peptides with significant XC and confirm the identification obtained with the doubly charged ion (peptide DTGLFGVYAVAK) or identify new peptides (IDAVDASTVK and VTEEDVIR). Each MS/MS spectrum correlated with several peptides but only the first hit is presented in **Table 2**. Delta Cn is the delta correlation value $(XC_1 - XC_2)/XC_1$ between the first and the second hit and should be greater than 0.1. The ions value is the number of experimental ions from the MS/MS spectrum matched with the theoretical ions predicted from the peptide sequence. For example, 18/22 says that the experimental MS/MS spectrum contained 18 of the 22 calculated ions. In the peptide sequences, M* corresponds to methionine sulfoxide; methionine oxidation (+16) is partial for the peptides ILVANTAMDTDK and VLFGGGVVPEMVMAK.

4. Notes

1. It is recommended to verify the efficiency (or inefficiency) of reduction and alkylation with a representative set of samples subjected or not to DTT/IAA. Check (1) the number of identified peptides, (2) the keratin contamination, and (3) the presence of carboxyamidomethylated cysteins (+ 57 Daltons).
2. The SpeedVac is known to induce difficulties in the solubilization of pellets and dried proteins. It is suspected to decrease peptide recovery. With a 30 µL extract, evaporation occurs after one night at room temperature or after 1 to 2 d at 4°C.

Table 2
Summary Output Example of "Protein Identification" With EST Database

File,Scans(s)	Sequence	MH+	Charge	XC	Delta Cn	Ions
#1 GPI:C:432431| Maize cDNA interbank contig						
dv02010302n045,1812-1817	DSFLDEGFILDK	1398.68	2	4.684	0.507	18/22
dv02010302n045,1202-1207	GASEHVLDEAR	1312.61	2	3.800	0.541	16/22
dv02010302n045,1319-1329	IHPM*TIIAGYR	1287.69	2	2.123	0.394	14/20
dv02010302n045,1357-1362	IIAHGNCFVNR	1413.74	2	3.480	0.416	18/22
dv02010302n045,1219-1222	ILVANTAMDTDK	1307.66	2	4.033	0.522	18/22
dv02010302n045,1309-1314	ILVANTAM*DTDK	1291.66	2	3.410	0.497	20/22
dv02010302n045, 1444-1447	SDLM*NIAM*TTLSSK	1543.74	2	3.000	0.375	16/26
dv02010302n045, 1209-1224	SLHIDNPAAK	1065.57	2	2.812	0.270	15/18
dv02010302n045,1317-1324	VEIDITCAPR	1173.59	2	3.846	0.333	17/18
dv02010302n045,1402-1407	VIEEIM*IGEDR	1319.66	2	4.052	0.574	16/20
dv02010302n045,1637-1642	VLFGGGVVPEM*VM*AK	1553.76	2	3.618	0.525	15/26
dv02010302n045,1749-1752	VLFGGGVVPEMVM*AK	1537.76	2	3.051	0.499	18/26
#2 GPI:C:435461| Maize cDNA interbank contig						
dv02010302n045,1657-1662	DTGLFGVYAVAK	1240.66	2	3.987	0.546	19/22
dv02010302n045,1664-1669	DTGLFGVYAVAK	1240.66	1	2.323	0.393	14/22
dv02010302n045,1212-1217	IDAVDASTVK	1018.54	1	1.945	0.388	13/18
dv02010302n045,1527-1532	IPTPELFAR	1043.59	2	2.699	0.234	14/16
dv02010302n045,1377-1372	LSTDPTTTNM*LVAK	1507.77	2	2.812	0.462	12/26
dv02010302n045,1439-1442	RIPTPELFAR	1199.69	2	2.965	0.365	13/18
dv02010302n045,1204-1242	VTEEDVIR	960.50	1	2.400	0.299	10/14

3. The liquid junction is robust enough to run hundreds of samples without any problems (except in case of clogging).
4. The electrode wire for the liquid junction must resist oxidation. Platinum wires used in the electrophoresis apparatus may be used.
5. Splitting is an efficient method to reduce the dead volumes coming from the pump, pump tubing, and mixer. On the other hand, the flow rate depends on the pressure induced by the split tube (generally 10 μm ID for nanoLC) and the column. Any change in these pressures will modify the flow rate. A recent improvement of the instruments was obtained with the active flow splitting system.
6. The full MS scan is a "low-resolution" MS scan on a large *m/z* range. The zoom MS scan is a "high-resolution" MS scan on a short m/z range (10 *m/z* centered on the parent ion). This allows determination of the charge and then the mass of the peptide. Here two ions are selected for MS/MS. The number of selected ions depends on the elution time of the peptides, the complexity of the peptide mixture, the speed of each scan, and the sensitivity of the instrument. With a complex mixture, remove the MS ZoomScans, and increase the number of selected ions (up to five). This increases the number of peptides analyzed.
7. The video camera mounted on the nanospray source lets the user see the tip and spray for easy tuning and troubleshooting. When the voltage is switched on after the LC pumps are started, the droplet must be ejected and any liquid must disappear from the tip.
8. A comprehensive tutorial on the IT and the effect of Q_z on low-mass cutoff can be found at http://www.abrf.org/ABRFNews/1996/September1996/sep96iontrap.html.
9. Dynamic exclusion is a list of ions corresponding to the masses of the parent ions detected during the few last seconds. This prevents repetitive MS/MS scan on the same parent ion. Moreover, this setting allows analysis of the less abundant ions during coelution.
10. The ESTs are DNA sequences (usually 200–500 nucleotides long) that are generated by systematically sequencing the ends of cDNA libraries. EST contigs are consensus sequences of EST clusters. Singlets are ESTs that were not found to be similar to any other EST. Usually the EST contigs and the EST singlets are grouped; the corresponding database is called tentative unique genes (TUG). TUG = TUC + TUS. TUC and TUS are short for, respectively, tentative unique contigs and tentative unique singlets.
11. See ftp://ftp.ebi.ac.uk/pub/databases/IPI/current/ for the IPI (International Protein Index), the new standard, ftp://ftp.ncbi.nih.gov/blast/db/FASTA/ for the most current databases, and http://www.plantgdb.org/download.php for plant sequence.
12. The selection of variable modification increases the size of the indexed database, increases the query time, and decreases the correlation score. No modification other than Met oxidation should be selected for a first-pass query, to limit false identifications.
13. Low contamination of keratin and trypsin peptides is not a problem in the case of MS/MS analysis. Difficulties only occur when their corresponding peaks are the

most intense. In this case, the minor peaks, which correspond to the peptides of interest, may be masked and not analyzed.

Acknowledgments

This work was supported by IFR 87, Intitut National de la Recherche Agronomique and Centre National de la Recherche Scientifique. The author would like to thanks Thierry Balliau and Marlene Davanture for their technical assistance.

References

1. Washburn, M. P., Wolters, D., and Yates, J. R. 3rd. (2001) Large-scale analysis of the yeast proteome by multidimensional protein identification technology. *Nat. Biotechnol.* **19,** 242–247.
2. Froehlich, J. E., Wilkerson, C. G., Ray, W. K., et al. (2003) Proteomic study of the *Arabidopsis thaliana* chloroplastic envelope membrane utilizing alternatives to traditional two-dimensional electrophoresis. *J. Proteome Res.* **2,** 413–425.
3. Lin, D., Tabb, D. L., Yates, J. R. III (2003) Large-scale protein identification using mass spectrometry. *Bioch. Biophys. Acta* **1646,** 1–10.
4. Ferro, M., Salvi, D., Brugiere, S., et al. (2003) Proteomics of the chloroplast envelope membranes from *Arabidopsis thaliana. Mol. Cell. Proteomics* **2,** 325–345.
5. Carter, C., Pan, S., Zouhar, J., Avila, E. L., Girke, T., and Raikhel, N. V. (2004) The vegetative vacuole proteome of *Arabidopsis thaliana* reveals predicted and unexpected proteins. *Plant Cell.* **16,** 3285–3303.
6. Friso, G., Giacomelli, L., Ytterberg, A. J., et al. (2004) In-depth analysis of the thylakoid membrane proteome of *Arabidopsis thaliana* chloroplasts: new proteins, new functions, and a plastid proteome database. *Plant Cell.* **16,** 478–499.
7. Abian, J., Oosterkamp, A. J., and GelpÌ, E. (1999) Comparison of conventional, narrow-bore and capillary liquid chromatography/mass spectrometry for electrospray ionization mass spectrometry: practical considerations. *J. Mass Spectrom.* **34,** 244–254.
8. Wilm, M. and Mann, M. (1996) Analytical properties of the nanoelectrospray ion source. *Anal. Chem.* **68,** 1–8.
9. Bodnar, W. M., Blackburn, R. K., Krise, J. M., and Moseley, M. A. (2003) Exploiting the complementary nature of LC/MALDI/MS/MS and LC/ESI/MS/MS for increased proteome coverage. *J. Am. Soc. Mass Spectrom.* **14,** 971–979.
10. Lim, H., Eng, J., Yates, J. R. 3rd, et al. (2003) Identification of 2D-gel proteins: a comparison of MALDI/TOF peptide mass mapping to mu LC-ESI tandem mass spectrometry. *J. Am. Soc. Mass Spectrom.* **14,** 957–970.
11. Schagger, H. and von Jagow, G. (1991) Blue native electrophoresis for isolation of membrane protein complexes in enzymatically active form. *Anal. Biochem.* **199,** 223–231.
12. Davis, M. T. and Lee, T. D. (1997) Variable flow liquid chromatography-tandem mass spectrometry and the comprehensive analysis of complex protein digest mixtures. *J. Am. Soc. Mass Spectrom.* **8,** 1059–1069.

13. Schlosser, A. and Volkmer-Engert, R. (2003) Volatile polydimethylcyclo-siloxanes in the ambient laboratory air identified as source of extreme background signals in nanoelectrospray mass spectrometry. *J. Mass. Spectrom.* **38,** 523–525.
14. Roepstorff, P. and Fohlman, J. (1984) Proposal for a common nomenclature for sequence ions in mass spectra of peptides. *Biomed. Mass Spectrom.* **11,** 601.
15. Yates, J. R. 3rd, Eng, J. K., and McCormack, A. L. (1995) Mining genomes: correlating tandem mass spectra of modified and unmodified peptides to sequences in nucleotide databases. *Anal. Chem.* **67,** 3202–3210.
16. Eng, J., McCormack, A. L., and Yates, J. R. III. (1994) An approach to correlate tandem mass spectral data of peptides with amino acid sequences in a protein database. *J. Am. Soc. Mass Spectrom.* **5,** 976–989.

21

Two-Dimensional Nanoflow Liquid Chromatography–Tandem Mass Spectrometry of Proteins Extracted from Rice Leaves and Roots

Linda Breci and Paul A. Haynes

Summary

In this chapter we present a detailed protocol for the large-scale identification of proteins present in rice leaf and root tissue samples using 2D liquid chromatography-tandem mass spectrometry of protein extracts. This is performed using biphasic (strong cation exchange/reversed phase) columns with integral electrospray emitters operating at nanoliter flow rates, a technique known by the acronym Mudpit (for multidimensional protein identification technique). The protocol involves harvesting of leaves and roots from rice plants, preparing protein extracts from the harvested tissues, preparing proteolytic digests of the extracted proteins, making a biphasic capillary column with an integral electrospray emitter, performing two-dimensional chromatographic separation of peptides with data-dependent tandem mass spectrometry, and the use of database searching of the acquired tandem mass spectra to identify peptides and proteins. This protocol is adaptable for use with a wide variety of plant materials and can be used to identify large numbers of proteins present in a specific tissue, organ, organelle, or other subcellular fraction. In addition to the detailed protocol, we also present the results of a representative experiment showing the identification of more than 1000 distinct proteins from rice leaf and root samples in two Mupdit experiments.

Key Words: Nanoflow liquid chromatography-tandem mass spectrometry; two-dimensional nanoflow chromatography; multidimensional protein identification technique (Mudpit); plant protein extraction; rice leaf; rice root.

1. Introduction

One of the most important advances in the field of proteomics in recent years has been the development of nanoflow 2D liquid chromatography-tandem mass spectrometry (LC-MS/MS) of complex peptide mixtures ***(1,2)***, commonly known as Mudpit (3) (an acronym derived from multidimensional

From: *Methods in Molecular Biology, vol. 335: Plant Proteomics: Methods and Protocols*
Edited by: H. Thiellement, M. Zivy, C. Damerval, and V. Méchin © Humana Press Inc., Totowa, NJ

peptide identification technique). In this approach a complex mixture of proteins is first digested into peptides with a protease such as trypsin to produce an even more complex mixture of peptides. This peptide mixture is then fractionated using two orthogonal dimensions of chromatographic separation, taking advantage of the fact that chromatographic theory dictates that the number of theoretical plates achieved in such a separation is the product, rather than the sum, of the theoretical plates achieved in each dimension of separation when considered in isolation ***(4,5)***.

The main strengths of this approach are that it is not biased against proteins with extremes of size, hydrophobicity, or isoelectric point, as 2D protein gel electrophoresis often is, and it can be used to generate a large number of protein identifications from a given sample in a relatively short time frame. The weaknesses of this approach are that it is technically challenging, is often not reproducible, and is consequently not well suited to analysis of subtle changes in protein expression between two samples. There have been some recent reports ***(6,7)*** using Mudpit analysis to quantify differential protein expression between two cellular states, by employing N^{15} metabolic labeling of organisms such as yeast and *E. coli*. This approach is necessarily limited to those organisms that can be metabolically labeled to completion with N^{15} and is therefore not well suited to analysis of plant tissues.

The term Mudpit was originally used to describe a highly specific experimental configuration, in which a biphasic column containing strong cation exchange and reversed-phase materials was packed into a single fused silica capillary with an integral nanoelectrospray emitter tip. In subsequent reports the term has been adapted to refer to almost any experimental system involving two dimensions of chromatographic separation. Many such systems involve numerous connections, dead volumes, and even fraction collection (with inherent losses), so their performance in terms of either chromatography or efficiency of peptide and protein identification is significantly worse than the original implementation. It is, of course, possible to perform two-dimensional chromatographic separations in an off-line dual column system, but, apart from the fact that it is incorrect to refer to this as a Mudpit analysis, these systems typically require significantly larger amounts of protein to begin with owing to losses inherent in fraction collecting and protein absorption onto the larger surfaces exposed in a larger bore chromatographic system.

Proteomics in plants is still at a relatively early stage of development compared with proteomics in other species. As an illustration, at the time of this writing, a simple search of the NCBI Pubmed publication database for the keyword "proteomics" retrieves 4704 articles, whereas a simple search with the keywords "plant proteomics" retrieves 260 articles. This means that approximately 95% of all research in proteomics is performed in species other than plants. Much of the research being published currently still relies on the use of 2D gel electrophoresis to resolve proteins prior to identification by MS techniques. However, there are an increasing number of reports using non-2D gel-based approaches such as off-line 2D-LC fractionation ***(8)***, sodium dodecyl sulfate-polyacrylamide gel electrophoresis (SDS-PAGE) gel fractionation ***(8–10)***, blue native gel fractionation ***(11)***, solvent partitioning fractionation ***(10)***, and Mudpit using biphasic nanoLC columns ***(12)***.

Several factors contribute to the fact that the adaptation and application of newer technologies has been slower in the plant proteomics field. First among these is the fact that there is still relatively little plant genome sequence information available. The complete genomes of both rice ***(13–15)*** and *Arabidopsis* ***(16–18)*** have been published, but they are the only plant genomes completed so far. This will obviously change in the near future as, for example, the current genome sequencing initiative in maize ***(19)*** reaches completion.

The other major issue in plant proteomics is that sample preparation of plant tissue is simply much more difficult than preparing samples of, for, example, bacterial cells or human physiological fluids for the same type of experiment. Plant tissue material typically contains high levels of cell wall proteins and carbohydrates for structural rigidity, chloroplast machinery proteins for photosynthesis, and the essential metabolic protein ribulose bisphosphate carboxylase (RuBisCo), estimated to be the single most abundant protein on the planet. The presence of a small number of very abundant proteins in the presence of a larger group of less abundant proteins represents a difficult analytical challenge, especially in Mudpit analysis, in which such proteins can occupy precious instrumental analysis time, leading to less than successful analysis of other peptides and proteins present in a complex mixture. The other physical problem presented by plant tissue material is that the techniques commonly used for extraction of protein from plant material are mostly based on grinding tissue in liquid nitrogen and precipitating protein with trichloroacetic acid and acetone ***(20)***. This produces a pale brown fibrous material that contains protein but also very high levels of undissolved structural components. It can be very difficult to extract sufficient protein from this material to use in a Mudpit experiment, but this can usually be overcome by combining repeated extractions prior to analysis.

Despite these caveats, Mudpit analysis still represents one of the best ways to produce a large amount of protein identification information from a given sample in a relatively short period. As more genome sequence information from plant species becomes available, it will become increasingly feasible to generate an initial expressed proteome catalog of any plant tissue material by using Mupdit analysis and cross-species identification of peptides by nanoLC-MS/MS. We have previously used this type of approach to characterize amyloplast proteins in wheat *(9)*, even though very little wheat genome sequence data was available, by taking advantage of the availability of the rice genome sequence *(15)*. It is important to note that this is only possible when using MS/MS approaches to identify individual peptides, as protein identification by matrix-assisted laser desorption ionization (MALDI) peptide mass fingerprinting has a low success rate for anything except the most well-known proteins *(21)*.

The other important point to consider is that plants have relatively large genomes, such as the more than 30,000 expected genes in rice, which means they have a very large potential array of expressed protein features. Even the best Mudpit analysis can only identify around 1000 to 1500 proteins in a single experiment, so a Mudpit experiment on plant tissue material should not be considered an exhaustive proteomic analysis. More protein expression and identification information can be obtained by multiplexing of orthogonal separations, such as protein fractionation by size or charge prior to mupdit analysis on each fraction, or protein fractionation by SDS-PAGE or isoelectric focusing *(22)* prior to nanoLC-MS/MS on each fraction, or an iterative system combining the latter with gas-phase fractionation in the mass spectrometer *(23,24)*, or any of the other innumerable combinations of these and other fractionation approaches.

In this report we demonstrate the use of a true capillary biphasic Mudpit system for the analysis of protein extracts prepared from the leaf and root of rice (*Oryza sativa*). We present a detailed protocol starting from plant seedlings and provide representative protein identification information and an overall protein identification summary for each tissue sample analyzed.

2. Materials

2.1. Instrumentation and Hardware

1. Tandem mass spectrometer: Thermo LCQ DecaXP Plus (Thermo Electron, San Jose, CA).
2 Nanoelectrospray ionization source: manufactured to specification in the University of Arizona Chemistry Department Machine Shop, also available in commercial versions from Thermo Electron, New Objectives (Woburn, MA), or Brechbuehler (Houston, TX).
3. Surveyor Autosampler (Thermo Electron).
4. Microbore HPLC Surveyor (Thermo Electron).
5. P2000 Laser Puller (Sutter Instrument, Novato, CA).
6. Helium pressure column-packing cell (Brechbuehler, Houston, TX).
7. Teflon ferrules for pressure cell, 1/16 in. to 0.4 mm (Chromatography Research Supplies, Louisville, KY).
8. SpeedVac vacuum centrifuge (Savant Instruments, Farmingdale, NY).
9. Rotating Shaker, LabQuake (Barnstead, Dubuque, IA).

2.3. HPLC Buffers and Column Packing Materials

1. Buffer A: 0.1% formic acid (98–100%, EMD Chemical, Gibbstown, NJ).
2. Buffer B: acetonitrile, 0.1% formic acid.
3. Buffer C: 95% 250 m*M* ammonium acetate, 5% acetonitrile, 0.1% formic acid.
4. Buffer D: 95% 1.5 *M* ammonium acetate, 5% acetonitrile, 0.1% formic acid.
5. Fused silica: 100 µm ID × 360 µm OD (Polymicro Technologies, Phoenix, AZ).
6. Reverse-phase packing material: Zorbax™ Eclipse® XDB-C18 5 µm (Agilent Technologies, Palo Alto, CA).
7. Cation exchange packing material: Poly Sulfoethyl A™ (PolyLC, Columbia, MD).
8. C18 Sep-Pak solid phase extraction cartridges (Waters, Bedford, MA), and a 10-mL disposable plastic Luer-lock syringe.
3. 9Spec-Plus PT C18 solid-phase extraction pipet tips (Varian, Lake Forest, CA).

2.4. Chemicals, Buffers, Solutions, and Enzymes

1. Milli-Q water (18 MΩ resistance).
2. Leaf resuspension buffer: 10% trichloroacetic acid (TCA), 0.07% (v/v) mercaptoethanol, in acetone.
3. EDTA wash solution: 0.07% (v/v) mercaptoethanol (Sigma, St. Louis, MO), 2 m*M* EDTA (Sigma, St. Louis, MO), in acetone (EMD Chemical).
4. HPLC Grade acetonitrile (EMD Chemical).
5. HPLC Grade methanol (EMD Chemical).
6. 0.1% Formic acid solution in water.
7. 0.1% Formic acid solution in 90% acetonitrile.
8. 100 m*M* Ammonium bicarbonate solution.
9. 100 m*M* Calcium chloride solution.
10. Proteomics grade Trypsin (Sigma): 1 µg/µL in 10 m*M* HCl.
11. Endoproteinase Lys-C (Roche Diagnostics, Indianapolis, IN): 0.2 µg/µL in 10 m*M* HCl.
12. Ethanol (EMD Chemical).
13. TCA (EMD Chemical).
14. Acetone (EMD Chemical).
15. Urea (Bio-Rad, Hercules, CA).

3. Methods

3.1. Leaf and Root Tissue Harvesting and Protein Precipitation

1. Cut green nonsenescent leaves from the middle of the plant, immediately place in a Ziploc plastic bag, and freeze. If there is no freezer readily available, place on ice until they can be stored at –20°C.
2. For root tissue, dig up plants and cut the roots off at least 1 in. below the start of the green stem. Shake the roots and dip rapidly into three successive large beakers of water to remove soil, then immediately place in a Ziploc plastic bag and freeze, or temporarily store on ice.

3. Weigh out 2.5 g of frozen leaf or root tissue with any stems and other extraneous materials removed. Cut the tissue into small pieces with chilled scissors, into a chilled ceramic mortar and pestle. Grind it to a fine powder in the presence of liquid nitrogen. More than one application of liquid nitrogen may be necessary to keep the tissue frozen.
4. To begin protein extraction, transfer the frozen leaf or root tissue powder to a 40-mL centrifuge tube, and resuspend the powder in 25 mL leaf resuspension buffer. Shake and mix thoroughly to ensure that all the powder has been resuspended. Let stand at –20°C for 45 min and shake again. Centrifuge the sample for 15 min at 35,000*g*.
5. Remove the supernatant with a glass pipet, trying not to disturb the pellet. Wash the pellet with the same volume of EDTA wash solution. Shake to disperse the pellet, let stand for 5 min on ice, and centrifuge again for 15 min at 35,000*g*. Repeat washings at least twice, or more if needed, until leaf tissue protein pellet is no longer green.
6. Lyophilize the pellet. The resulting dry powder should be pale brown and contains protein as well as other cell wall and fibrous materials.

3.2. Preparing Protein Sample from TCA/Acetone Powder

1. Weigh out 20 mg of the TCA/acetone powder for each sample (prepared in **Subheading 3.1.**) in 1.5-mL microcentrifuge tubes with screw caps.
2. Add 1 mL of 100 m*M* Tris-HCl (pH 8.5) to each tube, vortex for 1 min, and spin at 18,000*g* for 10 min. Remove and discard supernatant (*see* **Note 1**).
3. Pipet 500 µL of 8 *M* urea into each tube. Vortex for 5 min, place in a sonicating water bath for 10 min, and then place on a rotating shaker for 30 min.
4. Spin the pellet down at 18,000*g* for 10 min. The supernatant of approximately 400 µL now contains protein extracted from the leaf tissue, usually in the range 0.1 to 0.5 µg/µL (*see* **Note 2**). Remove the supernatant to another tube.
5. Repeat the 8 *M* urea extraction in **steps 3** and **4** with a second aliquot of 8 *M* urea, and combine this with the first supernatant, adjusting the volume used so that the combined supernatants total 600 µL.

3.3. Preparing Protease Digests of Extracted Protein

1. To the 600 µL of 8 *M* urea containing extracted protein, add 5 µL of endoproteinase Lys-C working solution, and digest at 37°C overnight on a rotating shaker (*see* **Note 3**).
2. Add to each sample: 2300 µL of 100 m*M* ammonium bicarbonate, 30 µL of 100 m*M* calcium chloride, 60 µL of acetonitrile, 10 µL of trypsin working solution. This gives a final mixture of 1.6 *M* urea, 76 m*M* ammonium bicarbonate, 1 m*M* calcium chloride, 2% (v/v) acetonitrile, 3.3 ng/µL trypsin.
3. Digest at 37°C for 24 to 48 h on a rotating shaker. Add 60 µL of formic acid to give a final concentration of 2%, and store at –20°C if necessary until further processing.

4. Equilibrate a C18 Sep-Pak for use with each sample, by rinsing with 15 mL of 0.1% formic acid in 90% acetonitrile, followed by four rinses with 20 mL of 0.1% formic acid in water.
5. Load the 3 mL of protease digest solution onto an equilibrated C18 Sep-Pak, collect the flowthrough, and slowly pass it over the C18 material twice more. Wash the C18 Sep-Pak with 20 mL of 0.1% formic acid in water to remove salts and other loosely bound materials. Elute the peptides from the C18 Sep-Pak with 3 mL of 0.1% formic acid in 90% acetonitrile.
6. Concentrate the eluted peptides down to a volume of approximately 500 μL in a SpeedVac vacuum centrifuge in Eppendorf centrifuge tubes. Centrifuge the samples at 16,000 rpm for 10 min and, if any pelleted material is visible, remove the supernatant to another tube and discard the pellet. Continue drying the samples down in the SpeedVac vacuum centrifuge to a volume of approximately 50 μL, and then dilute to 200 μL with 0.1% formic acid in water.
7. Equilibrate a Spec-Plus C18 solid phase extraction pipet tip for use with each sample, by rinsing with 300 μL of 0.1% formic acid in 90% acetonitrile, followed by 5 rinses with 300 μL of 0.1% formic acid in water.
8. Load the 200 μL of peptide solution onto an equilibrated Spec-Plus C18 solid phase extraction pipet tip, collect the flowthrough, and slowly pass it over the C18 material twice more (*see* **Note 4**). Wash the C18 material five times with 300 μL of 0.1% formic acid in water, and elute the peptides with 300 μL of 0.1% formic acid in 90% acetonitrile.
9. Concentrate the eluted peptides down to a volume of approximately 10 μL in a SpeedVac vacuum centrifuge in Eppendorf centrifuge tubes. This solution is then ready to load directly onto a biphasic Mudpit column.

3.4. Preparing a Two-Dimensional (SCX/RP) NanoHPLC Column with Integral Spray Emitter

1. The chromatography column is prepared by pulling a 25-cm piece of 100 μm I.D./360 μm O.D. fused silica capillary tubing in a laser puller to achieve a 3- to 8-μm tip at one end of the fused silica capillary. This acts as both the nanospray needle and the chromatography column, allowing the chromatography to be performed adjacent to the entrance to the mass spectrometer *(25)*.
2. A microcentrifuge tube containing the packing material suspended in methanol is seated into the pressure cell and the lid is secured. The blunt end of the capillary column is threaded through a Teflon ferrule in the lid to a depth consistent with the packing material, the ferrule is tightened, and the cell is pressurized to 500 psi, forcing the suspended material and methanol into the column. The methanol flows through the top of the column while the packing material is contained by the narrow 3- to 8-μm opening of the pulled capillary tip. The capillary is packed with 8 cm of reverse-phase (C-18) material followed by 4 cm of strong cation exchange (SCX) material.
3. The packed column is connected to the HPLC using a PEEK T-piece fitting, which allows for voltage to be applied upstream of the column using a gold wire

electrode (connected to the high-voltage power supply of the mass spectrometer) to form a liquid junction with the column eluent buffers.

4. Wash the column for 1 h with HPLC buffer B at a flow rate of 1 μL/min to ensure the mixed bed is thoroughly packed and settled.
5. Reequilibrate the column for 1 h with HPLC buffer A at a flow rate of 1 μL/min prior to use.

3.5. Two-Dimensional NanoHPLC Separation of Peptides

Orthogonal methods of separation are performed online, with continual elution of the HPLC buffers through the two-phase column and into the mass spectrometer to ensure no loss of sample during loading, salt fractionation steps, or column washing. The first dimension of peptide separation is by charge, and the second is by hydrophobicity.

1. The peptides are loaded onto the SCX bed by injection of 10 μL of sample from a sample well into the 10-μL autosampler injection loop (*see* **Note 5**).
   a. The peptides are transferred from the autosampler loop using 5% B isocratic buffer that passes through the loop for 25 min at a flow rate of 1 μL/min.
   b. The flow then continues at a reduced rate of 400 nL/min for 35 min to thoroughly transfer the sample and wash the column.
   c. Any peptides that pass through the cation exchange material will bind to the reverse-phase material. Any peptides that pass through both resins will be detected and fragmented in the mass spectrometer. The total run time is 70 min.
2. A reversed-phase gradient is applied that flows 5 to 50% buffer B at a rate of 400 nL/min for 60 min. Any remaining peptides are then removed from the column by application of a 5-min gradient from 50 to 98% buffer B for 5 min followed by a hold at 98% buffer B for 5 min. The column is then reequilibrated by a rapid (1 min) drop to 5% buffer B and an increase in flow rate to 1 μL/min that continues for 24 min. The total run time is 95 min (*see* **Note 6**).
3. Peptides are then transferred from the SCX bed onto the reverse-phase bed in a stepwise elution, by sending a 4-min plug of the 250 m*M* ammonium acetate buffer through the column in steps of increasing concentration, with each salt step followed by a reversed-phase gradient (as described in **step 2**) to resolve peptides transferred in the ion exchange step. Those peptides least tightly bound will transfer at lower concentrations of buffer C, whereas those tightly bound will require greater concentrations of buffers C or D.
   a. A flow rate of 400 μL/min is used to first wash with 5% buffer B for 5 min.
   b. A percentage (X%) of buffer C then flows through the column for 4 min, where $X = 10\%$ for the first salt step.
   c. The column is washed for 7 min with 5% buffer B and then a reverse-phase gradient flows 5 to 50% buffer B for 60 min.
   d. Any remaining peptides are then removed from the column by application of a 5-min gradient from 50 to 98% buffer B for 5 min followed by a hold at 98% buffer B for 4 min.

   e. The column is then reequilibrated by a rapid (1 min) drop to 5% buffer B and an increase in flow rate to 1 µL/min that continues for 24 min.
   f. Subsequent steps then follow sequentially with the percentage of buffer C changing (X = 20, 30, 40, 50, 60, 70, 80, 90, 100). Each of the 10 run times is 105 min.
4. After all 250 m*M* ammonium acetate salt steps have been performed, a final salt step with 50% buffer D is used in place of X% buffer C in the preceding description. This step ensures that all peptides strongly bound to the SCX will be transferred to the reverse-phase material. The reverse-phase gradient is also modified to include a rapid gradient development and an extended wash time with 98% B for 20 min to maximize recovery of strongly bound peptides. The total run time is 105 min. **Table 1** shows the chromatographic steps described in **Subheadings 3.** and **4.**

3.6. Data-Dependent Tandem Mass Spectrometry Using a Thermo Electron LCQ Deca XP Plus

1. Basic instrument parameters: Scan range for MS, m/z 400 to 1500; 5 microscans, injection time 500 ms; minimum MS signal for triggering to MS/MS 1E+05; isolation width for selection of precursors for fragmentation is 2.7 Daltons, default charge state of peptides selected for MS/MS = 2; applied electrospray voltage = 1.8 kV; dynamic exclusion is on for 7-min window, with repeat count = 2 and repeat duration = 0.5 min; exclusion mass widths are low = 0.8 amu and high = 1.5 amu; normalized collision energy = 35.0%, activation $Q = 0.25$, activation time = 30 msec; three most intense ions are selected for fragmentation, and fragmented in the order $n = 3, 2, 1$.
2. Using these parameters, the mass spectrometer continually acquires data throughout the 13 steps of HPLC separation described above, while it is programmed to acquire tandem mass spectra upon acquisition of a signal of predetermined intensity, an indication of peptides eluting from the column. Dynamic exclusion is used to ensure that an abundant peptide and its related isotopes are not fragmented repeatedly, thus maximizing the distribution of different peptide ions selected for fragmentation.

3.7. Protein Identification Using Database Searching

Peptides present in the protease-digested samples are identified by matching uninterpreted tandem mass spectra to protein sequence databases. Protein identifications are assembled from identification of peptides.

1. The entire set of tandem mass spectra collected during all 13 chromatographic steps are searched against an appropriate protein sequence database using one of several software programs that are available for this, such as Sequest (Thermo Electron) ***(26,27)***, Mascot (Matrix Sciences, London, UK) ***(28,29)***, or Xtandem (open source software, available from the Manitoba Centre for Proteomics at http://www.proteome.ca/opensource.html) ***(30,31)***.

Table 1
Chromatographic Steps Used in Mudpit Analysis

Step	Buffer C concentration (%)	Buffer D concentration (%)	Ammonium acetate concentration (m*M*)	Description
1	0	0	0	Initial isocratic load
2	0	0	0	Initial acetonitrile gradient
3	10	0	25	First salt step
4	20	0	50	Second salt step
5	30	0	75	Third salt step
6	40	0	100	Fourth salt step
7	50	0	125	Fifth salt step
8	60	0	150	Sixth salt step
9	70	0	175	Seventh salt step
10	80	0	200	Eighth salt step
11	90	0	225	Ninth salt step
12	100	0	250	Tenth salt step
13	0	50	750	High salt step, plus extended acetonitrile gradient

2. In most biological sample preparations there are some endogenous protease activities present, so some nontryptic peptides are produced during digestion. For this reason it is best not to specify identification of fully tryptic peptides only in search parameters, as this can cause significant loss of information.
3. Peptide identification data need to be sorted and filtered to remove as much extraneous information as possible. Unfortunately, there is no single definitive set of criteria that can be applied to ensure that all peptides appearing in the output from a database-searching program are correct and that no correctly assigned peptides have been excluded. A review of database searching parameters in the published literature will rapidly reveal that there is very little consensus on this issue. It is best to start with a set of published parameters and refine and optimize these using datasets of known protein digests prepared in the same laboratory using the same instrumentation. The accuracy of peptide assignments can also be improved by using postsearching statistical significance assessments ***(32,33)*** or comparison with repeated searching against a reversed protein sequence database ***(31,34,35)***.
4. One area of contention in the reporting of Mudpit data is the inclusion, or not, of protein identifications based on the identification of a single peptide. It is common to find that approximately half of all the protein identifications in a Mudpit experiment are based on single peptide identifications, so this is an important

area to consider. It is clear that some of these may well be correct, as there may only be one good tryptic peptide produced from a small protein, or there may be modifications masking the identification of other peptides, or there may be multiple tryptic peptides produced but only one actually selected for fragmentation. On the other hand, it is also clear that very large numbers of single peptide-based protein identifications can be made in most Mudpit datasets, and these numbers can be manipulated greatly by subtle changes to database searching parameters, so there is a high probability of a false-positive identification being made. It is best to report the number of proteins based on two or more peptide identifications and the number including single peptide-based identifications as two separate figures *(36)* (*see* **Subheading 3.8.**).

3.8. Results Example: Mudpit of Leaf and Root Tissue Prepared from Mature Rice Plants

3.8.1. Plant Growth and Harvesting

Rice plants (*Oryza sativa*, cv. Nipponbare), were grown from seed in a temperature-controlled greenhouse under a 12-h light 29°C/12-h dark 21°C regime. Humidity was maintained at 30%, and plants were grown in pots containing 50% Sunshine Soil Mix and 50% nitrohumus. Leaf and root samples were collected 50 d after germination and were pooled from multiple plants.

3.8.2. Protein Extraction, Sample Preparation and Mudpit Analysis

Harvested leaves and roots were ground to a fine powder using liquid nitrogen in a mortar and pestle. Protein extracts were prepared using TCA/acetone as described in **Subheading 3.1.**, protease digests of extracted protein were prepared as described in **Subheadings 3.2.** and **3.3.**, and the complex peptide mixtures were analyzed as described in **Subheadings 3.4.** to **3.7.**

3.8.3. Spectral Database Searching and Analysis

All tandem mass spectral data were searched against a database of rice protein sequences downloaded from publicly available resources at NCBI (www.ncbi.nlm.nih.gov), supplemented with an in-house contaminants file including trypsin, Lys-C, keratin, albumin, casein, and other common laboratory contaminants.

Data were searched using Xtandem and analyzed using the Global Proteome Machine website tools (www.thegpm.org). Xtandem search parameters included: fragment ion monoisotopic mass error 0.5 Daltons, parent ion monoisotopic mass error 2.0 Daltons plus or minus, spectrum dynamic range 100, spectrum total peaks 50, maximum valid expectation value > 0.1. Differential modification of methionine with 16 Daltons was allowed, to account for oxidation.

Table 2
Summary of Number of Proteins Identified in Mudpit Analysis of Rice Leaf and Root Protein Preparations

	Leaf (51,443 spectra)	Root (39,673 spectra)	Non-redundant total	Common to both leaf and root	Unique to leaf	Unique to root
Total proteins (two or more peptides)	329 (7245 peptides)	303 (3143 peptides)	549	83	246	220
Total proteins (two or more peptides or one high-scoring peptide)[a]	340 (7007 peptides)	332 (2937 peptides)	579	93	247	239
Total proteins (one or more peptide)[b]	578 (6996 peptides)	538 (2908 peptides)	1005	111	467	427

[a]A high-scoring single peptide is defined as one with a maximum expectation score of 10e–3.
[b]An identified single peptide is defined as one with a maximum expectation score of 0.1.

3.8.4. Results of Mupdit Analysis of Rice Leaf and Root

The results of Mupdit Analysis of rice leaf and root are shown in **Tables 2** to **4**. **Table 2** shows the total number of proteins identified in each experiment, separated into three categories: two or more peptides identified; two or more peptides identified or one peptide identified with an expectation value of < 10e-3; one or more peptides identified with an expectation value of < 10e-1. The number of proteins is further categorized to include those found to be common to both leaf and root, or unique to each tissue preparation. **Table 2** also includes information on the number of MS/MS spectra searched in each experiment and the number of peptides found in each category just described.

The results of these experiments clearly demonstrate several points, with the first being that using this approach it is possible to identify hundreds of proteins in a plant tissue in a single experiment. By searching all of the more than 91,000 MS/MS spectra acquired in these two experiments, and applying our minimal criteria, 578 proteins are identified in leaf and 538 in root, which includes 1005 nonredundant protein identifications. Using stricter criteria of two or more peptides identified, these numbers decrease to 329, 303, and 549, respectively. The number of proteins identified by one peptide constitutes 43% of the total, which is consistent with (or lower than) previously published

Table 3
First 25 Proteins Identified in Mudpit Analysis of Rice Leaf, Ordered by Protein Expectation Value

Protein identifier	Peptides	log(e) (protein)	p*I*	M. Wt. (kDa)
NP_039391.1| ribulose 1,5-bisphosphate carboxylase/oxygenase large chain [Oryza sativa	2140	–233.2	6.22	52.8
NP_920971.1| putative rbcL; RuBisCO large subunit from chromosome 10 chloroplast	2061	–244.9	6.45	52.9
NP_918587.1| putative 33kDa oxygen evolvingprotein of photosystem II [Oryza sativa	124	–59.3	6.08	34.8
NP_922436.1| ATPase alpha subunit [Oryza sativa (japonica cultivar-group)	108	–85	5.95	55.1
NP_039390.1| ATP synthase CF1 beta chain [Oryza sativa (japonica cultivar-group)	94	–66.8	5.47	54
NP_920964.1| putative ATPase alpha subunit from chromosome 10 chloroplast insertion	93	–70	5.88	55.7
NP_920970.1| putative atpB; ATPase beta subunit from chromosome 10 chloroplast insertion	91	–63.3	5.38	54
NP_917149.1| carbonic anhydrase [Oryza sativa (japonica cultivar-group)]	68	–62.5	8.41	29.1
XP_474199.1| OSJNBa0011F23.15 [Oryza sativa] plastidic glutamine synthetase	58	–49.4	5.96	46.6
XP_475868.1| putative ATP synthase beta chain [Oryza sativa (japonica cultivar-group)]	57	–62.2	5.95	58.9
NP_911136.1| probable photosystem II oxygen-evolving complex protein 2 precursor	49	–56.6	8.66	26.9
XP_470174.1| Putative catalase [Oryza sativa (japonica cultivar-group)]	48	–39	8.21	62.5
XP_478362.1| putative glycine-rich cell wall structural protein 1 precursor [Oryza sativa]	44	–157.7	9.02	26.9
XP_467296.1| phosphoribulokinase precursor [Oryza sativa (japonica cultivar-group)]	44	–27.4	5.68	44.8
XP_478627.1| putative oxygen-evolving enhancer protein 3-1, chloroplast precursor	43	–52.7	9.8	22.6
XP_472744.1| OSJNBa0036B21.24 [Oryza sativa] glyceraldehyde-3-phosphate dehydrogenase	41	–61.5	7.61	42.7
XP_474191.1| OSJNBa0011F23.7 [Oryza sativa] rubisco activase	37	–13	6	49.1
NP_916596.1| putative glycine dehydrogenase [Oryza sativa (japonica cultivar-group)]	34	–58.6	6.51	111.4
XP_475256.1| putative ATP synthase beta chain [Oryza sativa (japonica cultivar-group)]	34	–34.8	5.66	43
XP_475055.1| putative peptidylprolyl isomerase (EC 5.2.1.8) [Oryza sativa (japonica)]	33	–34.4	9.36	26.6
NP_039445.1| photosystem I subunit VII [Oryza sativa]	33	–21	6.52	8.9
NP_917525.1| putative chlorophyll a/b-binding protein 2 [Oryza sativa (japonica cultivar-group)]	33	–18.1	5.29	27.7
NP_915665.1| putative subtilisin-like protease [Oryza sativa (japonica cultivar-group)]	29	–32.1	8.72	78.8
XP_462797.1| putative triosephosphate isomerase [Oryza sativa (japonica cultivar-group)]	29	–31.2	5.38	27
XP_462851.1| B1146F03.16 [Oryza sativa)] chloroplast protein 12 (CP12) homolog	25	–16.5	4.58	12.8

Table 4
First 25 Proteins Identified in Mudpit Analysis of Rice Root, Ordered by Protein Expectation Value

Protein identifier	Peptides	log(e) (protein)	p*I*	M. Wt. (kDa)
NP_909830.1| sucrose-UDP glucosyltransferase 2 [Oryza sativa]	123	–118.7	5.94	92.8
XP_475868.1| putative ATP synthase beta chain [Oryza sativa (japonica cultivar-group)]	98	–100.8	5.95	58.9
XP_466582.1| putative glyceraldehyde-3-phosphate dehydrogenase (phosphorylating).	78	–60.3	7.69	36.5
XP_472949.1| OJ000223_09.15 [Oryza sativa] cytosolic glyceraldehyde-3-phosphate dehydrogenase	72	–59.2	6.34	36.7
NP_920013.1| putative enolase (2-phospho-D-glycerate hydroylase) [Oryza sativa (japonica)]	64	–87.1	5.16	45.9
XP_479895.1| glyceraldehyde 3-phosphate dehydrogenase, cytosolic [Oryza sativa (japonica)]	58	–39.2	6.61	36.4
XP_470632.1| Putative non-symbiotic hemoglobin 1 [Oryza sativa (japonica cultivar-group)]	58	–34.2	6.91	18.4
NP_039391.1| ribulose 1,5-bisphosphate carboxylase/oxygenase large chain [Oryza sativa]	57	–44.1	6.22	52.8
XP_465992.1| putative elongation factor 2 [Oryza sativa (japonica cultivar-group)]	54	–59.6	5.85	94
NP_912586.1| GLN1_ORYSA Glutamine synthetase root isozyme	53	–47.2	5.73	38.5
XP_483191.1| heat shock protein 82 [Oryza sativa (japonica cultivar-group)]	48	–67.5	4.99	80.1
XP_462797.1| putative triosephosphate isomerase [Oryza sativa (japonica cultivar-group)]	45	-47.9	5.38	27
XP_476473.1| putative pathogenesis-related protein [Oryza sativa (japonica cultivar-group)]	43	–19.5	4.37	19
XP_474202.1| OSJNBa0011F23.18 [Oryza sativa] putative serine/threonine kinase	43	–19.5	4.96	27.3
XP_469751.1| putative exoglucanase precursor [Oryza sativa]	38	–47.2	7.23	67.7
XP_472763.1| OSJNBa0072F16.20 [Oryza sativa] 14-3-3 like protein	36	–39.3	4.75	29.8
NP_920279.1| putative type-1 pathogenesis-related protein [Oryza sativa (japonica)]	36	–31.3	9.1	18.7
XP_475365.1| putative hsp70 [Oryza sativa (japonica cultivar-group)]	35	–45.6	5.1	70.8
XP_483167.1| putative caffeoyl-CoA O-methyltransferase 1 [Oryza sativa (japonica)]	34	–35.7	5.11	27.8
XP_469569.1| actin [Oryza sativa (japonica cultivar-group)]	32	–45.6	5.3	41.8
XP_470658.1| Putative ascorbate peroxidase [Oryza sativa (japonica cultivar-group)]	32	–34.1	5.42	27.1
XP_469508.1| putative 14-3-3 protein [Oryza sativa]	32	–34.1	4.81	29.2
NP_910962.1| copper/zinc-superoxide dismutase [Oryza sativa (japonica cultivar-group)]	31	–8.9	5.92	14.8
XP_470141.1| heat shock protein cognate 70 [Oryza sativa (japonica cultivar-group)]	28	–44.5	5.09	71.3
NP_908513.1| S-adenosyl methionine synthetase [Oryza sativa (japonica cultivar-group)]	27	–27.6	5.22	43.3

datasets in which that percentage has been specifically reported *(12,36–38)*. The number of proteins identified in the two experiments is very consistent, especially as reproducibility is not considered one of the strong points of the Mudpit technique *(39)*. However, the nonredundant total of protein identifications shows that there is only 10 to 15% overlap between the two datasets, which gives a good indication of the degree of tissue specificity of protein expression. One other very interesting numerical feature of these results is that although the number of proteins identified in leaf and root is similar, there are more than twice as many peptides identified in the leaf tissue. This is principally owing to the very large number of peptides identified from RuBisCo in the leaf tissue, as shown in **Table 3**.

Table 3 lists the first 25 proteins identified in the Mudpit analysis of rice leaf tissue, ordered by their Xtandem protein expectation value, and **Table 4** lists the first 25 proteins identified in the Mudpit analysis of rice root tissue, also ordered by their Xtandem protein expectation value. The proteins listed in **Tables 3** and **4** represent the most abundant proteins found in leaf and root tissue and are mainly proteins expected to be found in those tissues and expressed at a high level. The leaf tissue contains numerous isoforms of RuBisCo, ATP synthase, and photosystem complex proteins, as well as several other chloroplast-specific proteins. The root tissue contains numerous carbohydrate metabolism enzymes, such as sucrose-UDP glucosyl transferase 2, triosephosphate isomerase, ascorbate peroxidase, and glyceraldehyde-3-phosphate dehydrogenase. Interestingly, the root tissue also contains several highly expressed pathogenesis-related proteins, which are presumably important in mechanisms of response to biotic and abiotic stresses in the soil.

4. Notes

1. This washing step is essential to ensure that the pH of the extracted protein mixture is not too low owing to residual TCA. The presence of residual acid will produce poor results, as the proteases used will not function optimally at acidic pH.
2. The protein concentration in each sample can be quantified at this point using standard quantitative protein assay techniques on a separate protein extract aliquot. In the event that the extracted protein concentration is too low, multiple aliquots can be extracted, precipitated with TCA/acetone, and combined prior to further processing.
3. Endoproteinase Lys-C is active in 8 *M* urea, whereas trypsin is not. This step is optional, but the rationale for including it is that some digestion by Lys-C occurs, which helps to unfold some 3D structures and thus facilitate digestion by trypsin in the next step.
4. The Spec-plus C18 pipet tips can be run using gravity flow, but this often comes to a complete halt. Applying very gentle pressure to the top of the tip with a 3-mL plastic syringe can restart the flow.

5. It is not required to use an autosampler, but that is the system set up in our laboratory. The sample can also be transferred onto the biphasic column using a manual injector or directly via the pressure cell.
6. The initial C18 gradient often contains a large amount of material, especially for very complex protein mixtures. In such a case, lengthening the 5 to 50% B separation portion of the gradient to 90 or 120 min rather than the 60 min described here can produce a significant increase in the number of proteins identified.

References

1. Link, A. J., Eng, J., Schieltz, D. M., et al. (1999) Direct analysis of protein complexes using mass spectrometry. *Nat. Biotechnol.* **17,** 676–682.
2. Washburn, M. P., Wolters, D., and Yates, J. R. III (2001) Large-scale analysis of the yeast proteome by multidimensional protein identification technology. *Nat. Biotechnol.* **19,** 242–247.
3. Haynes, P. A. and Yates, J. R. III (2000) Proteome profiling-pitfalls and progress. *Yeast* **17,** 81–87.
4. Opiteck, G. J. and Jorgenson, J. W. (1997) Two-dimensional SEC/RPLC coupled to mass spectrometry for the analysis of peptides. *Anal. Chem.* **69,** 2283–2291.
5. Opiteck, G. J., Lewis, K. C., Jorgenson, J. W., and Anderegg, R. J. (1997) Comprehensive on-line LC/LC/MS of proteins. *Anal. Chem.* **69,** 1518–1524.
6. Washburn, M. P., Ulaszek, R., Deciu, C., Schieltz, D. M., and Yates, J. R. III (2002) Analysis of quantitative proteomic data generated via multidimensional protein identification technology. *Anal. Chem.* **74,** 1650–1657.
7. Wu, C. C., MacCoss, M. J., Howell, K. E., Matthews, D. E., and Yates, J. R. 3rd (2004) Metabolic labeling of mammalian organisms with stable isotopes for quantitative proteomic analysis. *Anal. Chem.* **76,** 4951–4959.
8. Froehlich, J. E., Wilkerson, C. G., Ray, W. K., et al. (2003) Proteomic study of the Arabidopsis thaliana chloroplastic envelope membrane utilizing alternatives to traditional two-dimensional electrophoresis. *J. Proteome Res.* **2,** 413–425.
9. Andon, N. L., Hollingworth, S., Koller, A., Greenland, A. J., Yates, J. R. III, and Haynes, P. A. (2002) Proteomic characterization of wheat amyloplasts using identification of proteins by tandem mass spectrometry. *Proteomics* **2,** 1156–1168.
10. Peltier, J. B., Ytterberg, A. J., Sun, Q., and van Wijk, K. J. (2004) New functions of the thylakoid membrane proteome of *Arabidopsis thaliana* revealed by a simple, fast, and versatile fractionation strategy. *J. Biol. Chem.* **279,** 49367–4983.
11. Heinemeyer, J., Eubel, H., Wehmhoner, D., Jansch, L., and Braun, H. P. (2004) Proteomic approach to characterize the supramolecular organization of photosystems in higher plants. *Phytochemistry* **65,** 1683–1692.
12. Koller, A., Washburn, M. P., Lange, B. M., et al. (2002) Proteomic survey of metabolic pathways in rice. *Proc. Natl. Acad. Sci. USA* **99,** 11969–11974.
13. Chen, M., Presting, G., Barbazuk, W. B., et al. (2002) An integrated physical and genetic map of the rice genome. *Plant Cell* **14,** 537–545.
14. Yu, J., Hu, S., Wang, J., et al. (2002) A draft sequence of the rice genome (*Oryza sativa* L. ssp. indica). *Science* **296,** 79–92.

15. Goff, S. A., Ricke, D., Lan, T. H., et al. (2002) A draft sequence of the rice genome (*Oryza sativa* L. ssp. japonica). *Science* **296,** 92–100.
16. Initiative, A. G. (2000) Analysis of the genome sequence of the flowering plant *Arabidopsis thaliana*. *Nature* **408,** 796–815.
17. Salanoubat, M., Lemcke, K., Rieger, M., et al.. (2000) Sequence and analysis of chromosome 3 of the plant *Arabidopsis thaliana*. *Nature* **408,** 820–822.
18. Tabata, S., Kaneko, T., Nakamura, Y., et al. (2000) Sequence and analysis of chromosome 5 of the plant *Arabidopsis thaliana*. *Nature* **408,** 823–826.
19. Lunde, C. F., Morrow, D. J., Roy, L. M., and Walbot, V. (2003) Progress in maize gene discovery: a project update. *Funct. Integr. Genomics* **3,** 25–32.
20. Damerval, C., de Vienne, D., Zivy, M., and Thiellement, H. (1986) Technical improvements in two-dimensional electrophoresis increase the level of genetic variation detected in wheat-seedling proteins. *Electrophoresis* **7,** 52–54.
21. Porubleva, L., Vander Velden, K., Kothari, S., Oliver, D. J., and Chitnis, P. R. (2001) The proteome of maize leaves: use of gene sequences and expressed sequence tag data for identification of proteins with peptide mass fingerprints. *Electrophoresis* **22,** 1724–1738.
22. Zuo, X., Echan, L., Hembach, P., et al. (2001) Towards global analysis of mammalian proteomes using sample prefractionation prior to narrow pH range two-dimensional gels and using one-dimensional gels for insoluble and large proteins. *Electrophoresis* **22,** 1603–1615.
23. Blonder, J., Hale, M. L., Lucas, D. A., et al. (2004) Proteomic analysis of detergent-resistant membrane rafts. *Electrophoresis* **25,** 1307–1318.
24. Guina, T., Wu, M., Miller, S. I., et al. (2003) Proteomic analysis of *Pseudomonas aeruginosa* grown under magnesium limitation. *J. Am. Soc. Mass Spectrom.* **14,** 742–751.
25. Gatlin, C. L., Kleemann, G. R., Hays, L. G., Link, A. J., and Yates, J. R. 3rd (1998) Protein identification at the low femtomole level from silver-stained gels using a new fritless electrospray interface for liquid chromatography-microspray and nanospray mass spectrometry. *Anal. Biochem.* **263,** 93–101.
26. Eng, J., McCormack, A. L., and Yates, J. R. III (1994) An approach to correlate tandem mass spectral data of peptides with amino acid sequences in a protein database. *J. Am. Mass Spectrom.* **5,** 976–989.
27. Yates, J. R. III, Eng, J. K., McCormack, A. L., and Schieltz, D. (1995) A method to correlate tandem mass spectra of modified peptides to amino acid sequences in the protein database. *Anal. Chem.* **67,** 1426–1436.
28. Perkins, D. N., Pappin, D. J., Creasy, D. M., and Cottrell, J. S. (1999) Probability-based protein identification by searching sequence databases using mass spectrometry data. *Electrophoresis* **20,** 3551–3567.
29. Creasy, D. M. and Cottrell, J. S. (2002) Error tolerant searching of uninterpreted tandem mass spectrometry data. *Proteomics* **2,** 1426–1434.
30. Craig, R. and Beavis, R. C. (2003) A method for reducing the time required to match protein sequences with tandem mass spectra. *Rapid Commun. Mass Spectrom.* **17,** 2310–2316.

31. Craig, R. and Beavis, R. C. (2004) TANDEM: matching proteins with tandem mass spectra. *Bioinformatics* **20,** 1466–1467.
32. Nesvizhskii, A. I., Keller, A., Kolker, E., and Aebersold, R. (2003) A statistical model for identifying proteins by tandem mass spectrometry. *Anal. Chem.* **75,** 4646–4658.
33. Sadygov, R. G., Liu, H., and Yates, J. R. (2004) Statistical models for protein validation using tandem mass spectral data and protein amino acid sequence databases. *Anal. Chem.* **76,** 1664–1671.
34. Peng, J., Elias, J. E., Thoreen, C. C., Licklider, L. J., and Gygi, S. P. (2003) Evaluation of multidimensional chromatography coupled with tandem mass spectrometry (LC/LC-MS/MS) for large-scale protein analysis: the yeast proteome. *J. Proteome Res.* **2,** 43–50.
35. Elias, J. E., Gibbons, F. D., King, O. D., Roth, F. P., and Gygi, S. P. (2004) Intensity-based protein identification by machine learning from a library of tandem mass spectra. *Nat. Biotechnol.* **22,** 214–219.
36. Breci, L., Hattrup, E., Keeler, M., Letarte, J., Johnson, R., and Haynes, P. A. (2005) Comprehensive proteomics in yeast using chromatographic fractionation, gas phase fractionation, protein gel electrophoresis, and isoelectric focusing. *Proteomics* **5,** 2018–2028.
37. Hall, N., Karras, M., Raine, J. D., et al. (2005) A comprehensive survey of the Plasmodium life cycle by genomic, transcriptomic, and proteomic analyses. *Science* **307,** 82–86.
38. Vitali, B., Wasinger, V., Brigidi, P., and Guilhaus, M. (2005) A proteomic view of *Bifidobacterium infantis* generated by multi-dimensional chromatography coupled with tandem mass spectrometry. *Proteomics* **5,** 1859–1867.
39. Durr, E., Yu, J., Krasinska, K. M., et al. (2004) Direct proteomic mapping of the lung microvascular endothelial cell surface in vivo and in cell culture. *Nat. Biotechnol.* **22,** 985–992.

22

Separation, Identification, and Profiling of Membrane Proteins by GFC/IEC/SDS-PAGE and MALDI TOF MS

Wojciech Szponarski, Frédéric Delom, Nicolas Sommerer, Michel Rossignol, and Rémy Gibrat

Summary

Membrane protein identification by matrix-assisted laser desorption/ionization-time of flight-mass spectrometry (MALDI-TOF-MS) requires that proteins be separated prior to MS analysis. After membrane solubilization with the nondenaturing detergent *n*-dodecyl-β-d-maltoside, proteins can be separated by ion-exchange chromatography (IEC) and further resolved by sodium dodecyl sulfate-polyacrylamide gel electrophoresis (SDS-PAGE). An additional separation step by gel filtration (GF) before IEC/SDS-PAGE can be required depending on the complexity of the membrane protein mixture. Staining of final SDS-PAGE gels allows one to establish simply the protein expression pattern of a membrane fraction and to profile responses. Moreover, in-gel digestion of hydrophobic integral proteins is valuable. Finally, the resolution capacity of this separation procedure allows identification of proteins by MALDI-TOF MS. The method is illustrated by application to plant and yeast plasma membrane and to plant vacuolar membrane.

Key Words: Membrane proteins; liquid chromatography; mass spectrometry; *Arabidopsis thaliana*; *Saccharomyces cerevisiae*.

1. Introduction

Achievement of sequencing and annotation of plant model genomes, as well as the rapid increase of various plants sequence data, allows to identify protein routinely from peptide mass fingerprints generated by matrix-assisted laser desorption ionization-time of flight-mass spectrometry (MALDI-TOF/MS) *(1,2)*. This MS method operates at high sensitivity, high throughput, and low cost but requires a preliminary protein separation step. Two-dimensional gel electrophoresis (2D-GE), the most popular method to separate cell-extracted proteins at a high resolution, is no well suited for a systematic resolution of membrane proteomes (***3–5***; and *see* Chapter 11) owing to precipitation of inte-

From: *Methods in Molecular Biology, vol. 335: Plant Proteomics: Methods and Protocols*
Edited by: H. Thiellement, M. Zivy, C. Damerval, and V. Méchin © Humana Press Inc., Totowa, NJ

gral proteins during the first isoelectric focusing (IEF) step *(6)*. By contrast, sodium dodecyl sulfate-polyacrylamide gel electrophoresis (SDS-PAGE) is efficient to resolve integral membrane proteins.

In the alternative approach detailed below, the IEF in 2D-GE is replaced by ion exchange chromatography (IEC). Gel filtration chromatography (GFC) constitutes a useful complementary step when sufficient amounts of membranes are available *(7)*. A final SDS-PAGE step allows resolution of further integral membrane proteins. Under these conditions, in-gel digestion of integral proteins and their analysis by MALDI-TOF/MS is successful for a variety of hydrophobic proteins *(7)*.

One aim of the method is to separate within a single experimental frame the different classes of proteins present in a membrane fraction, including the most hydrophobic, acidic, or basic ones, with a resolution compatible with simple analysis of protein expression by comparing stained gels and with simple protein fingerprinting by MALDI-TOF/MS. The protocols described here may not be directly applicable in detail to other materials. However, they apply to different materials and can guide the reader to optimize the principal steps.

2. Materials

1. Stripping buffer: 5 m*M* BTP-MES, pH 7.5, 0.2 *M* KI, 1 m*M* Na_2EDTA, 0.3 *M* mannitol, 20% (w/w) glycerol, 5 m*M* dithiothreitol (DTT), 0.5% (v/v) appropriate protease inhibitor cocktail (Sigma, for *Arabidopsis thaliana* or yeast), and 15 m*M* *n*-octyl-β-D-glucopyranoside.
2. Solubilization buffer: 5 m*M* BTP-MES, pH 7.5, 20% (w/w) glycerol, and 0.5% (v/v) appropriate protease inhibitor cocktail (Sigma, for *A. thaliana* or yeast).
3. Elution buffer: 3 m*M* BTP-MES, pH 7.5, 1.5 m*M* n-dodecyl β-d-maltoside (DM), and 20% glycerol (w/w).
4. Sample Laemmli buffer: 4% (w/v) lithium dodecyl sulfate, 20% (w/w) glycerol, 10% (v/v) 2-mercaptoethanol (2-ME), 0.004% (w/v) bromophenol blue, 0.125 *M* Tris-HCl, pH 6.8.
5. Gel segments washing buffer: 25 m*M* ammonium bicarbonate, acetonitrile 50% (v/v).
6. Digestion buffer: 15 m*M* ammonium bicarbonate.
7. Peptide extraction buffer: acetonitrile 60% (v/v), 0.15% (v/v) trifluoroacetic acid (TFA).

3. Methods

3.1. Membrane Stripping

Because isolated membranes are present as small vesicles with encapsulated soluble cytosolic proteins during the grinding step, it is preferable to eliminate these proteins before separation and identification of membrane proteins. For this purpose, membranes are incubated in the presence of detergent at subcriti-

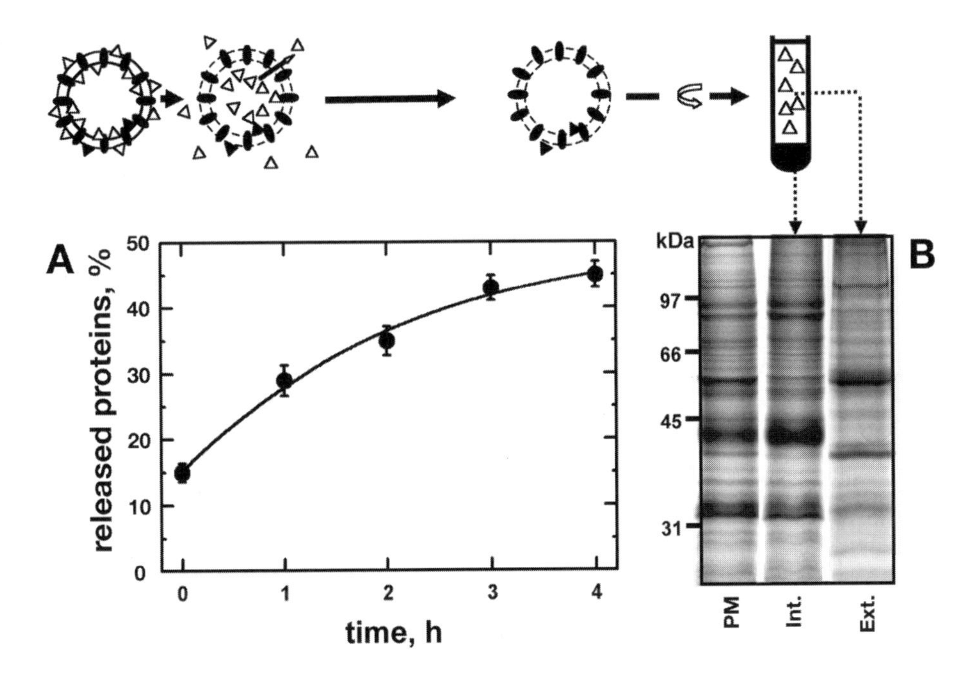

Fig. 1. Plasma membrane stripping. **(A)** Plasma membrane (PM) vesicles from *A. thaliana* are resuspended in stripping buffer, and protease inhibitor cocktail and incubated under agitation to allow for outward diffusion of loosely membrane-associated polar proteins (open triangles). At the indicated times, the suspension is centrifuged at 100,000*g* for 1 h, and proteins are assayed in both supernatant (extrinsic fraction) and resuspended pellet (intrinsic fraction, enriched in integral or tightly associated polar proteins, closed symbols). **(B)** SDS-PAGE of unstripped PM (PM), and intrinsic (Int) and extrinsic (Ext) protein fractions after 4 h of incubation.

cal micellar concentration to permeabilize the vesicles (*see* **Note 1**), allowing for slow outward diffusion of encapsulated proteins. The incubation buffer also contains a chaotropic salt and a divalent chelator to dissociate loosely bound peripheral membrane proteins. An example of plant plasma membrane (PM) stripping experiment, according to the protocol given here, is illustrated in **Fig. 1**.

1. Incubate the membrane fraction (0.5–5 mg at 0.2 mg/mL) under agitation at 4°C in stripping buffer.
2. Test incubations times are up to 4 h.
3. After incubation, centrifuge the membrane suspension at 100,000*g* for 1 h at 4°C.
4. Assay the proteins in both supernatant and pellet.
5. Analyze by SDS-PAGE the protein patterns of both extrinsic (supernatant) and intrinsic (pellet) fractions.

3.2. Protein Solubilization

Separation of integral membrane proteins requires that their native lipid environment be broken first, using detergents *(8,9)*. In the present approach, neutral or zwitterionic detergents are required because they do not modify the electrical charge of proteins, which can be therefore separated by IEC. Moreover, such detergents are in general mild, and they preserve protein native conformation and activity.

1. Resuspend stripped membranes (0.5–5 mg) in 1 mL of solubilization buffer at room temperature.
2. Add 1 mL of DM to the solubilization buffer at a DM concentration suitable to test DM/protein ratios (w/w) from 3 to 7 (*see* **Note 2**).
3. Vortex for 2 min and incubate at room temperature for 2 h under slow agitation.
4. Centrifuge at 190,000*g* for 1 h at room temperature.
5. Assay the proteins in both pellet and supernatant to determine the solubilization yield.
6. Analyze by SDS-PAGE the patterns of solubilized (supernatant) and insoluble (pellet) proteins.

Application of this procedure leads to SDS-PAGE patterns of solubilized proteins that are very similar to initial PM proteins (**Fig. 2**). About 85% of initial proteins are solubilized under these conditions.

3.3. Chromatography

3.3.1. IEC/SDS-PAGE Separation

Anion exchange chromatography (AEC; *see* **Note 3**) is performed at room temperature with a Mono Q HR 5/5 column (Pharmacia/Amersham Biosciences). Protein elution is monitored by absorbance at 280 nm, and 0.3-mL fractions are collected during all chromatographic steps. The presence of detergent in the sample and in elution buffers favors the formation of bubbles in the tubing system. Therefore, careful sonication (for 30 min) of buffers and moderate elution rates (0.5–1 mL/min) are recommended.

1. Equilibrate the column with 15 mL of elution buffer at room temperature.
2. Load freshly solubilized proteins (0.5–2 mg in 2 mL) and wash the column with 5 mL of elution buffer while collecting fractions (*see* **Note 4**).
3. Elute proteins with 10 to 15 mL of linear salt gradient (0–1 *M* NaCl in elution buffer).
4. Wash out the column with 5 mL of 1 *M* NaCl in elution buffer.
5. Precipitate proteins in collected fractions by addition of cold acetone (–20°C, 80% [v/v], final concentration) followed by overnight incubation at -20°C.
6. Spin down precipitated proteins at 4°C in a microfuge (17,000*g* for 30 min).

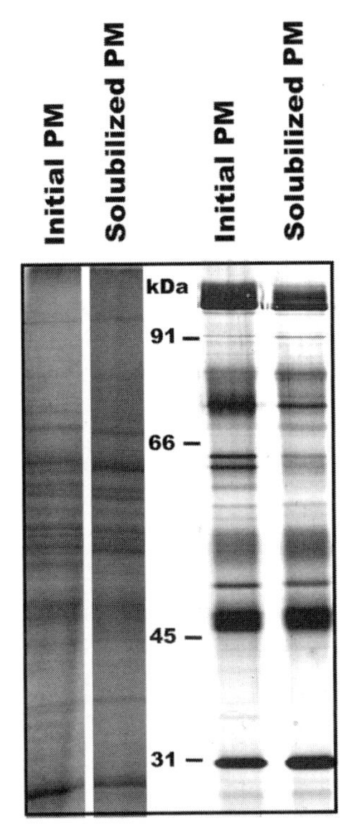

Fig. 2. Solubilization of stripped plasma membrane (PM) by *n*-dodecyl-β-D-maltoside. PM proteins are solubilized as described in **Subheading 3.2.** and resolved by SDS-PAGE. Gels was Coomassie blue (*A. thaliana*) or silver stained (*S. cerevisiae*).

7. Add 50 μL of sample Laemmli buffer and resuspend precipitated proteins by vortex (1 h) followed by sonication (15 min in a sonicator bath).
8. Separate proteins by SDS-PAGE.

Figure 3A illustrates representative IEC/SDS-PAGE separations starting from 0.5 mg of proteins solubilized from stripped yeast plasma membrane. A given band is typically observed in three to six lanes corresponding to successive fractions. MALDI-TOF/MS leads to the identification of one protein per band in 70% of the analyzed bands (about 100), and of mixtures of two or three proteins per band in 25 and 5% of bands, respectively. The reproducibility of the procedure is illustrated in **Fig. 3B** in a profiling experiment. Data analysis is facilitated by layering on successive SDS-PAGE lanes of the AEC fractions eluted at a same ionic strength but resulting from different treatments.

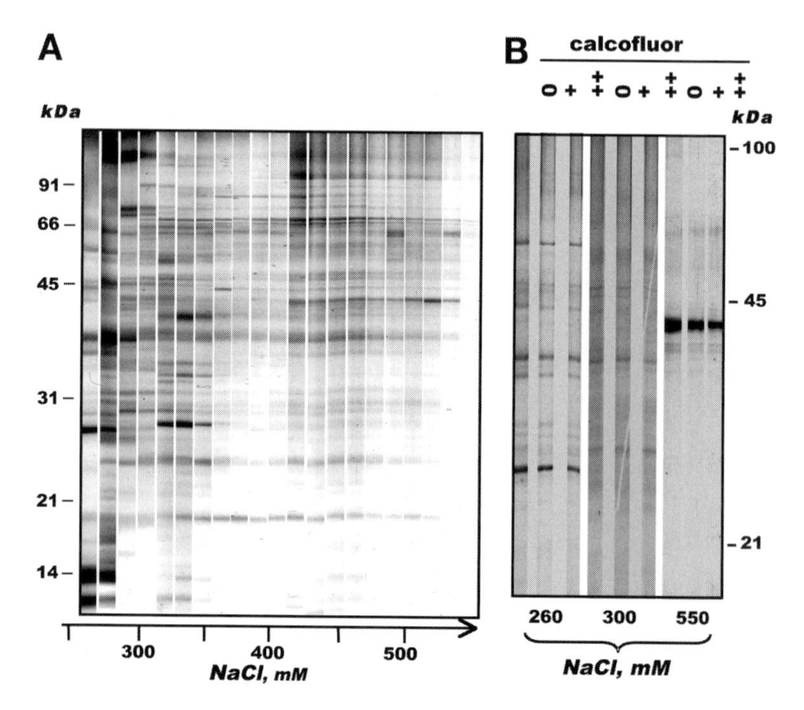

Fig. 3. IEC/SDS-PAGE separation of PM proteins from *S. cerevisiae*. (**A**) DM solubilized PM proteins (0.5 mg) are separated by IEC on a Mono Q HR 5/5 at a constant flow of about 0.5 mL/min according to a linear NaCl salt gradient and resolved further by SDS-PAGE. (**B**) examples of SDS-PAGE of IEC fractions eluted at 260, 300 or 550 m*M* NaCl but issued from control yeast cells (0) or cells treated with the antifungal calcofluor according to mild treatments which did not (+) or slightly (++) decreased their viability. The same fraction of eluted volume was layered onto each lane, and gels were Coomassie blue stained. Patterns of the three lane sets are essentially similar, with some exceptions. For example, protein fractions eluted at 550 m*M* NaCl display a well-stained band at 41 kDa whose intensity decreased upon calcofluor treatment. SDS-PAGE patterns before IEC separation are indistinguishable (data not shown).

3.3.2. GFC/AEC/SDS-PAGE Separation

If more complex protein mixtures (unstripped membranes for example) have to be resolved, two successive chromatographic steps are required. For this purpose, higher amounts (about 5 mg) of solubilized proteins must be available. This also allows analysis of proteins of lower abundance in final SDS-PAGE lanes. GFC is used in the first step because it dilutes the sample, which is further concentrated by AEC. In the following experiments, GFC of proteins solubilized from unstripped *A. thaliana* membranes is performed on a Superdex 200 HiLoad 16/60 (Pharmacia/Amersham Biosciences) (*see* **Note 5**).

1. Equilibrate the GFC column at room temperature with 180 mL of elution buffer at 1 mL/min elution rate.
2. Load freshly solubilized proteins (5–10 mg in 2 mL).
3. Elute proteins (0.5 mL/min) with 180 mL of elution buffer and collect 0.3 mL fractions.
4. Pool neighboring fractions if they contain less than 500 μg of proteins estimated by absorbance at 280 nm (*see* **Note 6**).
5. Isolate 100 μg aliquots of GFC fractions and precipitate proteins by addition of cold acetone (–20°C, 80% [v/v] final concentration) followed by overnight incubation at –20°C.
6. Spin down precipitated proteins at 4°C in a microfuge (17,000*g* for 30 min).
7. Add 50 μL of Laemmli sample buffer and resuspend precipitated proteins by vortex (1 h) followed by sonication (15 min in a sonicator bath).
8. Separate GFC protein fractions by SDS-PAGE using the aliquots just described.
9. Perform the second chromatographic step on a previous GFC fraction (containing at least 500 μg of proteins) by AEC on a Mono Q HR 5/5 (Pharmacia/Amersham Biosciences) according to the protocol described in the section above.

Application of the GFC/SDS-PAGE procedure to unstripped membranes leads to rather complex patterns, in particular in the 40- to 70-kDa range (**Fig. 4**). After GFC/AEC/SDS-PAGE separation, simpler SDS-PAGE patterns are obtained (**Fig. 5**), allowing for easier protein identification by MALDI-TOF/MS (fewer protein mixtures observed in excised bands).

3.4. In-Gel Protein Digestion by Trypsin

1. Cut out Coomassie blue-stained bands.
2. Wash the gel segments extensively in 2 mL of gel segment washing buffer (in 2 mL Eppendorf tubes) until complete elimination of the coloration. Generally, two 4-h washes are sufficient.
3. Dehydrate the segments by incubation in pure acetonitrile for 1 h.
4. Dry the segments completely in a SpeedVac (30 min).
5. Prepare a fresh solution of trypsin (0.02 μg/μL) in digestion buffer at 4°C to avoid enzyme autodigestion. Stock this solution in ice.
6. Rehydrate gel segments with trypsin solution by layering a droplet (15 μL) around the segments, and place the tube in ice. After about 30 min, only a thin layer of trypsin solution should be present around the segment. If the initial droplet is completely absorbed by the segment, form a thin layer by addition of 15 m*M* ammonium bicarbonate buffer. If the initial trypsin droplet is not completely absorbed by the segment, reduce the droplet to the thin layer (*see* **Note 7**).
7. Close the tubes containing gel segments and incubate for at least 5 h at 37°C. Overnight incubation is also possible.
8. Add 150 μL of peptide extraction buffer and incubate for 3 h at 20°C.
9. Remove gel segments and concentrate extracted peptides by a SpeedVac in 500-μL Eppendorf tubes to reduce the volume to about 5 μL. Never dry completely because this causes an irreversible adsorption of part of the peptides on the tube walls.

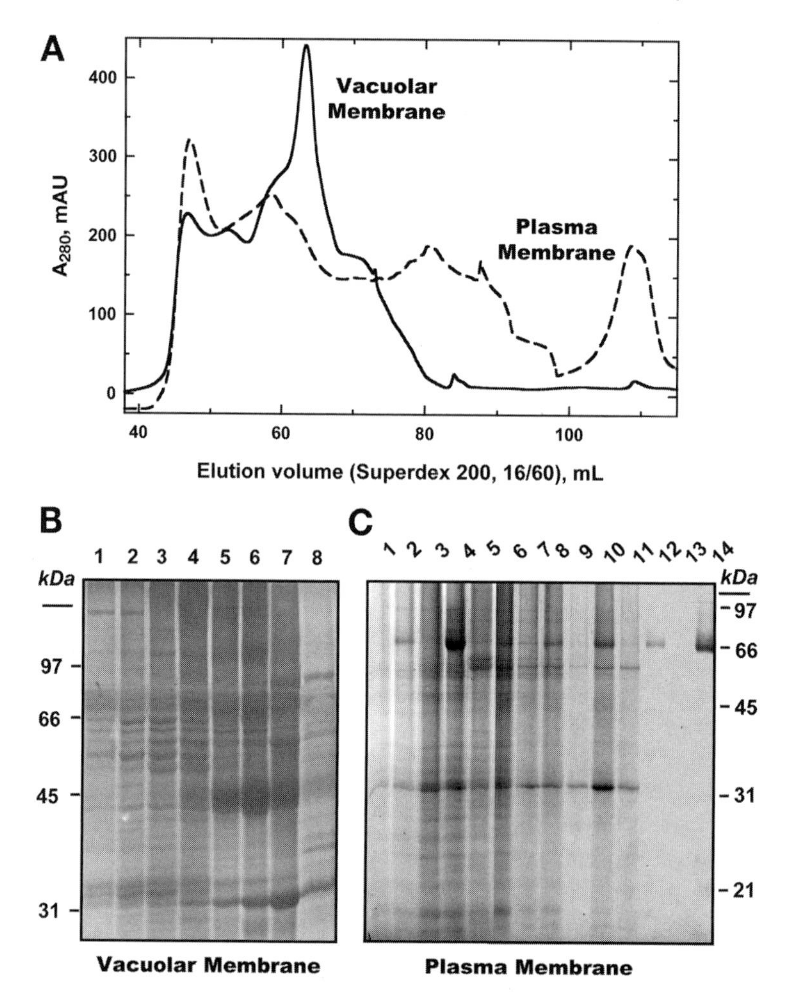

Fig. 4. First separation by GFC of DM-solubilized membrane proteins from *A. thaliana*. (**A**) GFC chromatograms obtained with a Superdex 200 16/60 (Amersham Bioscience) (solid line, vacuolar membrane proteins; dashed line, PM proteins). Fractions successively eluted are resolved further by SDS-PAGE (**B**: vacuolar membrane; **C**: PM). The same fraction of eluted volume was layered onto each lane, and gels were Coomassie blue stained.

3.5. MALDI-TOF/MS

The general procedure described in this book (*see* Chapter 19) is followed up to sample crystallization on the target. Briefly:

1. Prepare freshly α-cyano-4-hydroxycinnamic acid matrix at half-saturation in acetonitrile/water (1:1 [v/v]) acidified with 0.1% TFA.

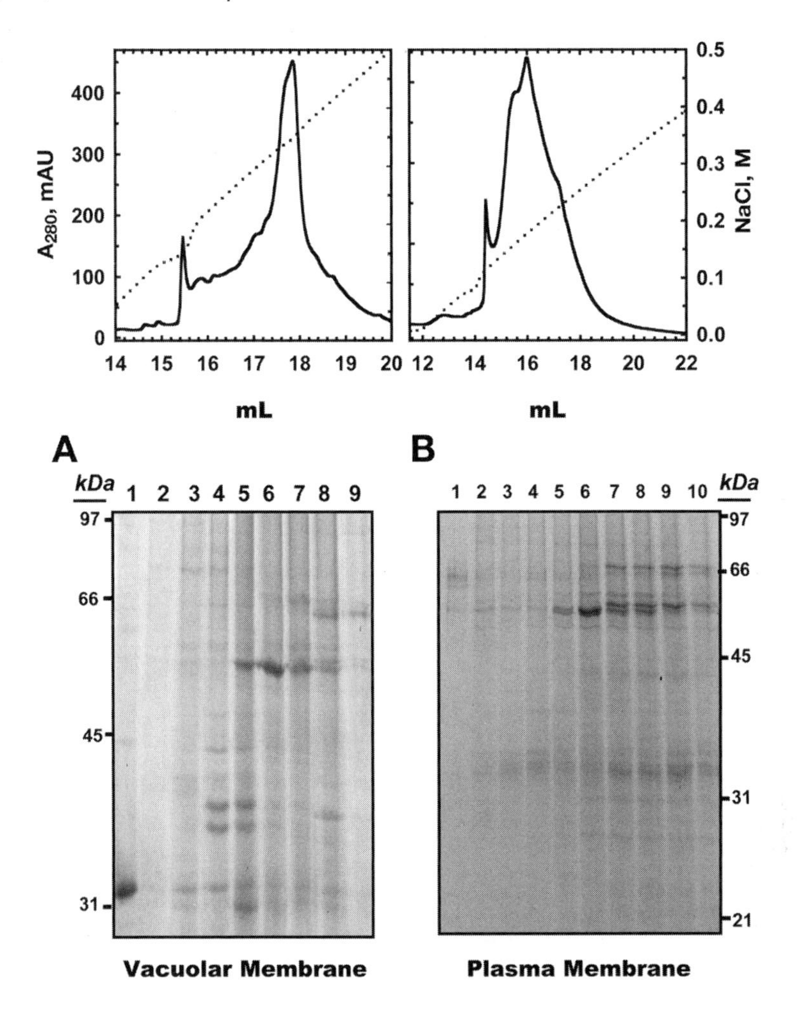

Fig. 5. Second separation by IEC of GFC fractions of membrane proteins from *A. thaliana*. GFC fractions in **Fig. 4** were separated by IEC on a Mono Q HR 5/5 (Amersham Biosciences) according to a linear NaCl gradient and further resolved by SDS-PAGE. IEC chromatograms (top) and SDS-PAGE patterns (bottom) of corresponding successive fractions are exemplified in **(A)** for the GFC fraction B2 (vacuolar membrane proteins) and in **(B)** for the fraction C5 (PM proteins) shown in **Fig. 4**. The same aliquot of eluted volume was layered onto each lane, and gels were Coomassie blue stained.

2. Mix 0.8 μL of each sample with 0.8 μL of the matrix, and spot the mixture immediately on the MALDI target.
3. Allow to dry and crystallize.
4. Rinse the target with ultrapure water and dry.

Spectra are summed until qualitative change are no longer observed (about 200–300 laser shots; *see* **Note 8**).

3.6. Database Search

Many different search engines are now available to identify proteins from an observed set of peptide mass. ProFound ***(10)*** explicitly allows for identification of mixtures (up to four different proteins), and mixtures are systematically assumed in our analyses (even if only one protein is finally identified univocally). Search parameters are as follows: appropriate taxonomic category (*A. thaliana* or *S. cerevisiae*), mass tolerance of 0.1 Dalton (to allow for a large matching of peptides outside the calibration range), one missed cleavage site allowed, protein MW range of 0–1000 kDa, p*I* range of 0 to 14, and no chemical modification of amino acids. To take into account the possible presence of protein mixtures, up to four different runs are performed by changing the supposed number (from one to four) of proteins present in the excised band. The run giving the best score is then conserved. For each identified protein, the list of experimental matching masses is used for a final Profound run in the version "single protein." Then, only proteins presenting a probability close to 1 and a *Z*-score higher than 1.65 are considered.

4. Notes

1. There is no ideal method for membrane stripping, allowing exclusive elimination of encapsulated proteins and of peripheral membrane proteins. The use of detergent at sub-critical micellar concentration to permeabilize the vesicles may lead to the unsuitable elimination of some integral membrane proteins. This merits to be checked by identification of eliminated proteins. On the other hand, if elimination of peripheral membrane proteins is not suitable, omit stripping and anticipate two successive chromatographic steps (GFC followed by AEC) as illustrated in **Figs. 4** and **5** for unstripped membranes *(7)*.
2. Membrane protein solubilization is a step deserving careful optimization. First, concerning the choice of detergent, DM has been increasingly and successfully used to solubilize many different proteins from different biological materials in a functional state *(8,9)*. It may therefore be recommended as the first detergent to be tested. Second, both the DM/protein ratio and the duration of the solubilization step under slow agitation must be optimized. In some cases, the solubilization yield can be increased further by vigorous vortexing of a DM-membrane protein mixture resuspended under argon in completely filled and sealed tubes containing 0.5 mm glass beads. Additional sonication in a bath sonicator can be also valuable.
3. The efficiency of different ion exchangers has to be tested in preliminary experiments on aliquots of solubilized proteins from the material of interest. In particular, both the capacity of columns to bind proteins and the capacity of the salt gradient to elute bound proteins must be compared. For the membranes used

here, cation exchangers irreversibly bound a significant fraction of proteins. For this reason, anion exchangers (allowing elution of 95% of loaded proteins) are preferred.

4. About 80% of the initial amount of solubilized proteins from plant or yeast PM bind to the Mono Q HR 5/5 column (Pharmacia/Amersham Biosciences) after loading at pH 7.5. Higher yields of protein binding (up to 100%) can be achieved by increasing the pH of the loading buffer up to 11.
5. Solubilization with mild detergents of a given membrane may form small micelles with individual proteins as well as large micelles containing supramolecular protein complexes. To estimate practically the size range of the micelles, a preliminary GFC must be performed on aliquots using a column offering a small loading capacity and a separation range as large as possible (like Superose 6 10/30, Pharmacia/Amersham Biosciences). This allows one to choose a more appropriate column after (offering a lower separation range, greater resolution, and greater loading capacity).
6. GFC fractions must contain sufficient amounts of proteins (typically, 0.5 mg proteins) to achieve the second separation by AEC and the last one by SDS-PAGE.
7. It is important to avoid autodigestion of trypsin because this precludes effective digestion of proteins in the sample and results in the dominating presence of trypsin peptides on MALDI-TOF spectra, suppressing signals from other peptides. For this purpose: (1) avoid excess of trypsin—the amounts proposed in the above protocol should be considered maximal, appropriate for 7 × 2-mm gel segments of medium-strong intensity after Coomassie blue staining; and (2) limit the volume of trypsin solution—only a thin layer should remain around the gel segment after its complete hydration and before incubation at 37°C. As mentioned in the text, never dry the solution containing the peptides completely during the concentration process.
8. After few tens of laser shots, the spectrum often changes qualitatively. This seems to be a property of peptides from membrane proteins and indicates that, after laser ablation of the matrix surface and after desorption of "easily flying" peptides, other peptides can be further desorbed and ionized. For this reason, it is fruitful to sum more spectra than usual, up to 300 laser shots.

Acknowledgments

Programs for plant and yeast membrane protein separation and identification were respectively supported by Génoplante (New Tools 19993663 and 2001027) and Aventis Pharma (FR00ANT043).

References

1. Godovac-Zimmermann, J. and Brown, L. R. (2001) Perspectives for mass spectrometry and functional proteomics. *Mass Spectrom. Rev.* **20,** 1–57.
2. Rappsilber, J., Moniatte, M., Nielsen, M. L., Podtelejnikov, A.V., and Mann, M. (2003) Experiences and perspectives of MALDI MS and MS/MS in proteomic research. *Int. J. Mass Spectr.* **226,** 223–237.

3. Santoni, V., Malloy, M., and Rabilloud, T. (2000) Membrane proteins and proteomics: un amour impossible? *Electrophoresis* **21,** 1054–1070.
4. Gygi, S., Corthals, G. L., Zhang, Y., Rochon, Y., and Aebersold, R. (2000) Evaluation of two-dimensional gel electrophoresis-based proteome analysis technology. *Proc. Natl. Acad. Sci. USA* **97,** 9390–9395.
5. Görg, A., Obermaier, C., Boguth, G., et al. (2000) The current state of two-dimensional electrophoresis with immobilized pH gradients. *Electrophoresis* **21,** 1037–1053.
6. Klein, C. Garcia-Rizo, C. Bisle, B., et al. (2005) The membrane proteome of *Halobacterium salinarum*. *Proteomics* **5,** 180–197.
7. Szponarski, W., Sommerer, N., Boyer, J. C., Rossignol M., and Gibrat, R. (2004) Large-scale characterization of integral proteins from *Arabidopsis* vacuolar membrane by two-dimensional liquid chromatography. *Proteomics* **4,** 397–406.
8. le Maire, M., Champeil, P., and Möller, J. V. (2000) Interaction of membrane proteins and lipids with solubilizing detergents. *Biochim. Biophys. Acta* **1508,** 86–111.
9. Seddon, A. M., Curnow, P., and Booth, P. J. (2004) Membrane proteins, lipids and detergents: not just a soap opera. *Biochim. Biophys. Acta* **1666,** 105–117.
10. Zhang, W. and Chait, B. T. (2000) ProFound: an expert system for protein identification using mass spectrometric peptide mapping information. *Anal. Chem.* **72,** 2482–2489; http://prowl.rockefeller.edu/profound_bin/WebProFound.exe.

23

The PROTICdb Database for 2-DE Proteomics

Olivier Langella, Michel Zivy, and Johann Joets

Summary

PROTICdb is a web-based database mainly designed to store and analyze plant proteome data obtained by 2D polyacrylamide gel electrophoresis (2D PAGE) and mass spectrometry (MS). The goals of PROTICdb are (1) to store, track, and query information related to proteomic experiments, i.e., from tissue sampling to protein identification and quantitative measurements; and (2) to integrate information from the user's own expertise and other sources into a knowledge base, used to support data interpretation (e.g., for the determination of allelic variants or products of posttranslational modifications). Data insertion into the relational database of PROTICdb is achieved either by uploading outputs from Mélanie, PDQuest, IM2d, ImageMaster(tm) 2D Platinum v5.0, Progenesis, Sequest, MS-Fit, and Mascot software, or by filling in web forms (experimental design and methods). 2D PAGE-annotated maps can be displayed, queried, and compared through the GelBrowser. Quantitative data can be easily exported in a tabulated format for statistical analyses with any third-party software. PROTICdb is based on the Oracle or the PostgreSQLDataBase Management System (DBMS) and is freely available upon request at http://cms.moulon.inra.fr/content/view/14/44/.

Key Words: Database; comparative proteomics; mass spectrometry; two-dimensional polyacrylamide gel electrophoresis; Gene Ontology.

1. Introduction

Two-dimensional polyacrylamide gel electrophoresis (2D PAGE) *(1)* and mass spectrometry (MS) are useful tools to catalog proteins found in a tissue (2D maps) *(2)* or to identify differentially expressed proteins *(3,4)*. A typical proteome experiment will generate several 2D gels as well as corresponding images, locations, and quantitation data about hundreds if not thousands of spots and mass spectra data. At the end, data have to be synthesized and compared to be turned into biological hypothesis, which is not an easy task con-

From: *Methods in Molecular Biology, vol. 335: Plant Proteomics: Methods and Protocols*
Edited by: H. Thiellement, M. Zivy, C. Damerval, and V. Méchin © Humana Press Inc., Totowa, NJ

sidering the various unrelated techniques and methods involved, as well as the amount of data produced.

The PROTICdb database was developed to handle all the data generated by a plant proteomic project from plant handling to 2D PAGE and MS outputs *(5)*. The database automatically builds links to external sequence databases and annotates proteins with Gene Ontology (GO) terms *(6)* in order to provide the user with a direct link to public functional information. As life scientist observations or conclusions about protein spots may be valuable for colleagues or subsequent data analysis, these can be stored in the knowledge base of PROTICdb. To date, it has been possible to create binary relationships between spots such as matching, allelism, and posttraductional modification. By inspecting these relationships, the database can infer new relationships, build spot networks, and spread identification data from a spot to other members of the same spot network, possibly completing the spot annotation of old experiments when a new relationship is added to the database.

PROTICdb is provided free of charge (the tool alone) at http://cms.moulon.inra.fr/content/view/14/44/ and is developed continuously. This chapter is based on the 1.0.7 version. A future version will include an advanced query tool with which the user can build a query iteratively. At each new step, results are available allowing the user to refine the query as needed.

2. Materials

PROTICdb client/server application includes a Database Management System (DBMS), a web server (the server side applications), and a user interface, usually a web browser (also referred to as the client).

2.1. Software Requirements

2.1.1. Server Side

A Unix/Linux server is required with Apache (version 1.3.33 or more, http://www.apache.org/), PHP (version 4.3.4 or more, http://www.php.net), and Perl (version 5.8.4 or more, http://www.perl.org). The database can be hosted on the same server or on a distant one. PROTICdb runs with either an Oracle8i or a Postgresql (version 7.4.7 or more, http://www.postgresql.org/) DBMS.

2.1.2. Client Side

The end-user may use any recent web browser compliant with CSS and JavaScript-enabled (Internet Explorer 5 [or more], Mozilla/Firefox (version 1.0 or more, http://www.mozilla.org), or Konqueror (version 3.3.2 or more, http://www.konqueror.org/). PROTICdb respects the XHTML 1.0 web standard (http://www.w3.org/).

The gel browser requires Java 1.4 or more (http://java.sun.com/) to be installed on the client computer.

2.2. Server Installation

Download the latest archive (http://cms.moulon.inra.fr/content/view/14/44/). Uncompress it and read the installation guidelines carefully (Protic/doc/install/pdf/install.pdf). At thc end of the installation process, the PROTICdb "diagnostic page" should be run (Protic/home/diagnose.php). This tool will help the user to fix setting errors related to Perl, PHP, and Apache configurations. PROTICdb support can be found at proticdb-support@moulon.inra.fr for any questions about installation or for bug reports.

3. Methods

3.1. PROTICdb Overview

To enter or browse data, each end-user needs a PROTICdb account (created by your administrator). The first step, for a new user, is to create a new project. Once created, the project is owned by the user, who can grant access to other users. To ensure a high level of data consistency, manual typing is limited as much as possible. Therefore, web forms contain combo boxes referring to a controlled vocabulary or to other data previously loaded into the database. When a new PROTICdb database is created, the controlled vocabulary library is empty, and the user implements it as needed. This will ensure a common and, if possible, standardized vocabulary. This is also important for query purposes, as this may reduce uses of synonymous terms as well as misspellings.

The web interface comprises three main areas: the main menu, the contextual menu, and the working space (**Fig. 1**). The user login is displayed below the PROTICdb logo along with the current project name (if one was selected).

Data must be loaded into a PROTICdb project. However, it is possible to create one or more projects and to organize data the way you need.

This training will be based on an exemplary dataset available at http://cms.moulon.inra.fr/content/view/14/44/.

3.2. Alimentation of the Database via Web Forms: Mélanie 2D Gel Image Analysis Output Uploading

For this part of the tutorial, the user will first have to create a project and then load all information related to a 2D gel from the plant specific for gel image acquisition and analysis, including plant culture conditions, protein extraction process, electrophoresis, and gel image analysis methods.

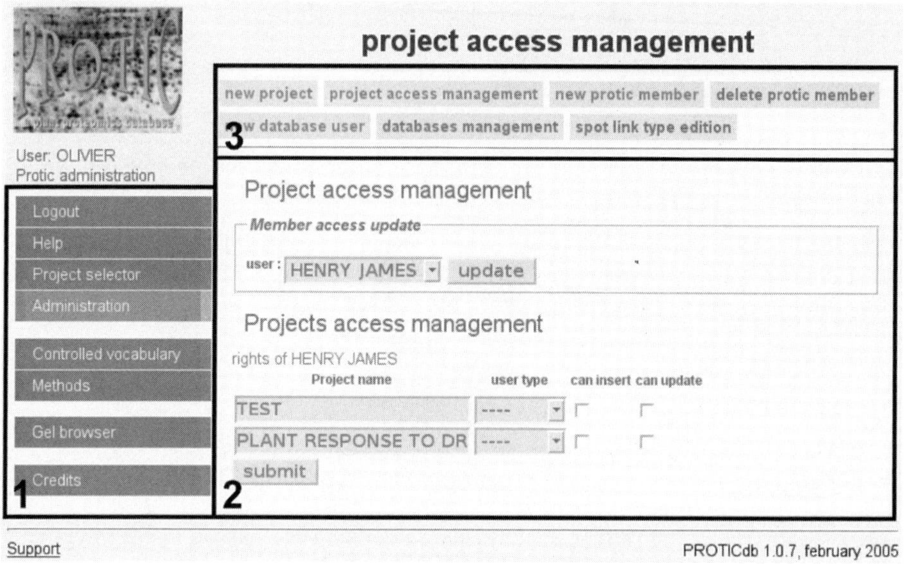

Fig. 1. General organization of the PROTICdb interface (1, main menu; 2, working space; 3, contextual menu).

3.2.1. Creation of a New Project

1. Go to the PROTICdb home page and log in (ask your PROTICdb administrator for a personal account).
2. Select the "administration" item on the main menu and then "new project" from the top contextual menu. The "new project" form is shown in **Fig. 2**.
3. Enter a name for the new project and add a text-free description and/or comment if needed.
4. Choose the key words in the list, or click on "new/update" to create or modify a key word (*see* **Note 1**). Initially, the database does not contain any key word. Thus new users must enter the first key words. This gives an opportunity to choose standardized terms or to try to build such standards at the laboratory level. Each key word may be entered along with a text-free description. Two "levels" of key words are proposed. This may help in classifying and mining projects. For instance, a first-level key word could be a general term like stress, and a second-level one could be more specific, like cold or drought.
5. Click on "go."

If an error occurs, fix it by following the instructions and submit the new project again.

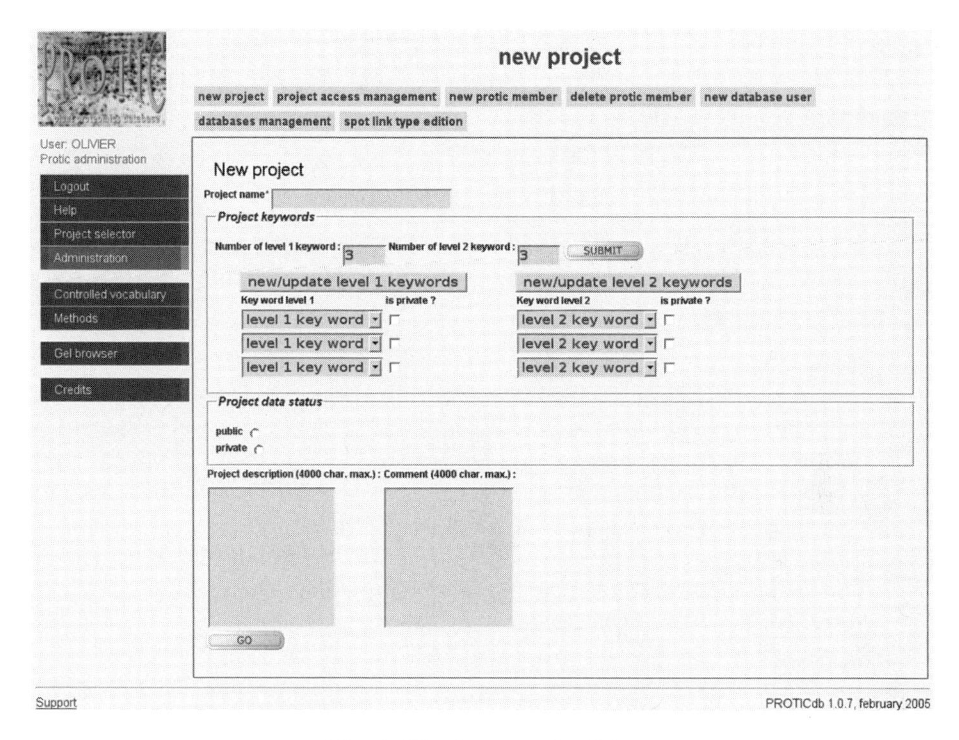

Fig. 2. The "new project" form.

3.2.2. Project Selector

Data must first be selected from the "project selector" in order to load them into the new project. This will also apply for future connections to PROTICdb.

1. Select the "project selector" item on the main menu.
2. Choose your project in the combo box.
3. Submit.

The selected project is now the current one for the database. Its name is displayed below the PROTICdb logo along with the logged user name. An overview of the current project (number of plants, protein extracts, gels, detections, and identifications) is indicated in the working space.

It can be seen that the main menu exhibits two new items: "form based feeding" and "file based feeding." These are the two gates for data submission to the database.

3.2.3. Plant Description to 2D Gel Image File Submission via Web Forms

To enter data from the plant to the image file, one way is to fill in several web forms. As already mentioned, these forms allow the user to choose words among lists when possible. If convenient terms are not listed, a new one may be created by selecting the "new" link (*see* **Note 2**). In this chapter, we will review the plant individual, plant sample, protein sample, gel, and gel image forms. Then we will upload a Mélanie output file.

1. Select the "form based feeding" item on the main menu.

The "new plant" form is displayed (*see* **Fig. 3**) in the working space (*see* **Note 3**). Inside a project, a plant must be associated with a "PROTICdb experiment" (*see* **Note 4**).

2. Enter a plant name (*see* **Note 4**).
   For this demo, there is no need to specify a particular plant name or other field. The demo will work with any name, controlled vocabulary term, or method name. However, the gel image name must correspond to a real 2D gel (*see* below).
3. Select or create (new buttons) genetic data as required. If a field is irrelevant, just leave it empty. The genotype is a term of the controlled vocabulary. If the convenient genotype is not listed, it may be created with the genotype "new" button as follows: once you are in the new genotype form, enter the scientific name (*see* **Note 1**) of the species and the genotype name (for example, *Zea mays* and F2). Click on submit; the new genotype will be recorded and the "new plant" genotype list updated. Continue by filling in the "new plant" form. Proceed in this way for each field by selecting the appropriate designations among the lists available in the form, including data about "cultivation" (*see* **Note 5**).
4. Create or select an experiment (*see* **Note 4**).
5. You can claim that your data are private or public. This feature will be implemented in future versions of PROTICdb, but you should consider it now.
6. A text-free comment may be added. This may be useful to enter data not proposed by the form or information that will be important for subsequent data analysis.
7. Submit the "new plant" (*see* **Note 6**) data.
   If an error occurs, fix it by following the instructions and submit the form again.
8. Enter data about plant sample harvest: click on the "plant sample" link of the contextual menu bar. Complete the form as described for the previous one. As the database has been updated after the "new plant" form submission, the corresponding plant can be selected in the plant field.
9. Complete the "gel" form. This last form is similar to the previous ones. A comigration electrophoresis experiment feature has been included in PROTICdb. The number of samples loaded must be entered in the "sample number" field. The form is updated with as many "protein sample" fields as needed (*see* **Note 7**).
10. Complete the "gel image" form and enter DV02121001 for the image name to be compliant with the demo dataset. The gel image file has to be in jpeg format. It is crucial that this image present the same resolution as the one used for spot detection (*see* **Note 8**). For the demo purpose, the image file DV02121001.jpg is available from the PROTICdbDemo.zip archive. This image will be necessary for later steps of this tutorial, so please make sure you have loaded it.

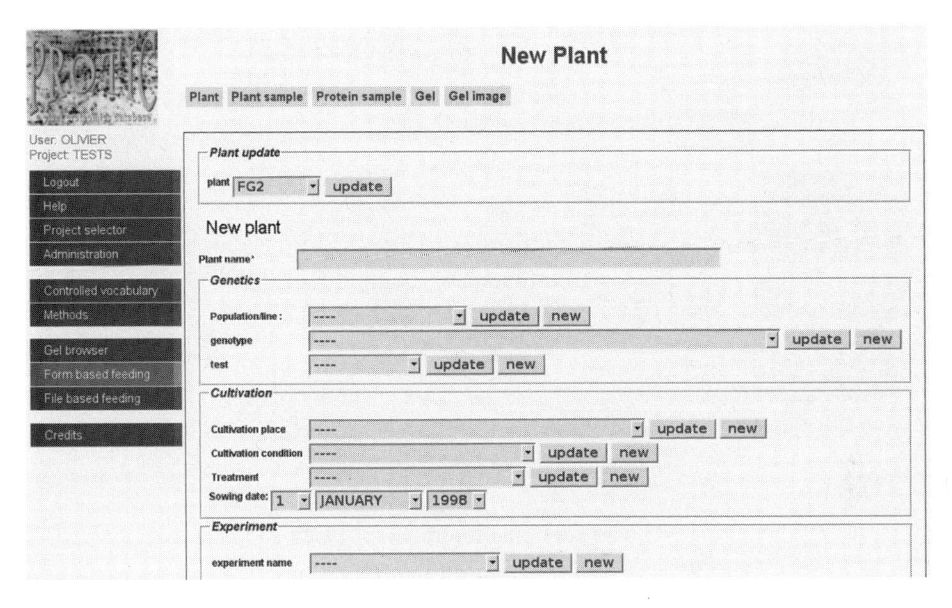

Fig. 3. The "new plant" form.

3.2.4. Mélanie Output File Uploading

1. From the Mélanie software, generate a detection report and export it to a tabulation-separated text format file (*see* **Note 9**). The file extension should be ".txt." For this demo, use the detections/DV02121001.txt file from the PROTICdb Demo.zip archive.
2. Select the main menu "form based feeding" → "spot detection (new/update)" (contextual menu bar). A combo box with a list of software appears.
3. Select the software "MELANIE" and submit.
   The form "spots detection file upload form" is displayed (**Fig. 4**).
4. Fill in this form as follows:
   a. Numbering system: this is a reference name you can choose. Usually, it corresponds to the name of the gel master.
   b. Detection software: the name of the software used to perform detection. Here, select "MELANIE."
   c. Detection method: the method used to detect spots. You can create or update one method if needed (*see* **Note 5**).
   d. Gel image: choose the gel on which spot detection was performed. Here, choose "DV02121001."
   e. Scaling method: the method used to normalize spot quantitative data.
   f. Master: indicates whether the gel is a master (reference) or not. For this demo, the gel DV02121001 must be considered as a master.
   g. Private: not activated at the moment. However, this field should be considered to anticipate a future policy of public release of data.

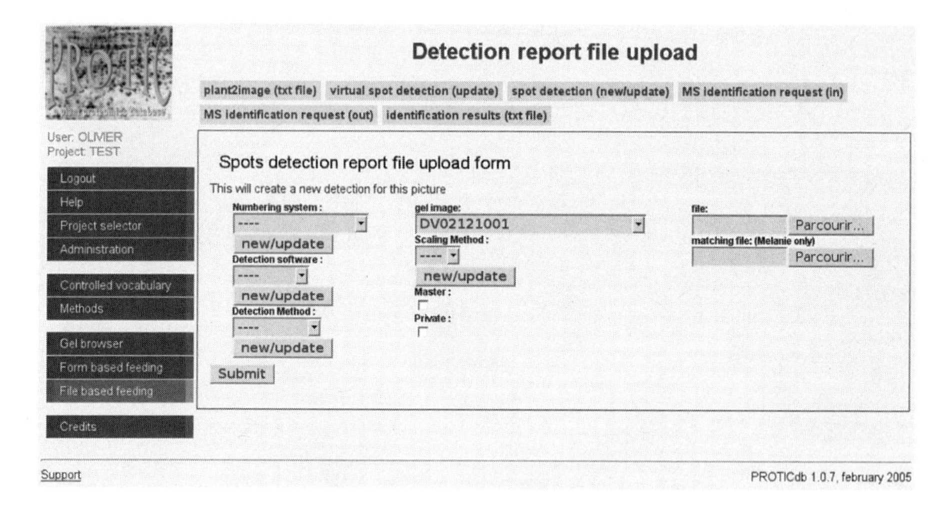

Fig. 4. The spot detection report file upload form.

h. File: click on this button to browse the local computer disk (or network distributed disk) and select the correct Mélanie output file. For this demo, select the "detections/DV02121001.txt" file, available from the PROTICdb Demo. zip archive.
i. Matching file: only used for Mélanie file uploading and required for a nonmaster gel. The Mélanie pair file contains the result of the matching experiment between a gel and the master gel. Since it is not relevant for the demo, leave it blank.

Once all the required fields have been filled, submit the form. Processing of this file by the database may take a long time; *do not interrupt* PROTICdb during this step. If an error occurs, correct it as suggested and submit the file again (*see* **Note 10**).

When the processing is done, a success/error notification is sent by e-mail to the current user. If errors occurred, none of the spot detection data are recorded. If so, just correct the errors as suggested and submit the file one more time.

The gel image, annotated with spot location and quantitative data, is now ready to be displayed by the Gel Browser (cf. **Subheading 3.5.**, Gel Browser).

3.3. File-Based Database Feeding: From Individual Plant to Mass Spectrometry

In this data submission example, we will see how to submit spot detection and mass spectrum identification outputs. First, we must create or choose a project (cf. **Subheading 3.2.**) and then feed the database with all information related to the plant, plant sample, protein extraction, gel, and image. These data will be useful to illustrate the second way to feed the database, namely, the file-based feeding method.

3.3.1. Uploading the "Plant2image" File

1. Go to the PROTICdb home page and log in.
2. Select a project from the main menu → "project selector" or create a new project.
3. Select the main-menu → "file based feeding" → "plant2image (txt file)" (contextual menu).
   The "plant to image file upload" and "gel image file upload" forms are displayed.
4. Choose or create an experiment in the "plant to image file upload" form (*see* **Note 4**).
5. From the "plant to image file upload" form, retrieve the "plant2image" file to upload using the browse button. For this demo, upload the plant2image.txt file available in the PROTICdb demo.zip archive.
   A plant2image template file is accessible following the link "here is the excel template file" (*see* **Note 11**). Each row of this file describes a sample and is divided into five sections: image, gel, protein sample preparation and gel loading, plant sample harvesting, and plant description. Once those sections have been filled, save the file by choosing the "tabulation-separated text" option.
6. Click on the "submit" button on the "Plant2image file upload" form. (Make sure not to click on the submit button of the second form.)
   The file is then processed by PROTICdb. This can be time consuming. Do not interrupt the loading process. Numerous tests are performed to ensure data consistency providing conformity to the controlled vocabulary. The method name from the submitted file must match a method already described in the database (new methods must be recorded beforehand; *see* **Note 5**). If tests failed, an error report message is sent by e-mail to the current user, and none of the data are recorded. In such a case, correct the file according to the error report suggestion and try to submit the entire file again.
   If the process is successful, a success report is also sent by e-mail.
7. From the form "gel image file upload" → "image file format," select "jpg."
   Browse your computer disk to select the gel image file corresponding to the data described in the plant2image file (*see* **Note 8**). Gel images in jpg format have to be uploaded one by one using this form. For the demo, upload the DV02041611.jpg and TB02052306.jpg files available in the "demo.zip" package.
8. Submit the "gel image file upload" form.

3.3.2. Mass Spectrometry Spot Identification Order

A single spot may be named with several names or numbers such as detection spot number, matching number, picking number. and so on. It is thus tedious to associate mass spectra data to the proper spot. To avoid errors, one of the PROTICdb workflows generates a file (referred to as an order) with a unique spot ID that is sent to the MS facility along with the excised spots. The

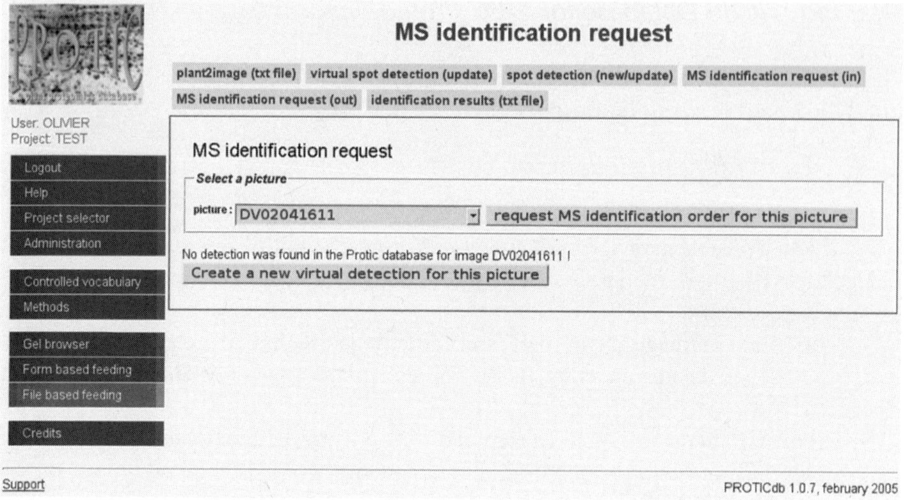

MS identification request

plant2image (txt file) | virtual spot detection (update) | spot detection (new/update) | MS identification request (in) | MS identification request (out) | identification results (txt file)

User: OLIVIER
Project: TEST

Logout
Help
Project selector
Administration

Controlled vocabulary
Methods

Gel browser
Form based feeding
File based feeding

Credits

MS identification request

Select a picture

picture : DV02041611 — request MS identification order for this picture

No detection was found in the Protic database for image DV02041611 !

Create a new virtual detection for this picture

Support

PROTICdb 1.0.7, february 2005

Fig. 5. MS identification request form (**step 1**).

MS facility can then return the result with the unique spot ID, hence avoiding ambiguous denominations. In this example, files describing the microplates containing the excised spots will be submitted to the database. On this microplate map, each excised spot is named with the "picked_spot" name previously defined by the user. This workflow supposes that the order is sent to the MS facility before the image analysis data (spot detection) are recorded in the database. That way, when the user will subsequently submit detection and matching data, he or she will simultaneously give the appropriate "picked_spot" designation associated to the spot detection number or the spot matching number. This step is crucial, as it allows the database to associate the "picked_spot" number to the matching number and then to ensure correct subsequent integration of MS data.

The output of the "Identification Order" creation step is a basic "Sequest" software-compliant file. It will contain a row for each sample of the microplate(s). This file is useful since it is directly readable by the "Sequest" software (*see* **Note 12**).

To create a new order, proceed as follows:

1. Select the main menu "file based feeding" →"MS identification request (in)" in the contextual menu.
2. Choose 'order spot identifications from one picture/detection' and submit.
3. Choose the gel image corresponding to the samples. (In this example, the spots located in the microplate derive from a single gel.) Then choose the "DV02041611" image.
4. Click on the button "request MS identification order for this picture."
   A new form is displayed (*see* **Fig. 5**, **step 1**).

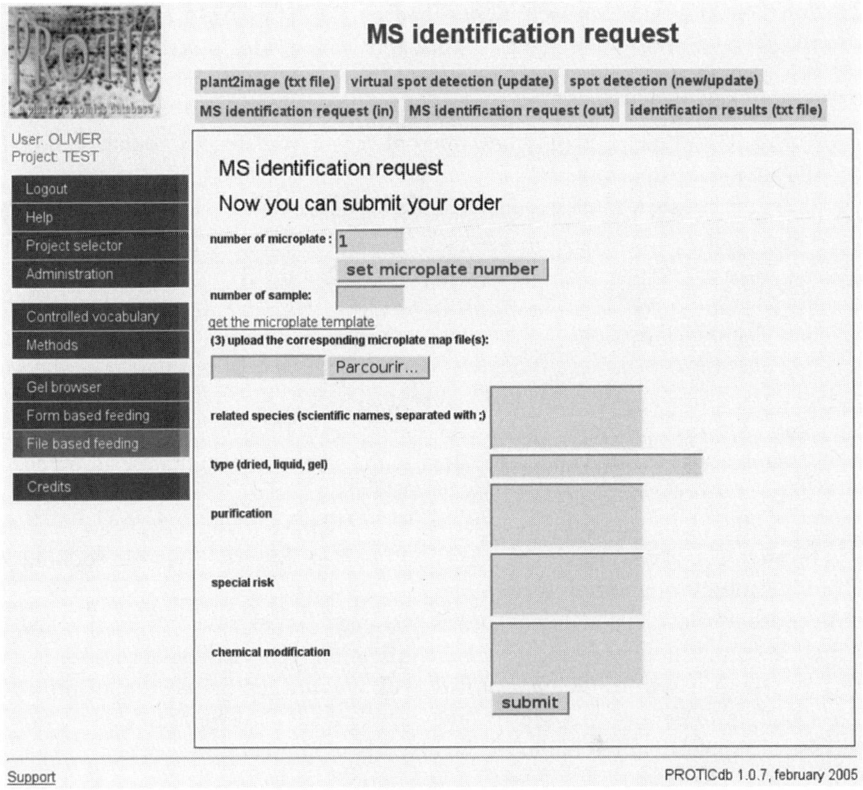

Fig. 6. MS identification request form (**step 2**).

5. Click on the button "create a new virtual detection for this picture."
   A virtual spot detection output is created (remember that the image analysis outputs were not submitted until now). The spots from the microplate are created in the database. These spots will be updated when the real spot detection experiment is finished and the output uploaded, and when the MS results will be fed to the database.
6. To load MS identification results, select the corresponding "virtual detections" created in **step 13**. Those "virtual detections" will be converted to "real detections" when the Mélanie output file has been submitted. A new form appears. Its title is "now you can submit your order" (*see* **Fig. 6**, **step 2**).
7. Enter the number of microplates used for this experiment, and click on "set microplate number."
   Click on "get the microplate template" and download the excel template file. In the microplate excel template file, enter the number of microplates used in this identification order, the mail address of the current PROTICdb user, and, in the grid (12 columns and 8 rows from A to H), the exact name of your samples: the

"picked_spot" number used in **Subheading 3.3.2.** (*see* **Note 1**). Save your Excel file to "tabulation-separated text" format (*see* **Note 11**). This map corresponds to the microplate used by your MS facility.
Prepare as many microplates files as needed.
For the demo, there is only one microplate: microplates/microplatedemo.txt in the "demo.zip" package.

8. Enter the number of samples loaded onto the microplate(s). Here, there are nine samples on the demo microplate.
9. Browse your local computer disk and select the microplate file(s).
For the demo, load the "microplatedemo.txt" file, ready to use and available in the "demo.zip" package.
Pay attention to the e-mail address in the microplate file; it has to be the same e-mail address as the PROTICdb current user one.
Enter "related species": genus and species according to the NCBI taxonomy (sequences from other species related to the sample are sometimes searched for protein identification). Here, enter "Oryza sativa."
10. Fill in the rest of the form as described and click on "Submit."
If errors occurred, an error message will appear. Modify your file according to this report and submit it again.
In case of success, a message "order n created !" is displayed. Keep this identification order number to access the identification report easily later.
An e-mail is sent to the current user with an attached Sequest input file corresponding to the order. Keep this message and save the Sequest input file. This file is supposed to be sent to the MS facility along with the spot microplate.

3.3.3 Spot Detection Output File Uploading

When using Mélanie, you need to annotate each spot with the corresponding "picked_spot" name (*see* **Subheading 3.3.2.**). To do so, a "picked_spot" column defined by the user must be created (*see* Mélanie documentation). Then, export the spot report in a tabulation-separated text format file (*see* **Note 9**).

Once the Mélanie detection file is saved, proceed as follows:

1. Select the main menu "file based feeding" item → "virtual spot detection (update)" (contextual menu).
2. Choose the image corresponding to the Mélanie output. Here, choose "DV02041611."
3. Submit.
A list of the previously created virtual spot detections for this gel image is displayed (*see* **Fig. 7**, **step 1**).
4. Choose the virtual detection to be updated (for the demo, only one virtual detection should be selectable).
5. Submit.
6. The "Spots detection file upload form" displayed is similar to the one described in **Subheading 3.2.4.** (*see* **Fig. 8**, **step 2**). Fill in this form as described in **Subheading 3.2.4.** The demo Mélanie output file DV02041611.txt can be found in

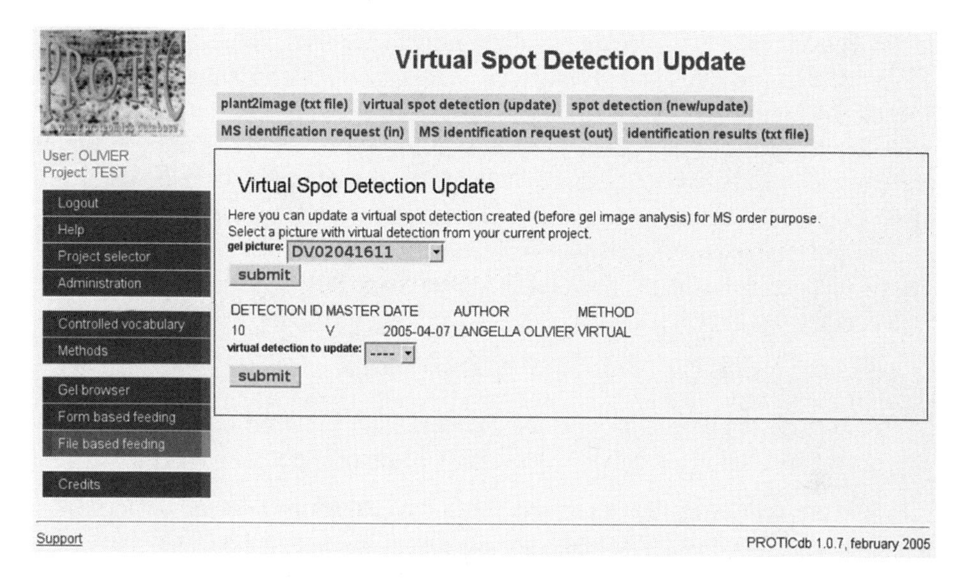

Fig. 7. Virtual spot detection update (**step 1**).

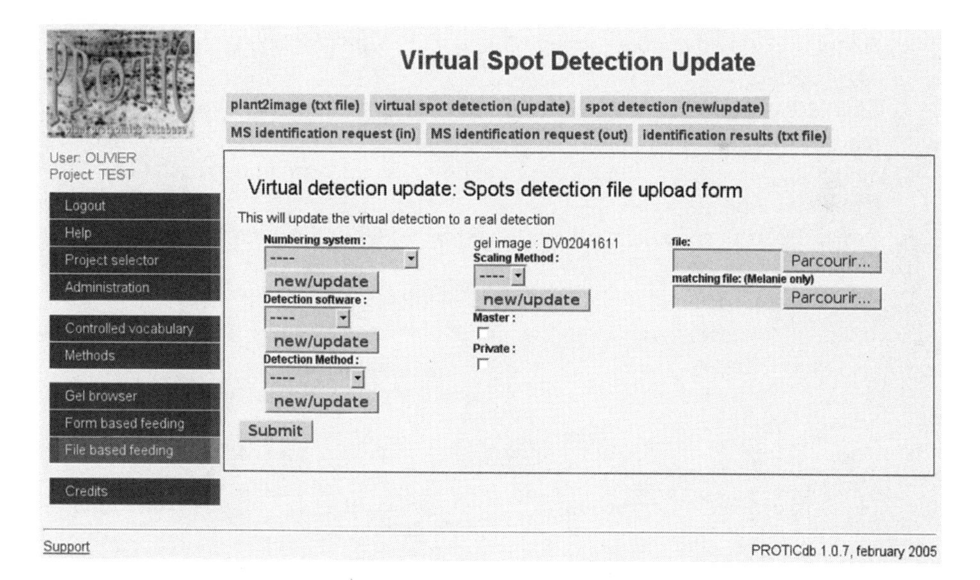

Fig. 8. Virtual spot detection update (**step 2**).

the "demo.zip" archive. This gel must be considered a master. Consequently, the "matching file" field has to be left empty. The private/public field is not used at the moment. However, this issue should have already been considered to anticipate a future policy of public release of data.

7. Submit.

Processing of this file by the database may take a long time; *do not interrupt* PROTICdb during this step. If an error occurs, correct it as suggested and submit the file again (*see* **Note 10**).

When the processing is done, a success/error notification is sent by e-mail to the current user. If errors occurred, none of the spot detection data are recorded. If so, correct the errors as suggested and submit the file one more time.

The gel image, annotated with spot location and quantitative data, is now ready to be displayed by the Gel Browser (*see* **Subheading 3.5.**, The Gel Browser).

3.3.4. MS Identification Report File Uploading

PROTICdb is compliant with the Sequest identification report if the latter has been exported in the tabulation-separated text format (*see* **Note 13**).

1. For the demo purpose, go to http://location_of/your/proticdb/home/demo.php. Click on "form to obtain the sequest identification demo file." The form "Get your sequest identification demo file" then appears. Upload the Sequest compliant file obtained in **Subheading 3.3.3.** and submit the form.
   An e-mail is send back to the current user along with the Sequest identification report file corresponding to the ordered spot identification created in **Subheading 3.3.3.**
   This last step ***(26)*** is for demo only and mimics the delivery of an identification report by the MS facility.
2. Select the main menu "File based feeding" → "identification results (txt file)" (contextual menu).
3. Select the software used to identify the samples. For the demo, select "Sequest."
4. Submit the form.
   A new "identification file upload" form is displayed (*see* **Fig. 9**).
5. Select or create the identification method (you may create a fictitious method for the demo purpose).
6. Select or create the database used for spot matching (you may create a fictitious database for the demo purpose).
7. Select the Sequest identification report file (received through e-mail in **step 26**) by clicking on "browse."
8. Submit.

Processing of this file by the database may be time consuming; *do not interrupt* PROTICdb during this step. If an error occurs, correct it as suggested and submit the file again (*see* **Note 10**).

The gel image, annotated with spot location and quantitative data and MS results, is now ready to be displayed by the Gel Browser (*see* **Subheading 3.5.**, The Gel Browser).

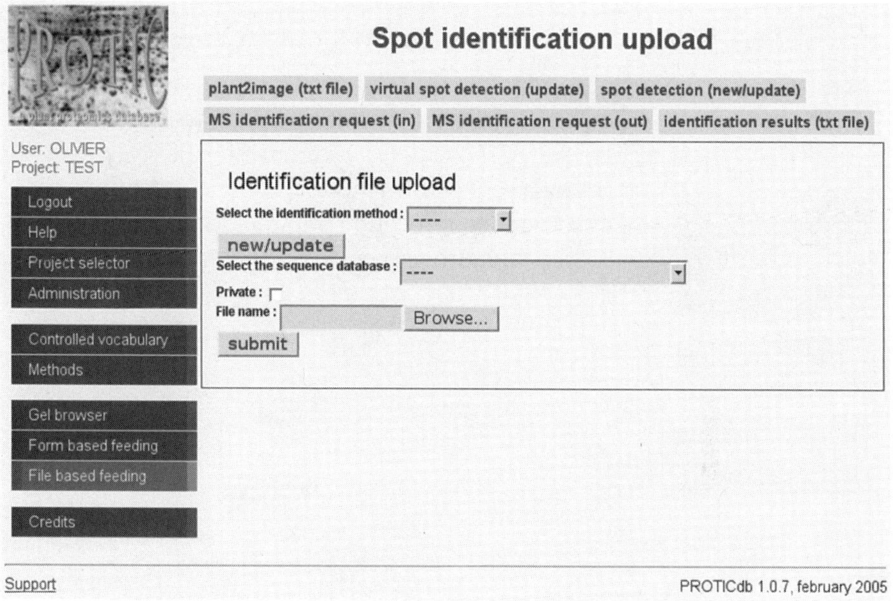

Spot identification upload

plant2image (txt file) | virtual spot detection (update) | spot detection (new/update)
MS identification request (in) | MS identification request (out) | identification results (txt file)

User: OLIVIER
Project: TEST

Logout
Help
Project selector
Administration

Controlled vocabulary
Methods

Gel browser
Form based feeding
File based feeding

Credits

Identification file upload
Select the identification method: ----
new/update
Select the sequence database: ----
Private:
File name: Browse...
submit

Support
PROTICdb 1.0.7, february 2005

Fig. 9. The spot identification report file upload form.

3.4. Sharing Data with Other Users

As a project owner in PROTICdb, you can easily share data with other PROTICdb users. You can grant access privileges and choose precisely what kind of privileges other users will be granted, such as read-only or full access. Only the PROTICdb administrator can create a new user account (*see* **Note 14**).

1. Go to the PROTICdb home page and log in.
2. Select the main menu "Administration" → "project access management" (contextual menu).
3. Select the user to whom you wish to update the rights (*see* **Fig. 10**, **step 1**).
4. Click on "update."
5. For each project you own, you can change the other user access rights (*see* **Fig. 11**, **step 2**):
   a. User type: owner or guest. An owner will have as many privileges as the current project's owner. A guest has limited access, read-only by default; more rights may be granted with "can insert" and "can update" options.
   b. Can insert: the user will be able to insert new data in this project.
   c. Can update: the user will be able to update existing data in this project.

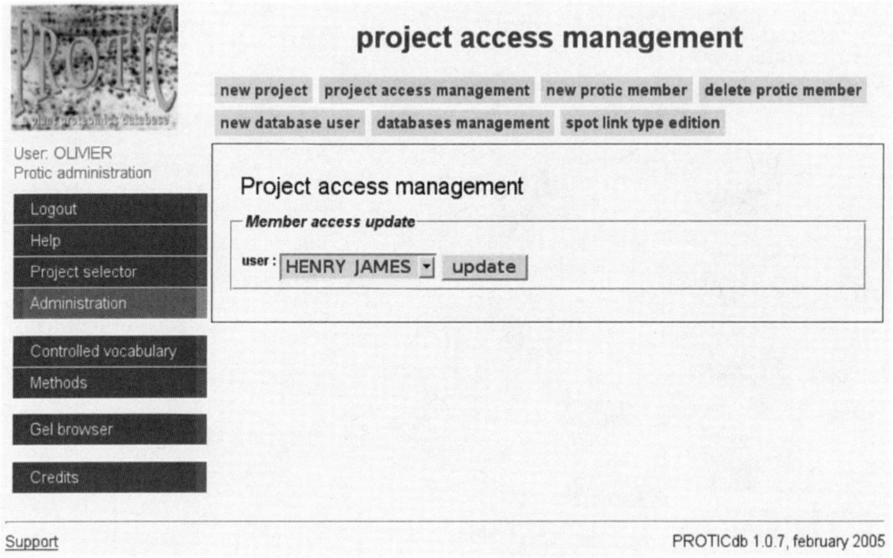

Fig. 10. The project access management form (**step 1**).

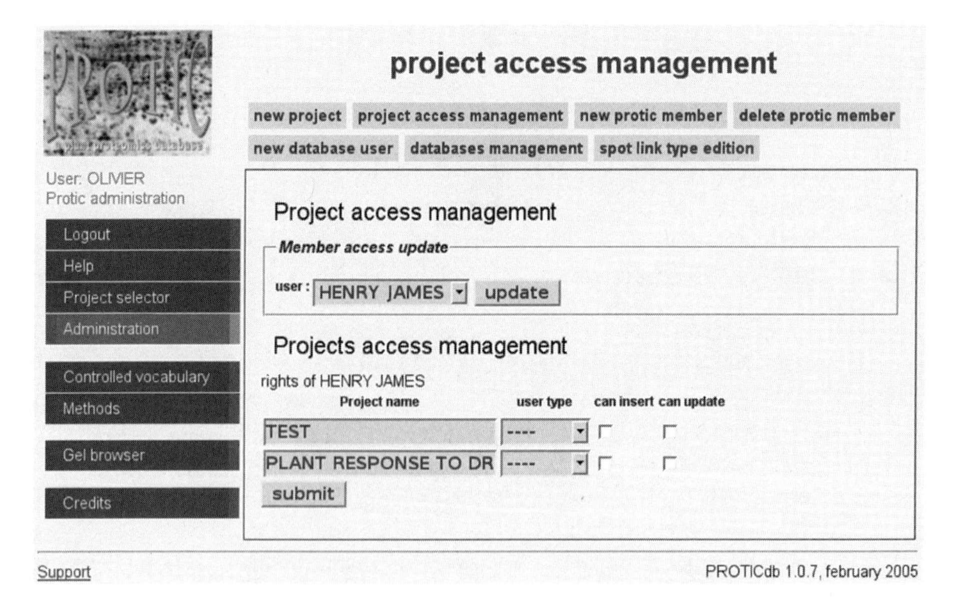

Fig. 11. The project access management form (**step 2**).

3.5. The Gel Browser

Gels can be accessed using the gel browser. Up to four gels can be displayed at a time. To make gel visual comparison easy, several zooming and image synchronization tools are available. For each protein spot, summarized or comprehensive data can be displayed, and links to external databases are provided. The gel browser also allows one to submit new data to the knowledge base.

These different points will be exemplified in the two following scenarios.

3.5.1. Which Spots on a New Gel Have Previously Been Identified?

One of the steps followed in a new gel analysis may be the selection of spots for mass spectra identification, without choosing spots that have previously been identified. The gel browser can be used to compare the new gel with up to three other master gels, displaying annotations. In this example, we will use the demo database at http://cms.moulon.inra.fr/content/view/14/44/ (we recommend installing a local version of the PROTICdb database and loading the demo dataset as explained previously), but it may also be conducted with your own dataset provided it contains at least a master gel and a gel that has not been analyzed yet (no identification or spot detection data; *see* **Note 15**).

1. Go to the PROTICdb home page and log in.
2. Launch the gel browser: main menu → Gel browser (*see* **Note 16**). The Project frame contains projects the user is authorized to browse. A description of the project, if available, can be found in the "project summary" item. Master gels are listed below the "project summary." Each master gel has a name and a numbering (*see* **Note 17**) system tag between parenthesis. Non-master gels are placed in the "non-master gel" subdirectory. The gel name is also followed by the numbering system tag. It is thus easy to determine to which master gel it has been matched. The "other gels" directory contains gels for which no spot detection or identification data are available, like newly produced gels.
3. Select the gel TB02052306 from the "projects" inner frame (in "other gels") and click on the "load" button (*see* **Notes 18** and **19**).
4. Repeat **step 3** for gel DV02041611(DLFMELANIE). There is no spot detection or identification information for the gel TB02052306. We assume we are interested in identifying several spots on this gel, and we have to decide whether mass spectra analysis is needed or not. We will compare TB02052306 with the available master gel, here DV02041611(DLFMELANIE), to find previously identified spots. The detected spots of DV02041611 are highlighted with blue crosses. The header of this gel indicates that 1820 spots were detected and that 9 are identified among them. To check which spots have been identified, we need to activate the Identification display mode.
5. In the menu bar choose Mode → ident or press key Alt i.
   Identified spots have a red cross. Let us take a closer look at these spots
6. Click anywhere on the DV02041611 window to make sure it has focus.

7. Select the Zoom level 3 by moving the cursor along the zoom slider. The displayed area of the gel is represented by the blue square in the top left thumbnail. This square can be selected and dragged over the thumbnail to change the area displayed.
8. Drag the thumbnail square of the DV02041611 gel on the right until five identified spots (red crosses) appear.
   By moving the mouse pointer over each spot, summarized information is displayed. This information includes the spot number, the numbering system (*see* **Note 17**), the pixel coordinates, a commentary, if available, and a functional annotation, if available (red crosses only). More comprehensive information can be displayed.
9. Fly over spot 342 (top left red spot of the area selected in **step 8**) to see the corresponding summary box.
10. Left click on the same spot (342) (*see* **Note 20**) to activate the Spot information box.
    Spot data are classified into four items: General, Relations, Identification, and Quantification. Each item may be developed to access the corresponding information. Identification includes MS details. (Validated is for future features.)
11. Develop the Identification and MS details and the next subdirectories.
    Each match obtained from mass spectra data analysis is reported. The first three items of a match are the sequence description, the global identification score (mass spectra matching score), and the access number of the sequence, the latter being linked to the corresponding database.
12. Double left click on the link "696997|entrez" of the spot 342 first match. A new browser will appear displaying the AAF34134 NCBI Entrez sequence report (*see* **Note 21**).
    When possible, corresponding Swiss-Prot key words and GO terms are provided. Again, direct links to the EBI GO Browser QuickGO are available.
    The next available information is about the peptides that led to the match. For each peptide, sequence and scores (Sequest or Mascot formats) are displayed.
    The quantification item provides access to spot area, volume, or peak height depending on what was submitted to the database.
    A spot can be located on a gel thanks to the spot finder.
13. Make sure that the gel DV02041611 window is active by left clicking into it.
14. From the menu bar, select Tools → Find or press key Ctrl f.
15. Type 1647 and click OK. The DV02041611 is zoomed and spot 1647 is highlighted by a yellow flashing circle. We are now able to proceed with DV02041611 and TB02052306 comparison.
16. Move the spot 1647 at the center of the image using the bottom right scrolling bar.
17. From the menu bar, select Tools → Synchronize or press key Ctrl s (*see* **Note 22**).
    The two gel images are now displaying the same area. This feature will not work properly if the images present too many differences. It may be necessary to use the scrolling bar to adjust the correspondence between images.
18. From the menu bar, select Tools and check the Link item or press key Ctrl l.

19. Drag the blue square of one of the two thumbnails.
    The two gel display areas are updated at the same time, conserving synchronization. However, as local gel phenomenon may lead to synchronization loss, the areas may be manually readjusted for one gel at a time using the scrolling bars.
20. Come back to the spot 1647 area (DV02041611). Follow **steps 13** to **17** again.
    It is easy in this case to find on TB02052306 the spot corresponding to spot 1647 on DV02041611. This spot does not need to be identified again and may not be excised for an MS experiment, except if the mass spectra outputs (accessible through the spot information box) are not considered of high quality by the user.

3.5.2. Alimentation of the Knowledge Base

The knowledge base was developed to store results obtained by experts when they are performing image gel comparison. The goal is to store binary relationships between spots from a single gel or from two master gels. Examples of such relationships are matching, physical interaction, allelism, posttransductional modification state, and so on. The user is free to create any type of relationship. Once two spots are linked, they share information like identification results. PROTICdb automatically builds spot networks by searching for transitive relationships within the knowledge base. For instance, if spot A is linked to spot B and spot B is linked to spot C, then the database suggests that spot A should be linked to spot C as well. It is then up to the user to validate the new information. If spot A is the only one identified, then the database will suggest the same identity for spots B and C. As the knowledge base is growing, more information is shared between spots, and this may result in a constant update of spot annotation level (even for old experiments), hence avoiding unnecessary MS analyses as well as allowing the building of new hypotheses.

1. Go to the PROTICdb home page and log in.
2. Launch the gel browser: main menu → Gel browser (*see* **Note 16**).
3. Select the master gel DV02041611(DLFMELANIE) from the "Projects" inner frame and click on the "Load" button (*see* **Notes 18** and **19**).
4. Repeat **step 3** for gel DV02121001(DLFM02121001).
5. Activate the Ident. display mode: menu bar → Mode → Ident. or press key Alt i.
6. Activate the DV02041611 gel window with a single left click into it.
7. Find spot 1647: menu bar → Tools → Find or press key Ctrl f and type in 1647, then click on OK.
   Spot 1647 is highlighted by a yellow flashing circle.
8. If necessary, move this spot to the center of the window using the scrolling bars.
9. From the menu bar, select Tools → Synchronize or press key Ctrl s (*see* **Note 22**). Despite the synchronization, the two displayed areas do not correspond to each other. The right area on gel DV02121001 is just above the displayed one.
10. Adjust the displayed area on gel DV02121001 with the scrolling bars so that it corresponds to the area displayed in the second gel. To do so, consider the red

spot of the gel DV02041611(DLFMELANIE) and the group of five other spots located to its left and try to find the corresponding ones on gel DV02121001. You may just have to move up with the right scrolling bar. If you fail to retrieve the correct area, then find spot number 1752 on gel DV02121001 (*see* **step 7**) and move it to the center of the window.

Now the two images are synchronized, and you find that the red spot (spot 1647 on gel DV02041611) can be matched to spot 1757 on gel DV02041611. You can also observe that this last spot is yet unidentified. Let us declare the matching relationship between these two spots.

The title bar of the application (at the very top of the main window) shows (you use display a lot; when possible use synonyms) the settings of the knowledge base alimentation tool: association mode is set OFF, and association type is set to manual matching. We first need to activate the association mode.

11. From the menu bar, select Association and then check the box Association mode or just press key Ctrl a.
    Now the Association mode is set ON. The new relationship to enter is a manual match, and it is already set up. You can check all the possible relationships available by default. This menu is configurable by the user. (*See* the spot link edition page from the PROTICdb Administration web forms.)
12. Activate the DV02041611 gel window with a single click into it.
13. Right click on the red spot (spot 1647 on gel DV02041611). The spot is then surrounded by a red circle.
14. Activate the DV02121001 gel window with a single click into it.
15. Right click on the matched spot (spot 1757 on gel DV02121001). The spot is then red circled and a confirmation box appears. You can add a text-free comment about this new relation and validate it by clicking OK.
16. Click OK. A box appears confirming that the new relation was successfully registered into the database.
17. From the menu bar, select Mode → Relation or press key Alt r. This time, only the spots included in at least one relationship are figured by a blue cross.
18. Place the mouse pointer over one of the two linked spots. The color of the cross of the designated spot switches to orange simultaneously for crosses on every spot linked to it (in this case, one matched spot on the other gel). Only spots included in a same network are displayed in orange when one of them is pointed at with the mouse (**Fig. 12**). You can verify that feature by declaring new relationships, thus enlarging the spot network.
19. Close gel DV02121001 and load it again (Alt p or as described in **steps 3** and **5**).
20. Find spot 1757 on gel DV02121001 (as described in **step 7**).
21. Synchronize the two gel displays (as described in **steps 8–10**).
22. Make sure that the ident display mode is activated (*see* **step 5**).
    The gel information summary appearing at the top right of each gel window reports the number of detected and identified spots. On gel DV02041611, the red crosses represent the nine spots that have been identified. No spot has been identified on gel DV02121001(DLFM02121001), although one spot is highlighted

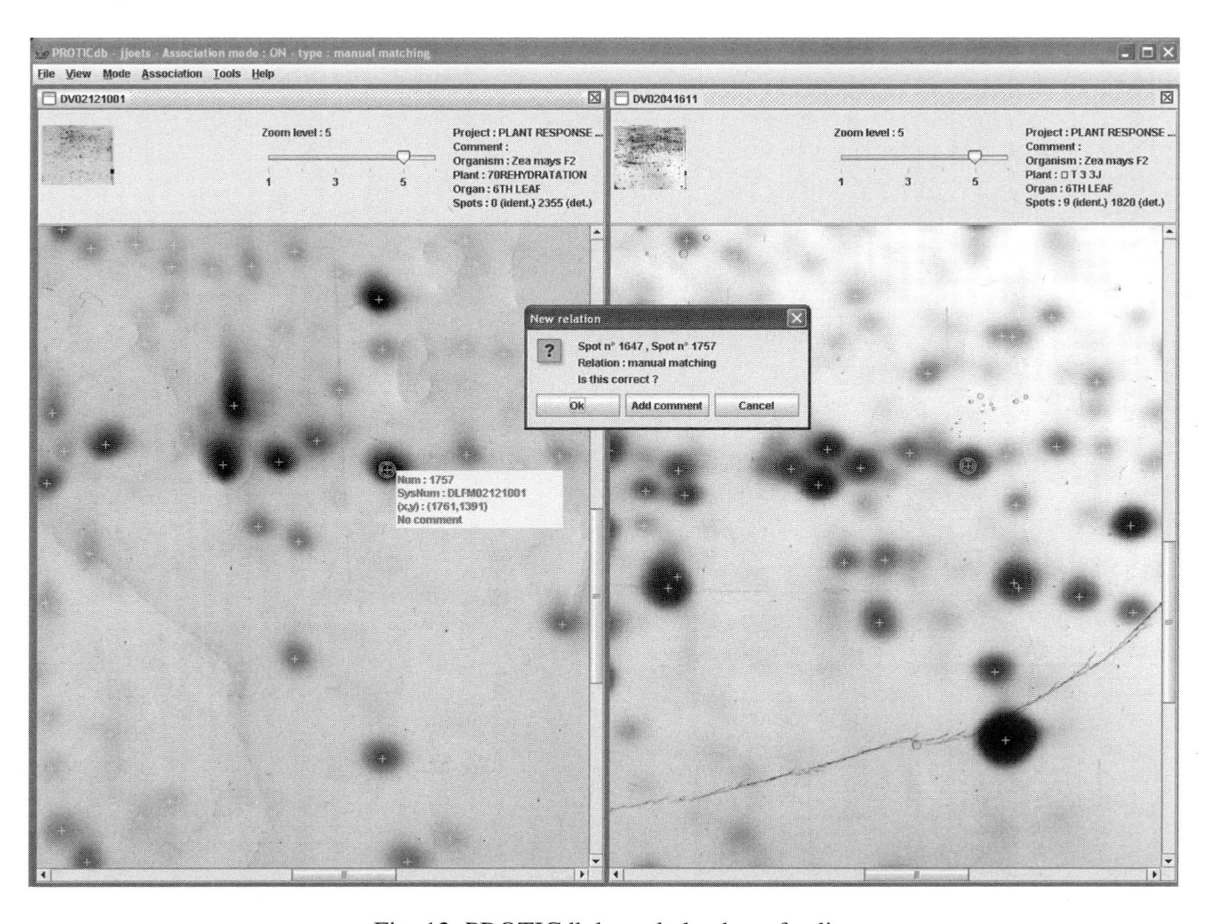

Fig. 12. PROTICdb knowledge base feeding.

by a red cross (spot 1386). These apparently conflicting data exemplify spot network implementation: spot 1757 on gel DV02121001 was identified (hence red crossed) because of a relationship with another identified spot (the relation previously created in this training). However, only spots that have been subjected to MS analyses are reported in the top right panel summary. Thus spot annotation is updated by PROTICdb via an automatic knowledge base analysis tool.

23. Place the mouse pointer over the red spot 1757 on gel DV02121001. The identification output is reported in blue text, meaning that this annotation is deduced from a relation with another spot. That particular relation is also reported just above the suggested identification.
24. Left click on the red spot 1757 on gel DV02121001.
    The Spot information inner box appears.
25. Develop the Relations and Identification directory. The relationship is reported along with the suggested identification. A warning message explains how this identification was obtained.
26. Compare with spot 1647 on gel DV02041611.

3.5.3 Quantitation Data Export

Spot quantity data (area, peak height, or volume) can be exported in tabulated a format for easy importation into any spreadsheet or statistics package.

1. Go to the PROTICdb home page and log in.
2. Launch the gel browser: main menu → Gel browser (*see* **Note 16**).
3. From the menu bar, choose Tools → Projects data export or press key Ctrl d. The data export box is opened. (There is no need to load any gel image prior to data exportation.)
4. Select the project containing the data you wish to export.
5. Select the numbering system (*see* **Note 17**).
   Comparing spots sharing the same number makes sense only within a given numbering system. Once the numbering system is selected, a table summarizing all available datasets is displayed. For each gel, the spot detection method name and the scaling method name are indicated. The user can export only data processed with the same method.
6. Select the data to be exported by checking the corresponding box and then click on OK.
7. Select area, volume, or peak height. You may also choose a missing data code. Click on OK.
8. Select the export output method: table, clipboard, or file. Spots are exported in rows and gels (conditions) in columns.

4. Notes

1. To avoid misspelling as much as possible and slight differences in software behavior dealing with strings, all words are uppercased for storing in PROTICdb (method names, controlled vocabulary, project keywords, and so on). Note that

this will not apply to species names, which must perfectly match the NCBI taxonomy format and therefore are case sensitive.

2. We encourage the user to check the lists carefully to avoid creation of synonymous terms. If a misspelling is detected at this step, it may be corrected by clicking the update button. Again, be careful: the update will be applied to the entire database. A first user of the database will have to create new terms.
3. Previously recorded plants may be selected to be updated at the top of the working space. This data "updating" tool is available from every other data submission form.
4. The name of a plant must be unique within a PROTICdb experiment. Thus, for instance, there may be two "plant 1" associated with a single project provided both these plants do not belong to the same experiment. A PROTICdb experiment may be viewed as the name of a group of unique plants.
5. Methods created in one project are available to all the other database projects. Methods can be created, browsed, and updated following the "methods" link of the main menu A method consists of a name, a description (required), and a comment. The private or public field will be implemented in future version of PROTICdb, but should already be considered.
6. If a mandatory field is not filled or if data are not entered in the required format, the form is rejected. Correct the form as suggested and submit it one more time.
7. Alternatively, you may consider a mix like a single sample (and enter it this way in the "Protein sample" form). In such a case, only one protein sample field is necessary even though this is a comigration experiment.
8. If the files are not initially in .jpeg format, it is necessary to convert them. The gel image files have to be at the same resolution (width and height in pixels) as the originals used by the detection software: this is strictly needed as coordinates of spots are given in pixels. To convert image files from tiff to jpeg, use your image manipulation software. XnView (http://www.xnview.com/) or The GIMP (http://www.gimp.org/) are available for free. A jpeg compression level up to 60% is compliant with efficient image visualization in the gel browser.
9. To generate a sport report file in Mélanie (or Image Master 5.0), go to menu "Reports" → "Spot Report" → choose columns → save as Text (*.txt), or see Mélanie documentation.
10. When submitting data files to PROTICdb (plant2image, detection results, identification results, and so on), several checkpoints are made to guarantee data consistency. Depending on the amount of data, this could be time consuming (seconds to minutes). Do not interrupt PROTICdb during data processing; you will be informed via e-mail or your browser of eventual problems. If so, correct your file and submit it again, entirely, until PROTICdb delivers a success message.
11. The template Excel files provided to submit plant2image data or microplate maps are in Microsoft Excel format, but you can use them with any spreadsheet software such as OpenOffice.org, StarOffice, ClarisWorks, and so on. Just save your spreadsheets as "text separated with tabulations" files before submitting them to PROTICdb.

12. To use this file with the Sequest tool (Xcalibur 1.3), choose "sequence setup" → "new" → "file" → "import sequence."
13. From the Sequest (Bioworks 3.1) software: display the identification results (protein and peptide list), right click, and select "export" → Excel. With Excel, save this spreadsheet as a "tabulation-separated text" format file.
14. PROTICdb new account creation should be restricted to the database administrator. A new account can be created as follows: "main menu" → "administration" → "new database user," fill the form and submit it. Next go to "new protic member" (contextual menu) to assign this new user a role in PROTICdb (guest, administrator, or user).
15. A minimal set of data about an image is needed to ensure its proper display and data consistency as well as to avoid loading without complete image information. The information required includes data about the experiment flowchart, from plant samples to 2D PAGE experiments and image numerization (jpeg format). The gel browser offer to display such gel images from the "non-analyzed gel" item of the corresponding project.
16. As stated in the Gel image visualization discussion, introducing the gel browser, an outgoing connection to the demo postgresql database (host: moulon.inra.fr port: 5432), is required. Such a connection may not be allowed by your personal or onsite firewall. Check this issue if a connection fails. Future versions of PROTICdb will be firewall friendly and will be as low bandwidth consuming as possible. The Gel Browser is a JAVA applet and requires version at least JAVA 1.4.
17. The numbering system is the name of a set of gels matched to the same master gel (referred to as a matchset in PDQuest software). Checking gel names allows one to quickly identify the gels presenting the same numbering system.
18. Loading a fourth gel image may fail because of a lack of memory. JAVA is configured to a default memory size of 64 Mo (maximal heap size). This parameter may be increased as follows (Windows-based computer): Control panel → JAVA plug-in control panel → JAVA tab → Java Applet Runtime Settings → view → double click into the blank JAVA runtime parameters and type in "-Xms128m -Xmx128m" (do not type the quotes). Then click on OK, Apply, and close both the JAVA Control and the Windows Control panels. Then close the Gel Browser and the browser and run them again. This will increase the memory size available for JAVA to 128 Mo. Do not enter a memory value exceeding your system's total physical memory size. For more information about JAVA memory tuning, see http://java.sun.com/docs/hotspot/index.html.
19. Be patient; depending on the server and Internet charge, loading may take up to 10 min. It is thus recommended to use a local database. (Loading time usually drops to less than 1 min.)
20. Two left clicks may be necessary if the gel DV02041611 window was not activated. In such a case, the first left click activates the window and the second one launches the spot information box.
21. Pop-up fighting tools may interfere with the linking process, so please disable such tools for the PROTICdb site.

22. It seems that, depending on the JAVA version installed, it may be necessary to use the synchronize command twice to obtain the correct result.

Acknowledgments

We are grateful to Delphine Vincent for a careful reading of the manuscript.

References

1. Görg, A., Weiss, and Dunn M. J. (2004) Current two-dimensional electrophoresis technology for proteomics. *Proteomics* **4,** 3665–4036.
2. Canovas, F. M., Dumas-Gaudot, E., Recorbet, G., Jorrin, J. , Mock H. P., and Rossignol M. (2004) Plant proteome analysis. *Proteomics* **4,** 285–298.
3. Zhu, H., Bilgin, M., and Snyder M. (2003) Proteomics. *Annu. Rev. Biochem.* **72,** 783–812.
4. Pandey, A. and Mann, M. (2000) proteomics to study genes and genomes. *Nature* **405,** 837–846.
5. Ferry-Dumazet H., Houel G., Montalent P., et al. (2005) PROTICdb: a web-based application to store, track, query and compare plant proteome data *Proteomics* **5,** 2069–2081.
6. Harris M. A., Clark J., Ireland A., et al. (2004) The Gene Ontology (GO) database and informatics resource. *Nucleic Acids Res.* **132,** D258–D261.

24

Identification of Phosphorylated Proteins

Maria V. Turkina and Alexander V. Vener

Summary

Reversible protein phosphorylation is crucially involved in all aspects of plant cell physiology. The highly challenging task of revealing and characterizing the dynamic protein phosphorylation networks in plants has only recently begun to become feasible, owing to application of dedicated proteomics and mass spectrometry techniques. The experimental methodology that identified most of the presently known proteins phosphorylated in vivo is based on protein cleavage with trypsin, following chromatographic enrichment of phosphorylated peptides and mass spectrometric fragmentation and sequencing of these phosphopeptides. This procedure is most efficient when it is limited to the tryptic digestion of proteins in distinct isolated fractions or compartments of plant cells. Immobilized metal affinity chromatography (IMAC) is most useful for phosphopeptide enrichment after methylation of the peptides in the complex protein digests. The following tandem mass spectrometry of the isolated phosphopeptides results in both identification of phosphorylated proteins and mapping of the in vivo phosphorylation sites. The relative quantitation of the extent of phosphorylation at individual protein modification sites may be accomplished by either stable isotope labeling technique or dedicated liquid chromatography-mass spectrometry measurements.

Key Words: IMAC; mass spectrometry; protein phosphorylation; phosphoproteins; phosphopeptides; peptide sequencing.

1. Introduction

Five major experimental approaches are used for identification of plant phosphorylated proteins: (1) labeling with the radioactive isotopes ^{32}P or ^{33}P *(1)*; (2) detection of the shift in the electrophoretic mobility of individual proteins *(2,3)*; (3) immunological analysis with phosphoamino acid-specific antibodies *(3,4)*; (4) measurement of the phosphorylation-induced increase in the mass of intact proteins by mass spectrometry *(5,6)*; and (5) identification and sequencing of the phosphorylated peptides obtained after proteolytic degradation of proteins

From: *Methods in Molecular Biology, vol. 335: Plant Proteomics: Methods and Protocols*
Edited by: H. Thiellement, M. Zivy, C. Damerval, and V. Méchin © Humana Press Inc., Totowa, NJ

and following phosphopeptide enrichment *(7–9)*. A number of experimental protocols for detection of phosphorylated proteins in plant thylakoid membranes by several of the techniques listed above have recently been described *(10)*. Thus, procedures 1 to 4, will not be discussed further in the present chapter, which is specifically focused on the most efficient and useful methodology, number 5.

This technique is based on the digestion of protein fractions by trypsin and further separation of peptides rather than proteins. To be successful, the proteolytic protein digestion should be done on distinct isolated membrane or soluble fractions of plant cell. The peptide mixtures obtained are enriched for the phosphopeptides by immobilized metal affinity chromatography (IMAC), and isolated phosphorylated peptides are sequenced by tandem mass spectrometry. This approach identifies both in vivo phosphorylated proteins and the exact sites of phosphorylation in protein sequences. The method has contributed most of the presently mapped protein phosphorylation sites in plants as well as other species *(11)*. The most important part of this methodology is the mass spectrometric sequencing of phosphorylated peptides. When the protein phosphorylation sites are mapped, the following relative quantitation of protein phosphorylation extent may be accomplished by two different experimental procedures, also described in these protocols. It is worth noting that successful application of the mass spectrometry-based phosphopeptide identification methods described here is usually limited to characterization of protein phosphorylation in plant species with sequenced genomes.

2. Materials

1. 25 m*M* NH_4HCO_3, 10 m*M* NaF. Prepare fresh solution.
2. Dithiothreitol (DTT).
3. Iodoacetamide, which is unstable in solution and therefore requires preparation immediately before use.
4. Sequencing-grade modified trypsin (Promega, Madison, WI).
5. GELoader tips (Eppendorf, Hamburg, Germany).
6. Chelating Sepharose Fast Flow (Amersham Biosciences, Uppsala, Sweden).
7. 0.1 *M* $FeCl_3$. Prepare fresh solution.
8. 20 m*M* Na_2HPO_4, 20% acetonitrile in water (v/v). Prepare fresh solution.
9. 0.1% (v/v) Acetic acid. Prepare fresh solution.
10. Acetonitrile isocratic grade for liquid chromatography.
11. $ZipTip_{C18}$ (Millipore, Bedford, MA, USA).
12. Trifluoroacetic acid (TFA) for spectroscopy.
13. Formic acid.
14. Acetyl chloride.
15. Methanol gradient grade for liquid chromatography, anhydrous.
16. Methanol-d_3 (d_3-methyl d-alcohol), anhydrous.

3. Methods

The methods described here outline the following procedures: digestion of proteins in membrane and soluble cellular fractions by trypsin (*see* **Subheading 3.1.**); enrichment of the phosphopeptides by IMAC (*see* **Subheading 3.2.**); identification and sequencing of the phosphorylated peptides by tandem mass spectrometry (*see* **Subheading 3.3.**); relative quantitation of phosphorylated peptides using stable isotope labeling (*see* **Subheading 3.4.**); and determination of protein phosphorylation ratios using liquid chromatography- mass spectrometry (LC-MS) (*see* **Subheading 3.5.**).

3.1. Trypsin Digestion of Cellular Subfractions

3.1.1. Trypsin Digestion of Membrane Fractions

For preparation of the surface exposed peptides from membranes:

1. Wash isolated membranes twice with 25 m*M* NH_4HCO_3 (*see* **Note 1**) containing 10 m*M* NaF (*see* **Note 2**).
2. Resuspend the membranes in the 25 m*M* NH_4HCO_3 with 10 m*M* NaF to obtain a total protein concentration of 10 to 15 mg/mL (or 2–3 mg of chlorophyll/mL for the photosynthetic membranes).
3. Add DTT to the membrane suspension to a final concentration of 1 m*M* and iodoacetamide to a final concentration of 3 m*M* to reduce and alkylate cysteine residues (*see* **Note 3**).
4. Incubate the suspension with sequencing-grade modified trypsin (5 μg trypsin/mg chlorophyll or 1 μg trypsin/mg protein; *see* **Note 4**) at 22°C for 3 h (*see* **Note 5**).
5. Freeze, thaw, and centrifuge the digestion products for 20 min at 15,000*g* or higher.

The supernatant containing peptides released from the membranes is used for mass spectrometric analyses directly or after enrichment for the phosphopeptides by IMAC.

3.1.2. Trypsin Digestion of Soluble Fractions

1. Resuspend proteins in the 25 m*M* NH_4HCO_3 containing 10 m*M* NaF to a protein concentration of 1 to 10 mg/mL.
2. Add DTT to a final concentration of 1 m*M* and iodoacetamide to the final concentration 3 m*M* to reduce and alkylate cysteine residues.
3. Incubate the suspension with sequencing-grade modified trypsin at 37°C for 24 h. The amount of trypsin added should correspond to 1 to 3% of the total protein amount. (Usually 2 μg of trypsin is added to 100 μg of protein.)

3.2. Enrichment of the Phosphopeptides by Immobilized Metal Affinity Chromatography

Enrichment of the phosphopeptides facilitates their subsequent identification and sequencing by mass spectrometry. Phosphopeptides are enriched by affinity chromatography on the columns with immobilized Fe(III) owing to the binding of negatively charged phospho groups to the Fe(III). Esterification of the free carboxylic groups under acidic conditions (MeOH/HCl) masks the negative charges of the carboxyl groups, which are responsible for the nonspecific binding of all acidic peptides. Thus, esterification, and particularly methylation, increases the specificity of enrichment for phosphopeptides ***(12)***.

3.2.1. Methylation of the Tryptic Peptides

1. Dry the tryptic peptides obtained from 300 μg of protein sample (50 μg of chlorophyll in the photosynthetic membranes) in the vacuum centrifuge.
2. Prepare 2 *M* methanolic HCl by dropwise addition of 80 μL of acetyl chloride to 500 μL of methanol with stirring. Add 300 μL of 2 *M* methanolic HCl to the peptides dried in the vacuum centrifuge. Incubate the mixture for 2 h at room temperature to allow proper esterification.
3. Dry the esterified peptides in the vacuum centrifuge.

3.2.2. IMAC

1. Make a microcolumn by packing GELoader tips (Eppendorf) with 8 μL of Chelating Sepharose Fast Flow (Amersham) beads suspension (1:1).
2. Wash the column two times with 20 μL 0.1% (v/v) acetic acid and charge with 100 μL of 0.1 *M* $FeCl_3$ (*see* **Note 6**).
3. Wash the column from the unbound $FeCl_3$ two times with 20 μL of 0.1% (v/v) acetic acid.
4. Dissolve the methylated tryptic peptides in 5 μL of methanol/water/acetonitrile (1:1:1, by vol.), load the solution onto the column, and incubate for 15 min.
5. Wash the column twice with 20 μL of 0.1% (v/v) acetic acid.
6. Wash the column twice with 20 μL of 0.1% (v/v) acetic acid, 20% acetonitrile.
7. Wash the column twice with 20 μL of 20% acetonitrile in water (v/v).
8. Elute the phosphopeptides with 40 μL of 20 m*M* of nonbuffered Na_2HPO_4, 20% acetonitrile.

The phosphopeptides enriched by IMAC could be directly analyzed by LC-MS. However, for electrospray ionization (ESI)-MS without prior LC, they should first be desalted on a $ZipTip_{C18}$ (Millipore).

3.2.3. Desalting of the Phosphorylated Peptides by Using $ZipTip_{C18}$

1. Dry the peptides eluted from IMAC in a vacuum centrifuge to remove acetonitrile, and dissolve the dried peptides in 10 μL 0.1% TFA in water.
2. Prewet the $ZipTip_{C18}$ with 50% acetonitrile in water using a pipet volume setting of 10 μL. Aspirate the solution into the ZipTip, and dispense to waste. Repeat three times.
3. Equilibrate the ZipTip with 0.1% TFA in water by aspirating solution into the tip and dispensing to waste. Repeat three times.
4. Bind the peptides to the ZipTip by aspirating and dispensing the peptide sample for 10 cycles for maximum binding of the peptides.

5. Wash the ZipTip with 0.1% TFA in water. Aspirate solution into tip and dispense to waste. Repeat five times.
6. Elute the desalted peptides into a clean tube by 10 μL of 50% acetonitrile in water (*see* **Note 7**). Aspirate and dispense the solvent through the ZipTip for 10 cycles before final pipeting into the tube.

3.3. Sequencing of Phosphopeptides Using Electrospray Ionization Mass Spectrometery

1. The mass spectrometric analysis is performed on a hybrid (quadrupole-time-of-flight) mass spectrometer equipped with a nanoelectrospray ion source.
2. The nanoelectrospray capillaries are loaded with 2 μL of peptide solutions after desalting on ZipTips (*see* **Subheading 3.2.2.**) in 50% acetonitrile in water with 1% formic acid (*see* **Note 8**).
3. The positive ionization mode full-scan spectra are recorded first.
4. Each peptide ion (*see* **Note 9**) present in the spectrum is subjected to collision-induced dissociation.
5. The fragmentation spectra allow identification of the phosphorylated peptides as well as determination of the phosphorylated residues by detection of characteristic neutral loss signals for phosphoric acid. Upon collision-induced fragmentation, the phosphoester bond between the phosphoryl group and a peptide is less stable than peptide bonds, which causes a prominent neutral loss of phosphoric acid H_3PO_4 (98 Daltons) from the positively charged peptide ions containing phosphorylated residues *(7,13)*. The neutral loss of the phosphoric acid occurs from the selected parent ion as well as from the fragment ions containing phosphorylated residue.
6. The spectra should be examined for the series of y (C-terminal) and b (N-terminal) fragment ions containing phosphoryl (80 Daltons added to the peptide fragment containing phosphorylated residue) and satellite fragment ions with the neutral loss of 98 Daltons (the peptide fragments that underwent the loss of phosphoric acid).
7. In parallel, the spectra are examined for the complementary ion signals originating from the peptide fragments that do not contain the phosphorylated residue. Such an interpretation of each spectrum usually allows complete peptide sequencing (even de novo sequencing) with identification of the phosphorylation sites.
8. Peptide identification can also be performed by submitting the experimental fragmentation spectra to a database search using the Mascot server (http://www.matrixscience.com/). The settings of the Mascot MS/MS Ion Search allow methylation and phosphorylation of the peptide to be included in the search.

3.4. Relative Quantitation of Phosphorylated Peptides

The intensity of ionic species detected by mass spectrometry depends on the nature of a particular peptide and its ionization properties. Consequently, a quantitative comparison of different peptides is impossible. However, stable isotope labeling in combination with mass spectrometry allows relative quantitation of protein phosphorylation in two distinct protein samples. The scheme for this type of experiment is shown in **Fig. 1A**. Peptides obtained after tryptic digestion of equal amounts of protein samples from the same tissue or cellular fraction isolated from plants under two distinct conditions are labeled in parallel by esterification with methanolic HCl using methanol-d_0 for one sample and methanol-d_3 for the second sample to yield light and heavy labeled peptides, respectively ***(14,15)***. The two samples are then combined, and the enrichment for phosphopeptides by IMAC is made. The ratio between the light and heavy forms for each phosphorylated peptide is detected by mass spectrometric measurement. Each peptide in the mass spectrum is represented by the light and the heavy form (**Fig. 1B** and **C**). The ratio of the total peak areas for light and heavy isotope-labeled peptides corresponds to difference in the phosphorylation at the particular site in the protein in two distinct conditions.

1. Dry in the vacuum centrifuge two samples of tryptic peptides originating from 150 μg of protein from the same cellular fraction obtained from plants in two distinct conditions.
2. Prepare two mixtures of light and heavy 2 *M* methanolic HCl by dropwise addition of 80 μL of acetyl chloride either to 500 μL of methanol-d_0 or to 500 μL of deuterated methanol-d_3 with stirring. Add 200 μL of 2 *M* methanolic-d_0 HCl to the dried peptide sample 1 and 200 μL of 2 *M* methanolic-d_3 HCl to the dried peptide sample 2. Incubate the mixtures for 2 h at room temperature to allow proper esterification.
3. Combine samples 1 and 2, and dry them in the vacuum centrifuge.
4. Dissolve the dried peptides in 5 μL of methanol/water/acetonitrile (1:1:1, by vol.) and perform the IMAC procedure as described above (*see* **Subheading 3.2.2.**).

Fig. 1. (*opposite page*) Relative quantitation of phosphorylated peptides. (**A**) Scheme of stable isotope labeling and phosphopeptide enrichment for quantitative analysis of phosphopeptides originating from the same protein fraction obtained from plants in two different conditions. (**B** and **C**) The resulting full-scan spectra for the mixture of the peptides obtained from the thylakoid membranes of *Arabidopsis thaliana* in two different conditions and treated according to the scheme shown above. (**B**) Ions 738.4 and 741.4 represent the light (condition 1) and the heavy form (condition 2) of the methylated phosphopeptide *Ac*-tIALGK (m/z = 724.4) from the N-terminus of the D2 protein (the lower case t designates phosphothreonine in the peptide sequence). The ratio of the total peak areas for the peptides labeled with light and heavy methanol is 3:1. Ions 852.4 and 858.4 represent the light (condition 1) and the heavy form (condition 2) of the phosphopeptide *Ac*-tAILER (m/z = 824.4) from the N-terminus of the D1 protein. The C-terminus and carboxylic group of glutamic acid were methylated in this peptide during esterification (**A**). The ratio between the light and heavy isotopes is 2:1. (**C**) The spectrum corresponding to the same samples from the two different conditions; however, the peptides corresponding to conditions 1 and 2 were labeled during esterification in a way opposite to the samples in (**B**). The ratio between the light and heavy isotopes is 1:3 for the *Ac*-tIALGK and 1:2 for *Ac*-tAILER.

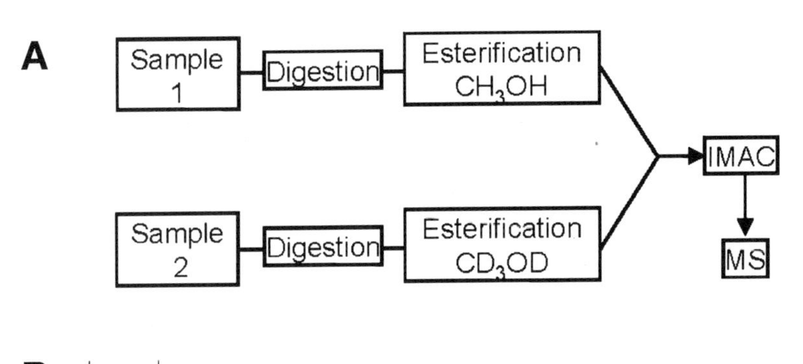
A
Sample 1
Digestion
Esterification CH_3OH
Sample 2
Digestion
Esterification CD_3OD
IMAC
MS

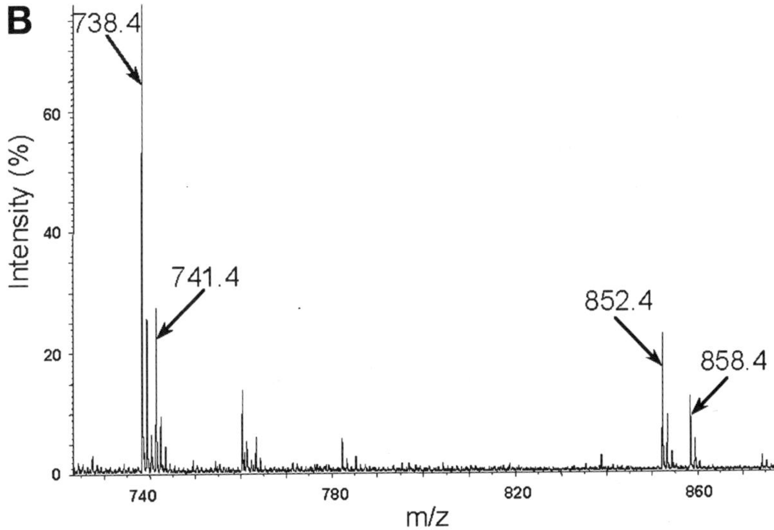
B
738.4
741.4
852.4
858.4
Intensity (%)
0
20
40
60
740
780
820
860
m/z

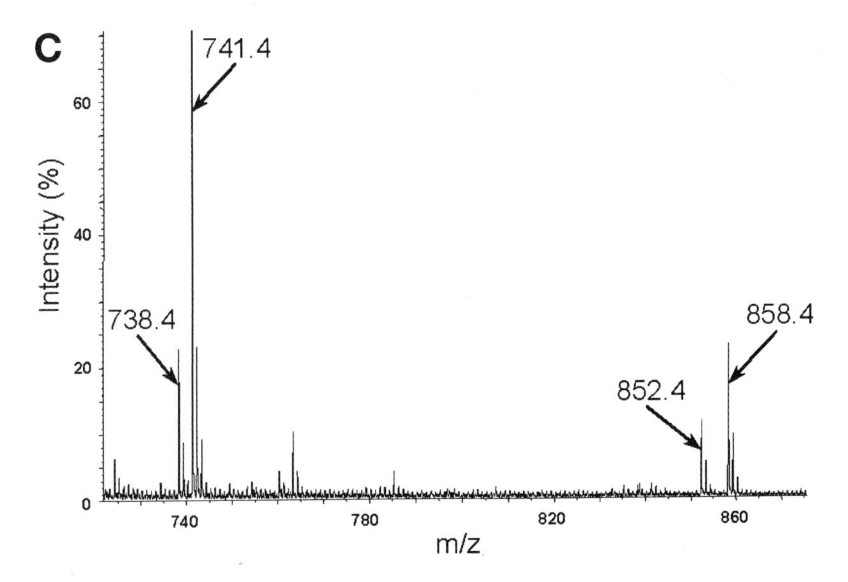
C
741.4
738.4
858.4
852.4
Intensity (%)
0
20
40
60
740
780
820
860
m/z

To make the internal control, an additional experiment can be made in parallel. For this experiment the two samples are labeled in the opposite way, i.e., if in the first case sample 1 was labeled with the light isotope and sample 2 with the heavy one, then in the second experimental setup sample 1 is methylated with the heavy methanol and sample 2 with the light methanol. If the experimental handling of both protein samples has been correct and equal, then resulting mass spectra for the "direct" (**Fig. 1B**) and "opposite" (**Fig. 1C**) labeling experiments give the same ratio for phosphorylation of each peptide in the samples originating from the plants in two distinct conditions.

3.5. Determination of Protein Phosphorylation Ratios Using LC-MS

After the masses of phosphorylated peptides present in the peptide mixtures from membrane (*see* **Subheading 3.1.1.**) or soluble (*see* **Subheading 3.1.2.**) cellular fractions have been established, the stoichiometry of phosphorylation at each initial site can be determined by LC-MS. The method is based on measurement of the ratios of phosphorylated to corresponding nonphosphorylated peptides present in the mixtures *(7)* (*see* **Note 10**).

1. For this experiment, 20 to 40 μL of the total peptide mixture obtained after protein digestion with trypsin (*see* **Subheadings 3.1.1.** and **3.1.2.**) are subjected to LC-MS. The peptides are separated on 5 μm C18 MetaChem 150 × 1.0-mm column at a flow rate of 20 μL/min.
2. A gradient of 0.1% (v/v) formic acid in water (A) and 0.1% (v/v) formic acid in acetonitrile (B) is distributed as follows: 0% B for the first 3 min; 0 to 20% B from 3 to 20 min; 20 to 70% B from 20 to 105 min; 70 to 99% B from 105 to 115 min.
3. Online detection of the peptides is performed using a triple quadrupole mass spectrometer with a standard ionspray source. The instrument settings recommended for a particular mass spectrometer are used.
4. Two parallel mass spectrometric measurements are continuously made during the whole LC-MS run: (1) in positive ionization mode, to obtain the spectra of all peptides present in the fractions; and (2) in negative ionization mode with the detection of diagnostic phosphoryl ions at m/z = –79, to map the fractions containing phosphorylated peptides (*see* **Note 11**).
5. The results of both parallel experiments (**Fig. 2A** and **B**) are saved in the electronic computer files, which are analyzed as described in **Fig. 2**.

Fig. 2. (*opposite page*) Determination of the phosphorylation levels for membrane proteins according to Vener et al. *(7)*. Thylakoid membranes isolated from *Arabidopsis thaliana* were treated with trypsin (*see* **Subheading 3.1.1.**). The peptides released from the membrane proteins contained both phosphorylated and corresponding nonphosphorylated peptides in ratios equivalent to the extent of in vivo modification at each individual site. These peptides were subjected to LC-MS with two parallel MS experiments: in positive ionization mode **(A)** and in negative ionization mode **(B)** to map the elution times of phosphorylated peptides by detection of diagnostic phosphoryl

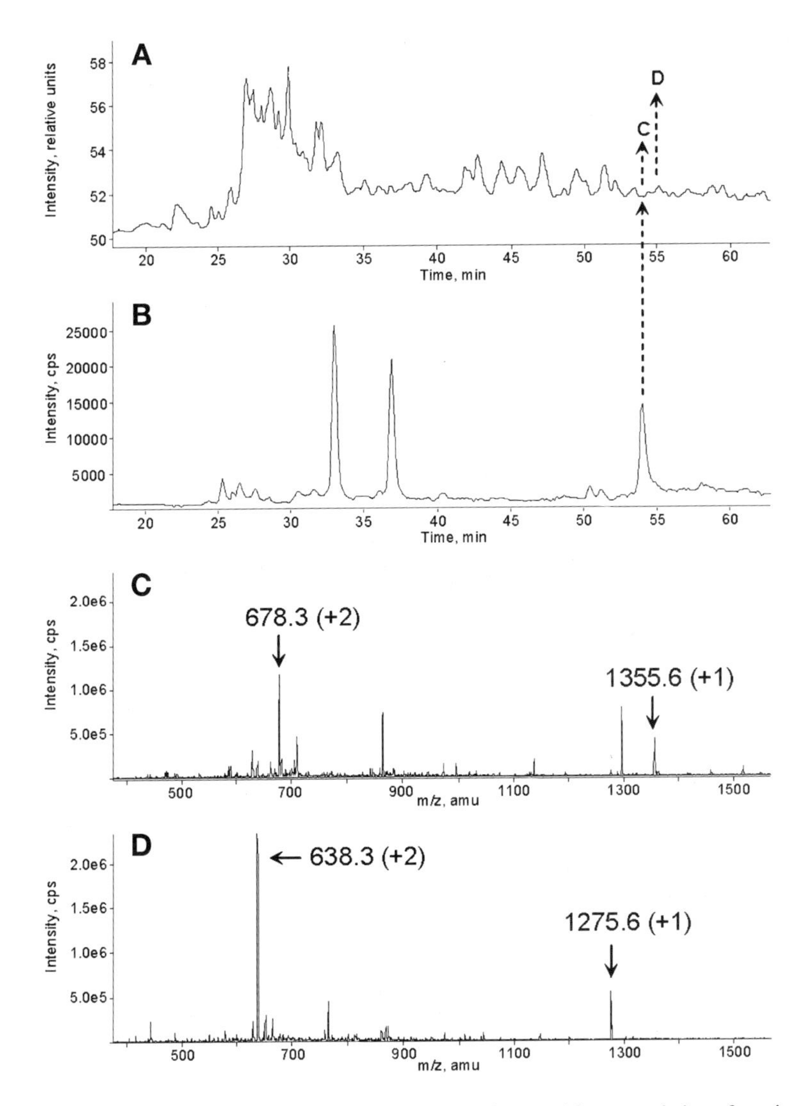

Fig. 2. (*continued*) ions (m/z = –79). The phosphopeptide-containing fractions are revealed by the peaks in the negative ionization experiment **(B)**. The positive ionization spectra in corresponding fractions (the dashed line from **B** to **A**) are examined for the presence of phosphopeptides with known molecular masses, which have been determined in the prior experiments on phosphorylation site mapping. For instance, the fraction marked "C" in **(A)** has a spectrum shown in **(C)**. Spectrum **(C)** contains signals of monoprotonated and doubly protonated phosphopeptide corresponding to the m/z values indicated in the spectrum. Examination of the spectra in the fractions eluted 1 to 2 min after the phosphopeptide (marked "D" in **A**) reveals the signals of the corresponding nonphosphorylated peptide, as marked by m/z values for singly and doubly charged ions in **(D)**. The peak areas for each phosphopeptide are extracted electronically from the logged LC-MS file, as well as the areas for the corresponding nonphosphorylated peptide. The ratio of phosphopeptide to nonphosphopeptide determines the extent of phosphorylation for each individual peptide (modification site).

6. First, the elution positions for the phosphopeptides with the known m/z are determined.
7. Then the signals of the corresponding nonphosphorylated forms of these peptides are revealed.
8. The ratio of the sum of the peak intensities for all ionization states of each phosphopeptide to that of the corresponding nonphosphorylated peptide gives the stoichiometry of phosphorylation for the particular protein.
9. Because both phospho and nonphospho forms are detected simultaneously in the same sample, the stoichiometry of phosphorylation can be measured for individual proteins from the particular membrane or soluble fraction sample.

4. Notes

1. The resulting supernatant of tryptic peptides in 25 m*M* ammonium bicarbonate (NH_4HCO_3) is compatible with direct analyses by mass spectrometry. To remove the excess high salts in buffers used during the particular membrane preparation, the membrane washing with 25 m*M* NH_4HCO_3 may be done more than twice.
2. To avoid protein dephosphorylation by protein phosphatases, add 10 m*M* NaF to the washing solution of 25 m*M* NH_4HCO_3. For the same reason membrane washing by resuspension and centrifugation should be done at 4°C.
3. Carbamidomethylation prevents cysteine residues from forming unexpected modifications, by primary oxidation.
4. It is important to make the final resuspension of the membranes up to a high protein (chlorophyll) concentration (10–15 mg/mL of protein or up to 3 mg of chlorophyll/ml) for good proteolytic digestion and for a high concentration of the resultant peptides for direct mass spectrometric analyses.
5. It is important to perform the proteolytic treatment of photosynthetic membranes at 22 to 25°C but not at 37°C, because a number of photosystem II phosphoproteins are rapidly dephosphorylated by the heat-shock-activated membrane protein phosphatase at 37°C ***(16)***.
6. The $FeCl_3$ solution requires preparation immediately before use.
7. Do not add TFA to the elution solution. TFA suppresses ionization of the peptides during ESI-MS.
8. The peptides should be ionized as positively charged, protonated ions. To increase positive ionization, the peptide solutions are acidified by the addition of formic acid to a final concentration of 1% prior to the analyses.
9. Both singly and multiply charged ion peaks can represent the peptides. The chemical noise in ESI is composed mostly of singly charged ions. A high signal-to-noise ratio for the peptide ions allows their rapid localization in the spectrum. Multiply charged peptide ions at the noise level can also be found because of the isotopic pattern for multiply charged ions, which is denser than that of the singly charged chemical background ions.
10. The intensity of ionic species detected by mass spectrometry depends on the nature of a particular peptide and its ionization properties. Consequently, quantitative comparisons of different peptides are generally not allowed. However,

because phosphorylation of a peptide adds just 80 m/z (HPO_3) and does not change the ionization state of the peptide under acidic conditions, quantitation of the ratio of a phosphopeptide to a corresponding nonphosphorylated peptide is possible using LC-MS in the positive ionization mode *(7)*.

11. Two parallel mass spectrometric experiments are performed according to the following settings. The positive mode ion scan in the m/z range from 320 to 1800 is made for 5.2 s, followed by a 0.7-s pause for polarity switching; then a 1.5-s single ion monitoring of the negative 79 m/z ion is performed. Then another 0.7-s pause to return to positive-ion mode is done. This cycle of positive/negative scanning is constantly repeated during the whole LC-MS run.

Acknowledgments

This work was supported by grants from the Swedish Research Council for Environment, Agriculture, and Spatial Planning (Formas), the Nordiskt Kontaktorgan för Jordbruksforskning (NKJ), and the Graduate Research School in Genomics and Bioinformatics (FGB).

References

1. Michel, H. P. and Bennett, J. (1987) Identification of the phosphorylation site of an 8.3 kDa protein from photosystem II of spinach. *FEBS Lett.* **212,** 103–108.
2. Elich, T. D., Edelman, M., and Mattoo, A. K. (1992) Identification, characterization, and resolution of the in vivo phosphorylated form of the D1 photosystem II reaction center protein. *J. Biol. Chem.* **267,** 3523–3529.
3. Rintamäki, E., Salonen, M., Suoranta, U. M., Carlberg, I., Andersson, B., and Aro, E.-M. (1997) Phosphorylation of light-harvesting complex II and photosystem II core proteins shows different irradiance-dependent regulation in vivo. Application of phosphothreonine antibodies to analysis of thylakoid phosphoproteins. *J. Biol. Chem.* **272,** 30476–30482.
4. Rintamäki, E. and Aro, E.-M. (2001) Phosphorylation of photosystem II proteins, in *Regulation of Photosynthesis* (Aro, E.-M. and Andersson, B., eds.), Kluwer Academic Publishers, Dordrecht, Nederlands, pp. 395–418.
5. Gomez, S. M., Nishio, J. N., Faull, K. F., and Whitelegge, J. P. (2002) The chloroplast grana proteome defined by intact mass measurements from liquid chromatography mass spectrometry. *Mol. Cell. Proteomics* **1,** 46–59.
6. Carlberg, I., Hansson, M., Kieselbach, T., Schroder, W. P., Andersson, B., and Vener, A. V. (2003) A novel plant protein undergoing light-induced phosphorylation and release from the photosynthetic thylakoid membranes. *Proc. Natl. Acad. Sci. USA* **100,** 757–762.
7. Vener, A. V., Harms, A., Sussman, M. R., and Vierstra, R. D. (2001) Mass spectrometric resolution of reversible protein phosphorylation in photosynthetic membranes of *Arabidopsis thaliana. J. Biol. Chem.* **276,** 6959–6966.
8. Hansson, M. and Vener, A. V. (2003) Identification of three previously unknown in vivo protein phosphorylation sites in thylakoid membranes of *Arabidopsis thaliana. Mol. Cell. Proteomics* **2,** 550–559.

9. Nuhse, T. S., Stensballe, A., Jensen, O. N., and Peck, S. C. (2003) Large-scale analysis of in vivo phosphorylated membrane proteins by immobilized metal ion affinity chromatography and mass spectrometry. *Mol. Cell. Proteomics* **2,** 1234–1243.
10. Aro, E.-M., Rokka, A., and Vener, A. V. (2004) Determination of phosphoproteins in higher plant thylakoids. *Methods Mol. Biol.* **274,** 271–285.
11. Loyet, K. M., Stults, J. T., and Arnott, D. (2005) Mass spectrometric contributions to the practice of phosphorylation site mapping through 2003: a literature review. *Mol. Cell. Proteomics* **4,** 235–245.
12. Ficarro, S. B., McCleland, M. L., Stukenberg, P. T., et al. (2002) Phosphoproteome analysis by mass spectrometry and its application to *Saccharomyces cerevisiae*. *Nat. Biotechnol.* **20,** 301–305.
13. Shou, W., Verma, R., Annan, R. S., et al. (2002) Mapping phosphorylation sites in proteins by mass spectrometry. *Methods Enzymol.* **351,** 279–296.
14. He, T., Alving, K., Field, B., et al. (2004) Quantitation of phosphopeptides using affinity chromatography and stable isotope labeling. *J. Am. Soc. Mass Spectrom.* **15,** 363–373.
15. Ficarro, S., Chertihin, O., Westbrook, V. A., et al. (2003) Phosphoproteome analysis of capacitated human sperm. Evidence of tyrosine phosphorylation of a kinase-anchoring protein 3 and valosin-containing protein/p97 during capacitation. *J. Biol. Chem.* **278,** 11,579–11,589.
16. Rokka, A., Aro, E.-M., Herrmann, R. G., Andersson, B., and Vener, A. V. (2000) Dephosphorylation of photosystem II reaction center proteins in plant photosynthetic membranes as an immediate response to abrupt elevation of temperature. *Plant Physiol.* **123,** 1525–1536.

25

Plant Proteomics and Glycosylation

Anne-Catherine Fitchette, Olivia Tran Dinh, Loïc Faye, and Muriel Bardor

Summary

In plant cells, as in other eucaryotic cells, glycosylation is one of the most studied posttranslational events. It can be of two types, *N*- or *O*-glycosylation, depending on the linkage involved between the protein backbone and the oligosaccharide moiety. In this review, we present different methods, commonly used in our laboratory, to study the glycosylation of plant proteins. These approaches rely on blot detection with glycan-specific probes, as well as specific deglycosylation of the glycoproteins, followed by mass spectrometry analysis. Such experiments not only allow determination of whether the protein is a glycoprotein, but also how and where it is glycosylated. The last part of this chapter is dedicated to the specific purification and identification of glycoprotein populations in plant cells, so-called glycoproteomics.

Key Words: *N*-glycosylation; *O*-glycosylation; *O*-*N*-acetylglucosamine; lectin; anti-glycan antibodies; β-elimination; endoglycosidase H; peptide-*N*-glycosidase F; peptide-*N*-glycosidase A; glycoproteomics; MALDI-TOF; LC-MS/MS.

1. Introduction

In plants, as in other eukaryotes, glycosylation is mostly encountered on secreted proteins, although some forms of glycosylation can also be found on several cytosolic or nuclear proteins. It can be of two types: *N*- or *O*-glycosylation, depending on the linkage involved between oligosaccharides and the protein backbone. In plants, *N*-glycosylation is the most studied event.

1.1. N-Glycosylation

In plant cells, as in other eukaryotic cells, *N*-glycosylation occurs exclusively on proteins that enter the secretory pathway. It starts with the cotranslational transfer, in the endoplasmic reticulum (ER), of an oligosaccharide precursor, $Glc_3Man_9GlcNAc_2$, onto specific asparagine residues constitutive of

From: *Methods in Molecular Biology, vol. 335: Plant Proteomics: Methods and Protocols*
Edited by: H. Thiellement, M. Zivy, C. Damerval, and V. Méchin © Humana Press Inc., Totowa, NJ

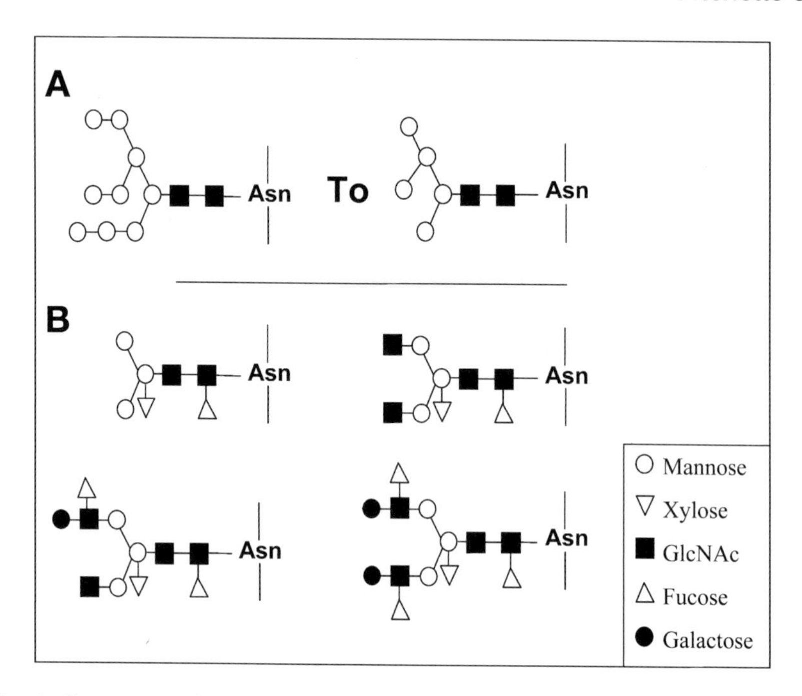

Fig. 1. Structures of *N*-linked oligosaccharides in plants. High-mannose type *N*-glycans are composed of the core $Man_3GlcNAc_2$ substituted by two to six mannose residues (**A**). Complex type *N*-glycans present a β1-2 xylose and and/or α1-3 fucose linked to the core. Some of them also carry terminal antennae consisting of a single GlcNAc or a Galβ1-3(Fucα1-4)GlcNAc Lewis a trisaccharide (**B**).

N-glycosylation sequences present on the nascent protein (Asn-*X*-Ser/Thr, where *X* can be any amino acid except proline and aspartic acid). This event takes place as soon as the newly synthesized protein enters the ER lumen. Then, further processing of the $Glc_3Man_9GlcNAc_2$ precursor occurs along the secretory pathway as the glycoprotein moves from the ER and through the Golgi apparatus to its final destination. If the glycan side chain is accessible to the processing enzymes, such as glycosidases and glycosyltransferases present in the ER and Golgi apparatus, the precursor can be converted successively to high-mannose-type *N*-glycans, ranging from $Man_9GlcNAc_2$ to $Man_5GlcNAc_2$ (**Fig. 1A**), and to complex-type *N*-glycans. Plant complex *N*-linked glycans present an α1-3-fucose attached to the proximal glucosamine residue of the core, a β1-2-xylose residue linked to the core β-mannose and/or Gal β1-3(Fucα1-4)GlcNAc oligosaccharide terminal sequence, linked to their terminal antennae, called Lewis a (**Fig. 1B**) *(1)*.

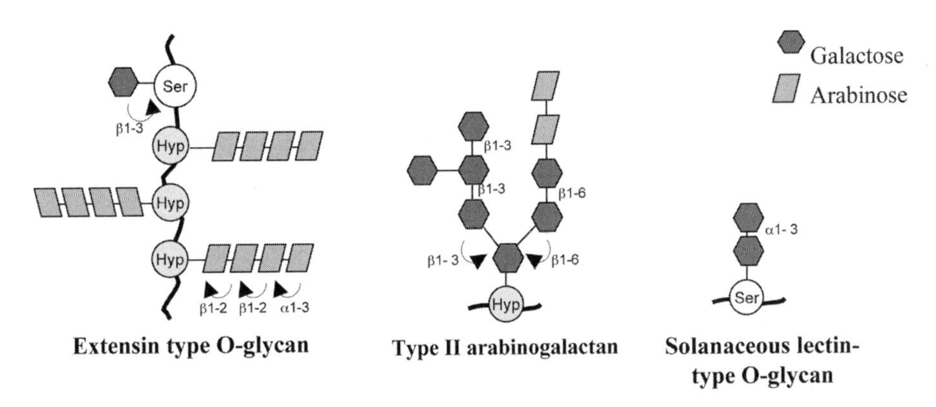

Fig. 2. Structures of *O*-glycans linked to plant glycoproteins. (Adapted from **ref. *4*.**)

1.2. O-Glycosylation

O-glycosylation can be of two types in plant cells, i.e., born by secreted proteins or by cytosolic/nuclear proteins. The nature of the *O*-linked glycans differs according to the location of the proteins.

1.2.1. O-Glycosylation of Secreted Proteins

Little is known about *O*-glycosylation of secreted proteins in plants. A few papers have reported a first class of *O*-glycans that are linked to some Ser or Thr residues of plant glycoproteins. Indeed, several glycoproteins, like cell wall extensin and vacuolar sweet potato sporamin, have been described as glycoproteins containing Ser residues *O*-glycosylated with one single galactose *(2,3)*. The characterization of rice glutelin, a protein transported to the vacuole, tends to demonstrate the presence of mammalian mucin-type glycosylation. This type of *O*-glycosylation in plants is still a matter of debate *(4)*.

The second type of secreted protein *O*-glycosylation requires the hydroxylation of proline residues from glycoproteins and has been found in plants on hydroxyprolin-rich glycoproteins (HRGPs). HRGPs represent the major surface glycoproteins in plants. These HRGPs, essentially located in the cell wall or at the outer surface of plasma membrane, represent three goups of proteins: extensins, arabinogalactan proteins (AGPs), and solanaceous lectins (**Fig. 2**). Extensins and solanaceous lectins contain repeated sequences consisting of serine-$(hydroxyproline)_4$ in which the serine can be *O*-galactosylated, and the hydroxyprolines can be substituted by arabinosyl chains *(5)*. This arabinosylation processes occur at the Golgi apparatus level *(6)*. The AGP family is composed of extracellular and plasma membrane proteins that are hyperglycosylated. Their protein moiety is rich in hydroxyproline (hyp), serine, alanine, threonine, and glycine and contains Ala-hyp repeats. Conversion of selected proline to hyp occurs either in the ER and/or in the Golgi apparatus by the action of a plant-specific prolyl-4-hydroxylase enzyme *(4,7)*. AGPs are *O*-glycosylated in a stepwise manner by addition on their hyp of polysaccharide chains consisting of β1-3-galactose substituted by β1-6-galactose side chains, which are in turn branched with arabinose or other less abundant monosaccharide (**Fig. 2**) *(8)*.

1.2.2. The O-GlcNAc Modification

Another *O*-glycosylation, well described in mammals, is the *O*-GlcNAcylation of cytosolic and nuclear proteins. This addition of *O*-GlcNAc on protein is quite ubiquitous among eukaryotes and occurs in a β linkage onto Ser or Thr. There is no well-defined consensus sequence for *O*-GlcNAcylation, although the sequence Tyr-Ser-Pro-Thr*-Ser-Pro-Ser*, where the amino acids* are the glycosylated ones, represents a typical GlcNAc attachment site *(9)*. *O*-GlcNAcylation is a highly dynamic process and often acts in a reciprocal manner to *O*-phosphorylation *(10)*. It is driven by an *O*-GlcNAc transferase and an *O*-GlcNAcase that adds and removes, respectively, the sugar residue on/from proteins *(10)*. This glycosylation event is involved in numerous cellular processes, such as protein nuclear import, transcription, attachment of cytoskeleton to membranes, or protein synthesis *(9)*, although its precise role is far from understood.

In plants, little information is available concerning *O*-GlcNAcylation. Two enzymes similar to the *O*-GlcNActransferase, named SPINDLY and SEC, have been cloned in *Arabidopsis thaliana* *(11–13)*. They both present an *O*-GlcNAc transferase activity in vitro, but their natural substrates have not been defined yet. On the other hand, some nuclear proteins have been described as glycoproteins bearing *O*-linked glycans with terminal GlcNAc *(14,15)*. Whether the latter are the substrates of SPINDLY and SEC is a question that remains to be answered.

1.3. Heterogeneity of Glycosylation

The *N*-glycosylation pattern of a mature secreted glycoprotein in eukaryotic cells results both from cotranslational transfer in the ER of an oligosaccharide and from its processing in the ER and Golgi apparatus. Factors like polypeptide folding can influence the availability of potential glycosylation sites in the ER *(16)*. As a consequence, when multiple *N*-glycosylation sites are present in a protein sequence, some sites might be glycosylated inefficiently and some others not used at all *(17–19)*.

In parallel, the processing of glycan side chains is related to their accessibility to the maturation enzymes. Glycans that are located on the protein surface will mature into complex-type *N*-glycans, whereas oligosaccharides that are buried in the protein structure remain unmodified high-manose glycans ***(17)***. This is also true for *O*-linked glycans. Hence, identification of the glycosylation of a plant glycoprotein not only requires the identification of the glycan-protein linkage and the elucidation of the oligosaccharide structures attached to the glycoprotein, but also necessitates determination of their distribution/location on potential glycosylation sites of the protein backbone.

In this context, the aim of this chapter is to introduce thc plant proteomist to several tools and methods for determining whether the proteins identified are glycoproteins, to which category they belong, and, eventually, which part(s) of the protein backbone is (are) wearing the glycan side chain. The last paragraph of this chapter will be dedicated to the specific identification of glycoproteins in plant glycoproteomics.

2. Materials

2.1. Is My Protein a Glycoprotein?

1. DIG Glycan Detection Kit (Roche Applied Science, Mannheim, Germany) (store at 2–8°C).

2.2. How Is My Protein Glycosylated?

2.2.1. Detection of Glycans by Western Blotting

2.2.1.1. Affinodetection with Lectins

1. TTBS buffer: 20 m*M* Tris-HCl, pH 7.4, containing 0.5 *M* NaCl and 0.1% Tween-20.
2. Lectin-biotin: GNA (*Galanthus nivalis* agglutinin)-biotin (Vector Labs/Abcys, France) (store at 2–8°C), and WGA (*Triticum vulgaris* agglutinin)- biotin (Sigma-Aldrich, Lyon, France) (store at 2–8°C).
3. Streptavidin-peroxidase conjugate (Amersham Biosciences/GE, Healthcare, Uppsala, Sweden) (store at 2–8°C).
4. TBS buffer : 20 m*M* Tris-HCl, pH 7.4, containing 0.5 *M* NaCl.
5. 4-Chloro-1-naphtol (Bio-Rad, Hercules, CA) (store at –20°C).
6. Ribonuclease B from bovine pancreas (Sigma Aldrich) (store at 2–8°C).
7. Methyl α-D-mannopyranoside (Sigma Aldrich) (store at room temperature).
8. α-A-crystallin from bovine lens (Sigma-Aldrich) (store at –70°C).
9. Ovalbumin (Sigma-Aldrich) (store at 2–8°C).
10. TTBS* buffer: 20 m*M* Tris-HCl, pH 7.4, containing 0.5 *M* NaCl, 1 m*M* $CaCl_2$, 1 m*M* $MgCl_2$, and 0.1% Tween-20.
11. Concanavalin A (Sigma-Aldrich) (store at 5–8°C).
12. Horseradish peroxidase (HRP; Sigma-Aldrich) (store at 2–8°C).
13. Soybean agglutinin (Sigma-Aldrich) (store at 2–8°C).

2.2.1.2. Immunodetection with Specific Antibodies

1. TBS buffer : 20 m*M* Tris-HCl, pH 7.4, containing 0.5 *M* NaCl.
2. Gelatin (Bio-Rad).
3. Anti-HRP rabbit immunserum (Sigma Aldrich) (store at 2–8°C).

4. Anti-*Apis mellifera* venom (honeybee venom; Sigma Aldrich) (store at 2–8°C).
5. Mouse anti-human Lewis a monoclonal antibody (Calbiochem, Meudon, France) (store at –20°C).
6. TTBS buffer: 20 m*M* Tris-HCl, pH 7.4, containing 0.5 *M* NaCl and 0.1% Tween-20.
7. HRP-conjugated goat antibodies directed at mouse polyvalent immunoglobulins (Sigma Aldrich) (store at –20°C).
8. HRP-conjugated goat antibodies directed at rabbit immunoglobulins (Bio-Rad) (store at 2–8°C).
9. Phospholipase A2 from honeybee venom (Sigma Aldrich) (store at 2–8°C).
10. Bean phytohemagglutinin L (PHA-L; Sigma Aldrich) (store at 2–8°C).
11. Recombinant avidin expressed in maize (Sigma Aldrich) (store at 2–8°C).

2.2.2. Glycan Analysis by Is Monosaccharide Composition

1. Teflon-faced screw cap: kit flacons 2 mL (Interchim, Montluçon, France).
2. 2 m*M* Inositol solution in water (store at 4°C).
3. Methanolic-HCl 3 *N* Kit (Supelco, Bellefonte, PA) (store at 2–8°C). Open one ampule of 3 *N* methanolic-HCl and dilute it with methanol, in order to have a 1 *N* methanol-HCl solution (*see* **Note 1**).
4. Acetic acid, anhydrous 98% A.C.S. reagent (Sigma-Aldrich, Aldrich product) (store at room temperature).
5. Pyridine, anhydrous 99.8% (Sigma-Aldrich) (store at room temperature).
6. Silylation reagent: HMDS + TMCS + pyridine, 3:1:9 (Sylon HTP) kit (Supelco) (store at 2–8°C; *see* **Note 1**).
7. Cyclohexane Rectapur (VWR, Fontenay sous Bois, France) (store at room temperature).
8. Gas chromatography apparatus equipped with a silica column CP-SIL 5 CB (25 m × 0.25 mm).

2.2.3. Analysis of the Protein after Glycan Release

2.2.3.1. Chemical Treatment for Glycan Release

1. Glass reaction tube (Pyrex culture tubes) equipped with Teflon-faced screw cap (Bibby Sterilin Stone, Staffs, England).
2. Sodium borohydride solution: sodium borohydride freshly prepared at 40 mg/mL, in a 100 m*M* NaOH solution.
3. Acetic acid, anhydrous 98% A.C.S. reagent (Sigma-Aldrich, Aldrich product) (store at room temperature).

2.2.3.2 Enzymatic Treatment by *N*- or *O*-Glycosidases

1. 1% Sodium dodecyl sulfate (SDS).
2. 150 m*M* Sodium acetate buffer, pH 5.7.
3. Endo H: endoglycosidase H from *Streptomyces plicatus*, recombinant from *E. coli* (Roche Applied Science) (store at 2–8°C; freezing inactivates the protein).

4. 0.1 *M* Tris-HCl buffer, pH 7.5, containing 0.1% SDS.
5. 0.1 *M* Tris-HCl buffer, pH 7.5, containing 0.5% Nonidet P40.
6. Nonidet P40 (called now IGEPAL; Sigma-Aldrich).
7. PNGase F: peptide-*N*-glycosidase F from *Flavobacterium meningosepticum* (Roche Applied Science) (store between –15°C and –25°C).
8. 10 m*M* HCl solution, pH 2.2.
9. C18 column: Bond Elut C18 column (Varian, Palo Alto, CA) (store at room temperature).
10. Acetonitrile, minimum 99.5% (Sigma-Aldrich).
11. 100 m*M* Sodium acetate buffer, pH 5.5.
12. Pepsin A from porcin stomach mucosa (Sigma-Aldrich).
13. PNGase A: peptide-*N*-glycosidase A from sweet almond (Roche Applied Science) (store between –15°C and –25°C).
14. Enzymatic deglycosylation kit (Prozyme, San Leandro, CA) (store at 4°C).

2.3. Where Is My Protein Glycosylated?

1. 50 m*M* Ammonium bicarbonate, pH 8.0.
2. Trypsin: TPCK-treated trypsin from bovine pancreas (Sigma-Aldrich, Fluka product) (store at -20°C).
3. Chymotrypsin: TLCK-treated chymotrypsin from bovine pancreas (Sigma Aldrich) (store at -20°C).
4. C18 column: Merck C18 Spherisorb ODS 2 column (5 µm, 4.6 × 250 mm, VWR)
5. Solvent A: 90:10 water/acetonitrile containing 0.1% trifluoroacetic acid (TFA).
6. Solvent B: 10:90 water/acetonitrile containing 0.1% TFA.

2.4. How Can the Whole Glycoproteome Be Analyzed?

2.4.1. Identification of Glycoproteins with High-Mannose N-Glycans

1. TBS buffer: 20 m*M* Tris-HCl, pH 7.4, containing 0.5 *M* NaCl.
2. Bradford Assay: Bio-Rad protein assay (store at 4°C).
3. 0.20-µm Membrane: Filtropur S 0.2 (Sarstedt, Nümbrecht, Germany).
4. Concanavalin A-Sepharose: Con A Sepharose™ 4B (Amersham Biosciences) (store at 4°C).
5. TBS*: 20 m*M* Tris-HCl buffer, pH 7.4, containing 0.5 *M* NaCl, 1 m*M* $CaCl_2$, and 1 m*M* $MgCl_2$.
6. Poly-Prep® column (Bio-Rad).
7. TTBS*: 20 m*M* Tris-HCl buffer, pH 7.4, containing 0.5 *M* NaCl, 1 m*M* $CaCl_2$, 1 m*M* $MgCl_2$, and 0.1% Tween-20.
8. Methyl α-D-mannopyranoside (Sigma Aldrich) (store at room temperature).

2.4.2. Identification of O-GlcNAcylated Glycoproteins

1. Miracloth (Calbiochem, Meudon, France).
2. WGA buffer: 20 m*M* Tris-HCl, pH 7.8, containing 150 m*M* KCl, 2 m*M* $CaCl_2$, 10 m*M* $MgCl_2$, 1 m*M* ditiothreitol (DTT), and protease inhibitors.

3. Protease inhibitor cocktail tablets (Roche Applied Science) (store at 4°C).
4. Amicon Bioseparation YM-10 (Millipore, Saint Quentin en Yvelines, France).
5. Nonidet P40 (Sigma-Aldrich).
5. PNGase F: peptide-*N*-glycosidase F from *Flavobacterium meningosepticum* (Roche Applied Science) (store between –15°C and –25°C).
6. WGA-agarose (Sigma-Aldrich) (store at 2–8°C).
7. GlcNAc (Sigma-Aldrich).

3. Methods

3.1. Is My Protein a Glycoprotein?

Only one approach is available to answer the question globally: is my protein a glycoprotein? It necessitates the use of a general sugar detection kit that allows the detection and quantification of glycoproteins on blot. All the reagents necessary for such a determination are provided in a detection kit, the DIG Glycan Detection Kit commercialized by Roche Applied Science, which also includes the manufacturer's instructions. It should be noted that the only information the experimenter will get using this kit is that the proteins do bear at least one oligosaccharide moiety. No information will be provided about the type of glycan involved or its linkage to the protein backbone.

3.2. How Is My Protein Glycosylated?

Several strategies can be initiated to determine the nature of the glycan borne by a glycoprotein. The first one involves the use of probes specific for glycoprotein oligosaccharide moieties in Western blotting experiments. These probes are of two types, lectins and antibodies. This strategy requires neither glycoprotein sample purification nor glycan cleavage or separation (*see* **Subheading 3.2.1.**). Another strategy is to analyze the linked glycan directly by its monosaccharide composition, after the glycoprotein isolation (see **Subheading 3.2.2.**). Alternatively, the glycan can be selectively cleaved from the purified glycoprotein by chemical or enzymatical treatment. The glycoprotein is then analyzed by electrophoresis or mass spectrometry prior to and after deglycosylation (*see* **Subheading 3.2.3.**). The selective cleavage will inform the experimenter about the nature of the linked glycans. In parallel, the protein mass difference observed by mass spectrometry will give information on the structure of the linked glycans.

3.2.1. Detection of Glycans by Western Blotting

The detection of glycans borne by glycoproteins can be performed after Western blotting of either 1D or 2D electrophoresis gels. The probes used for this type of detection are mostly specific for *N*-linked glycans.

3.2.1.1. Affinodetection with Lectins

Lectins are plant proteins able to bind oligosaccharide moieties specifically. They have been characterized on the basis of their affinity for mammalian glycoproteins. Only a few of them are available for the detection of plant glycans. They are presented in **Table 1** (*see* **Note 2**).

These lectins are available commercially in a biotinylated form. The recognition event between the lectin and the oligosaccharide is visualized on a blot using a streptavidin-peroxidase conjugate. An exception is the lectin concanavalin A, which is able to bind directly to peroxidase. The two protocols are detailed here.

The lectin-biotin/streptavidin-peroxidase procedure:

1. Separate the proteins to be studied by 1D or 2D electrophoresis and transfer onto a nitrocellulose membrane.
2. Saturate the blot with TTBS buffer for 1 h (*see* **Note 3**).
3. Incubate the blot in TTBS containing the lectin-biotin complex (0.1 mg/20 mL) for 2 h at room temperature.
4. Wash four times in TTBS, for 15 min each.
5. Incubate the blot with a streptavidin-peroxidase conjugate diluted 1:1000 in TTBS for 1 h at room temperature.
6. Wash the blot in TTBS (4 × 15 min) and once in TBS prior to development.
7. Prepare extemporaneously the peroxidase development mixture by dissolving, in a beaker, 30 mg of 4-chloro-1-naphtol in 10 mL cold methanol (–20°C), and mixing in another beaker 30 µL H_2O_2 with 50 mL TBS. Mix the two solutions just prior to pouring onto the membrane, and shake gently to optimize the development reaction.
8. Stop the reaction by discarding the development mixture and rinsing the blot several times with distilled water. Dry the membrane and store between 2 sheets of filter paper.
9. Suggested controls for the experiment are presented in **Note 4**.

The concanavalin A (ConA)-peroxidase procedure (modified from **ref. *20***):

1. Separate the proteins to be studied by 1D or 2D electrophoresis, and transfer onto a nitrocellulose membrane.
2. Saturate the blot for 1 h in TTBS at room temperature.
3. Incubate the membrane in TTBS* containing ConA (25 µg/mL) for 2 h at room temperature.
4. Wash the membrane four times for 15 min each in TTBS.
5. Incubate the membrane in TTBS containing horseradish peroxidase (HRP, 50 µg/mL) for 1 h at room temperature.
6. Rinse the blot four times for 15 min each in TTBS and once in TBS for 15 min.
7. Develop the peroxidase reaction as described in the lectin-biotin/streptavidin-peroxidase procedure.
8. Controls have to be performed as presented in **Note 4**.

Table 1
Lectins Used for *N*- and *O*-Linked Glycan Detection in Plant Cell Extracts in Western Blotting

Lectin	Specificity	Detected oligosaccharides in plants	Suggested positive controls	Sugar inhibitor	Ref.
Concanavalin A from *Canavalia ensiformis*	αD: mannose αD: glucose	Precursor structures: M9G3 to M9G1 High-mannose structures: M9 to M5	Soybean agglutinin Ovalbumin Ribonuclease B	Methyl α-D-mannopyranoside	***20***
Galanthus nivalis agglutinin (GNA)	Terminal α1→3 mannose	High-mannose structures: M8 to M5	Ribonuclease B	Methyl α-D-mannopyranoside	***42,43***
Wheat germ agglutinin (WGA) from *Triticum aestivum*	Terminal GlcNAc Internal chitobiose units	All *N*-linked glycans *O*-linked glycans with terminal GlcNAc	α-A-crystalline from bovine eye lens (*O*-GlcNAc) Ovalbumin (*N*-glycans)	GlcNAc Chitotriose Chitobiose	***44***

3.2.1.2. Immunodetection with Specific Antibodies

We have shown that the β1-2 xylose and the α1-3 fucose epitopes of plant *N*-linked complex glycans are highly immunogenic in rabbits *(21)*. As a consequence, sera prepared against glycoproteins containing complex glycans generally contain antibodies directed at the β1-2 xylose and/or the α1-3 fucose. Some commercially available immunsera present the same characteristics as our home-made probes. For instance, immunsera directed against HRP can be used as a probe specific for plant complex N-glycans with α1-3 fucose and β1-2 xylose residues. More specific is the immunserum raised against honeybee venom proteins, which can be used as a specific probe for α1-3 fucose-containing plant complex glycans *(21)*. We also raised in our laboratory rabbit antibodies specific for the terminal antennae of complex *N*-glycans, the trisaccharide Lewis a *(22)*. Anti-Lewis a monoclonal antibodies are commercially available, although they are very expensive.

AGPs and extensins can be detected using monoclonal antibodies that are now commercially available (http://www.plantprobes.co.uk/). As we do not use these antibodies in our lab, the experimenter should refer to the initial publications describing the specificities of these antibodies for further details (*see* **Note 5**).

1. Separate the proteins by 1D or 2D electrophoresis and transfer onto a nitrocellulose membrane.
2. Saturate the blot with 3% gelatin prepared in TBS for 1 h at room temperature.
3. Incubate the blot in TBS containing 1% gelatin and the immunserum at a convenient dilution for 2 h at room temperature. We used the commercial immunsera described above in **Subheading 2.2.1.2.** at the following dilutions:
   a. Anti-HRP rabbit immunserum, 1:300.
   b. Anti-honeybee venom rabbit immunserum, 1:200.
   c. Anti-Lewis a mouse monoclonal antibody, 1:100.
4. Wash the blot with TTBS buffer four times for 15 min each.
5. Incubate the blot in TBS containing 1% gelatin and the suitable conjugate, i.e., a goat anti-rabbit IgG conjugate coupled to HRP at a dilution of 1:2000, or HRP conjugated goat antibodies directed at mouse polyvalent immunoglobulins at a dilution of 1:500, for 1 h at room temperature.
6. Wash the blot four times with TTBS for 15 min each and once in TBS prior to development.
7. Develop the peroxidase reaction as described for the lectin/biotin-streptavidin/peroxidase procedure (*see* **Subheading 3.2.1.1.**).
8. For controls, *see* **Notes 6** and **7**.

3.2.2. Glycan Analysis by its Monosaccharide Composition

The analysis of the glycan monosaccharide composition is carried out by hydrolysis with methanol-HCl, followed by derivatization and analysis of the resulting monosaccharide(s) using gas chromatography. The monosaccharide composition obtained from this experiment gives preliminary data on the type of sugar linked to the protein (*O*- and/or *N*-linked oligosaccharide).

1. Freeze-dry the sample containing 1 mg of purified protein into a glass reaction tube equipped with a Teflon-faced screw cap (*see* **Note 8**).
2. Add 5 to 10 µL of a 2 m*M* inositol stock solution to the protein under analysis (*see* **Note 9**). Then freeze-dry the sample again prior to methanolysis.
3. Add 500 µL of 1 *N* methanol-HCl to the sample, close the tube tightly with the Teflon-faced screw cap, and place the tube overnight at 80°C.
4. Dry the sample under nitrogen atmosphere after methanolysis while heating the sample at 40°C. **Caution:** Work in a hood because of the methanol toxicity.
5. Wash with 250 µL of methanol. Dry under nitrogen atmosphere as in **step 4** and repeat this washing step twice.
6. Resuspend the sample in 250 µL of methanol. Then add 25 µL of acetic acid, 25 µL of pyridine, homogenize, and incubate for 6 h at room temperature. Dry under nitrogen atmosphere as in **step 4**.
7. Add 250 µL of silylation reagent and incubate at 80°C during 20 min (*see* **Note 1**). Air-dry.
8. Wash with 1 mL of cyclohexane. Air-dry and resuspend in 200 µL of cyclohexane, homogenize, and centrifuge before transferring 100 µL of the derivatized sample in a reaction vial with screw cap.
9. Set the gas chromatography apparatus to 1.4 bars with helium. Prior to injection, heat the injector and FID detector to 250°C and 280°C, respectively. Set the helium pressure to 20 psi and its flow rate to 3 mL/min. Equilibrate the column temperature to 120°C prior to the injection of 5 µL of the derivatized sample (equivalent to 25 µg of protein). Eluate the derivatized monosaccharides by a temperature gradient: (1) +10°C per minute until 160°C; (2) +1.5°C per minute until 235°C; (3) 235°C for 4 minutes; (4) +20°C per minute until 280°C; and, finally, (5) 280°C for 3 min.

3.2.3. Analysis of the Protein after Glycan Release

3.2.3.1. Chemical Treatment for Glycan Release

O-linked glycans can be selectively cleaved from glycoproteins by reductive amination (*see* **Note 10**). The deglycosylated protein will be then analyzed by 1D SDS-PAGE and its migration will be compared with the migration of its glycosylated form. An increase in its electrophoretical mobility is the proof of *O*-glycan occurrence on the protein (*see* **Note 11**).

1. Freeze-dry the purified protein in a glass reaction tube equipped with a Teflon-faced screw cap.
2. Dissolve it into 500 µL of a sodium borohydride solution. Close the glass tube tightly and incubate it overnight at 37°C.

3. Add acetic acid drop by drop to stop the reductive amination reaction until the gas production finishes. Add then 500 µL of 10% acetic acid dissolved in methanol. Air-dry under a hood. Repeat the washing step three times. This will coevaporate excess of borates.
4. After treatment, the deglycosylated protein is precipitated overnight at –20°C, by 4 vol of ethanol, and dissolved in a suitable buffer for 1D SDS-PAGE. In parallel, the glycans contained in the ethanol supernatant can be recovered and analyzed for monosaccharide composition (*see* **Subheading 3.2.2.**).

3.2.3.2. Enzymatic Treatment by N- or O-Glycosidases

Further information can be deduced by treating the purified glycoprotein with glycosidases. Endoglycosidase H (Endo H) is only able to release high-mannose type *N*-glycans from plant glycoproteins by hydrolyzing the glycosidic bond between the two GlcNAc residues on the core of the *N*-glycan (**Fig. 3**). Peptide *N*-glycosidases (PNGases) hydrolyze the bond between the Asn of the peptide backbone and the proximal GlcNAc of the oligosaccharide part, for both high-mannose type *N*-glycans and complex type *N*-glycans. However, PNGase F, which is widely used in the analysis of mammalian glycoproteins, is active on high-mannose and complex plant *N*-glycans, except those presenting an α1-3 fucose residue linked to the proximal GlcNAc. PNGase A is able to release all types of plant *N*-glycans (**Fig. 3**), but it is almost only efficient on glycopeptides and necessitates the proteolytic digestion of the glycoprotein prior to deglycosylation ***(23,24)***. The deglycosylation protein or peptides can be analyzed for (1) an increased electrophoretic mobility, (2) the loss of glycoprotein reactivity on blots with glycan-specific probes after Endo H or PNGase F treatment, or (3) mass spectrometry analysis after Endo H, PNGase F, or PNGase A treatment. The same approach and strategies can be performed by using *O*-glycosidases but are rarely used in our lab and will not be detailed here (*see* **refs. *25*** and ***26*** for details).

Deglycosylation with endoglycosidase H:

1. Prior to enzyme deglycosylation with Endo H, denature the purified protein by heating it to 100°C for 5 min in the presence of 1% SDS (w/v).
2. Dilute the sample fivefold with 150 m*M* sodium acetate, pH 5.7, and add then 10 mU Endo H. Incubate the mixture at 37°C for at least 6 h.
3. After the deglycosylation reaction, add an equal volume of twice concentrated electrophoresis sample buffer if the result of the digestion is followed by electrophoresis gel, affino-, or immunodetection (*see* **Subheading 3.2.1.**). The sample can also be desalted and analyzed by mass spectrometry.

Deglycosylation with PNGase F:

1. Dissolve the purified protein to be digested in 0.1 m*M* Tris-HCl, pH 7.5, containing 0.1% SDS.

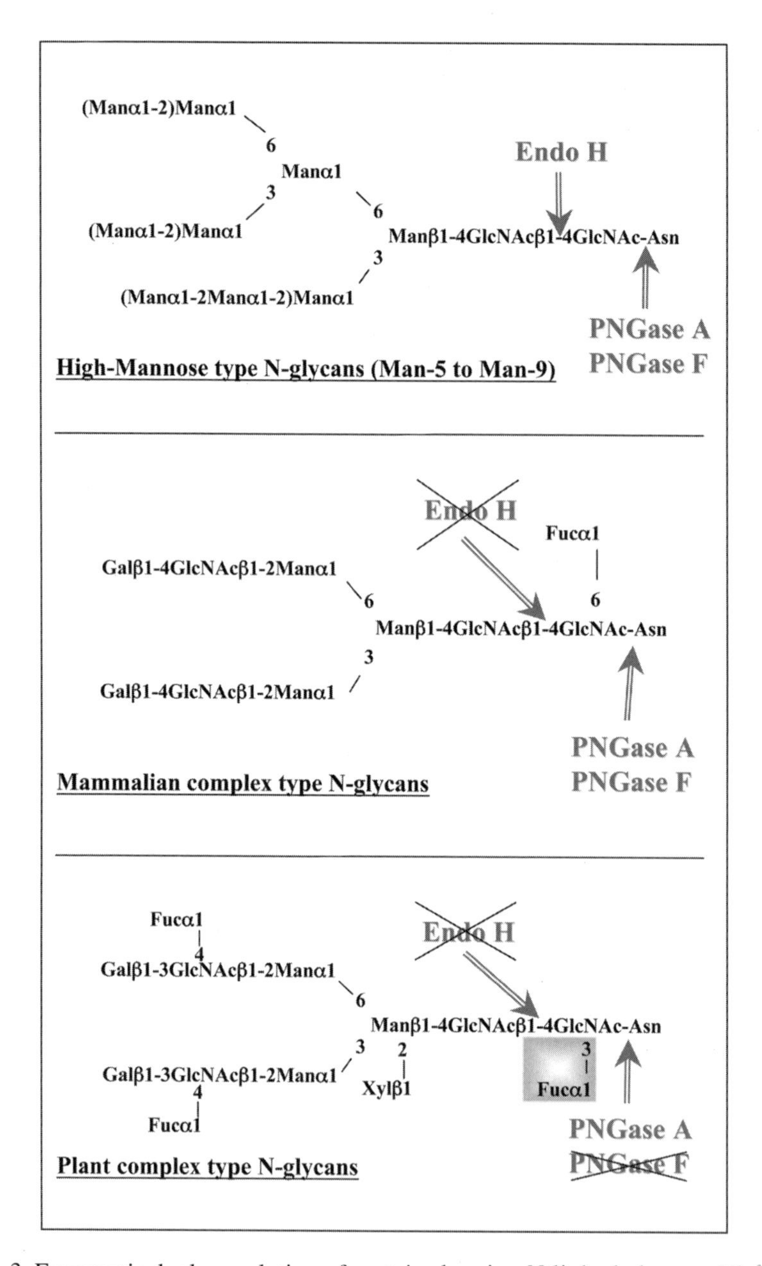

Fig. 3. Enzymatic deglycosylation of proteins bearing *N*-linked glycans. High-mannose type *N*-glycans can be cleaved by either endoglycosidase H (Endo H), peptide *N*-glycosidase A (PNGase A), or peptide *N*-glycosidase F (PNGase F). Mammalian complex type *N*-glycans can be removed using PNGase A and PNGase F. Plant complex type *N*-glycans can only be released by PNGase A, owing to the presence of α1-3 fucose linked to the core. It should be noted that PNGase A mostly acts on glycopeptides and is not efficient on a complete glycoprotein.

2. Heat the sample for 5 min at 100°C to denature the protein. Let it cool down at room temperature and add an equal volume of 0.1 m*M* Tris-HCl, pH 7.5, containing 0.5% Nonidet P-40 (*see* **Note 12**).
3. Incubate the sample with PNGase F (1U enzyme for 100 μg protein) at 37°C during 24 h.
4. After digestion, precipitate the deglycosylated protein overnight at –20°C, by 4 vol of ethanol.
5. Centrifuge the sample. Recover the deglycosylated protein in the pellet and dissolve it in a suitable buffer for gel electrophoresis or mass spectrometry analysis.

Deglycosylation with PNGase A:

1. Dissolve 100 μg of purified protein in 500 μL of 10 m*M* HCl, pH 2.2, and add 10 μg of pepsin. Incubate the sample for 24 h at 37°C and add again the same amount of pepsin. Continue the digestion for additional 24 h. Stop the reaction by heating for 5 min at 100°C.
2. Cool the sample down and take 10% of the solution to purify the peptide and glycopeptide mixture on a C18 column according to the following steps.
3. First rinse the C18 column with 5 mL of acetonitrile followed by 5 mL of water. Apply the diluted sample to 1 mL final volume with distilled water. Rinse the column with 5 mL of water to remove salts from the sample, and elute the peptides that bind to the column by using 5 mL of acetonitrile. Concentrate the peptide sample by evaporating the acetonitrile. Analyze the mass of the peptides and glycopeptides by MALDI-TOF mass spectrometry, to get the glycopeptide masses prior to PNGase A treatment.
4. Freeze-dry the sample left over and dissolve it in 500 μL of 100 m*M* sodium acetate, pH 5.5.
5. Incubate at 37°C for 18 h with 0.1 mU PNGase A. After digestion, freeze-dry the sample and separate the peptides from the released oligosaccharides using a C18 column, as explained above in **steps 2** and **3**. Analyze the deglycosylated peptides by MALDI-TOF mass spectrometry. The comparison of the two spectra, obtained before and after the deglycosylation treatment, provides a mass difference for some ions owing to the removal of *N*-linked glycans. This mass difference, in addition to the enzyme used for the deglycosylation procedure and the knowledge of plant glycosylation, indicate which type of glycan is *N*-linked to the protein and also give an idea of the structure of this glycan (**Table 2**).

All the reagents necessary for such a determination are included in a detection kit, the Enzymatic deglycosylation Kit commercialized by Prozyme. It should be noted that, using this kit according to the manufacturer's conditions, the experimenter can obtain information about the type of glycan (*N*- and *O*-linked) borne by the glycoprotein under study.

3.3. Where Is My Protein Glycosylated?

This approach will allow the experimenter to get information about both the distribution of the glycans on the protein backbone and the structures of the

Table 2
Structures and Theorical (M+Na)+ Values of High-Mannose Type *N*-Glycans (Man5 to Man9) and Complex Type *N*-Glycans Isolated from Plants[a]

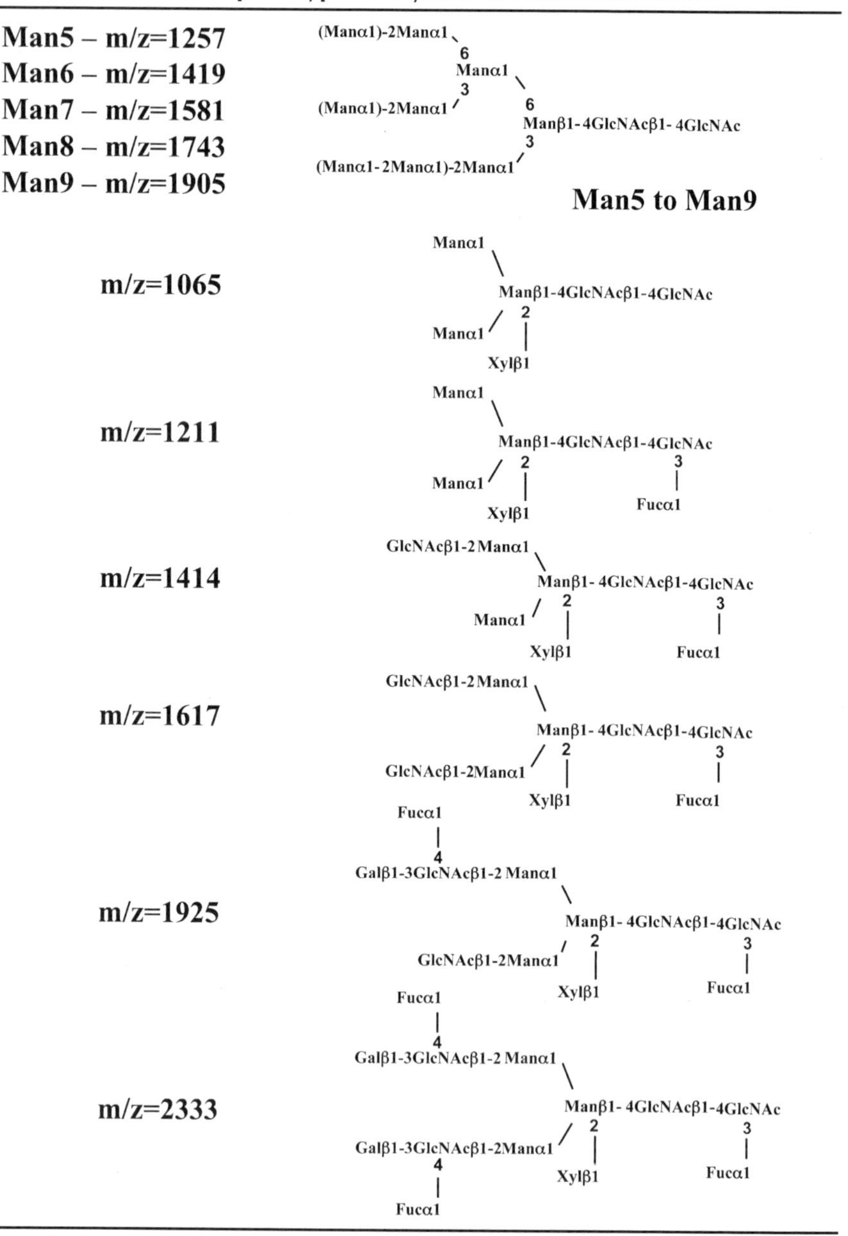

[a] Values indicated on the left correspond to the m/z (mass/charge) ratio calculated for the ions that can be observed when *N*-glycans are analyzed by MALDI-TOF mass spectrometry. These masses represent sodium adducts of the oligosaccharides.

glycans themselves. These experiments can be performed on a purified glycoprotein or on a protein isolated from a 2D or 1D electrophoresis gel. First, the protein is digested by endoproteases, and the resulting peptide and glycopeptide mixture is separated by high-performance liquid chromatography (HPLC). The presence of glycans within the collected fractions is analyzed by sugar composition (*see* **Subheading 3.2.2.**). The glycopeptide fractions are then analyzed by MALDI-TOF mass spectrometry prior to and after PNGase A digestion (*see* **Subheading 3.2.3.2.**) ***(18,19)***.

1. Dissolve 1 mg of purified protein in 500 μL of 50 m*M* ammonium bicarbonate, pH 8.0, and heat for 3 min at 100°C.
2. Add 50 μg of TPCK-treated trypsin from bovine pancreas and carry out the digestion for 2 h, at 37°C.
3. Add a second aliquot of 50 μg of the enzyme and incubate for additional 2 h at 37°C.
4. To perform a double digestion, dissolve 50 μg of TLCK-treated chymotrypsin in 50 m*M* ammonium bicarbonate, pH 8.0, and add them to the trypsin-digested sample. Incubate the sample at 37°C for 2 h.
5. Incubate for 5 min at 100°C to stop the protease digestion.
6. Separate peptides and glycopeptides generated by the trypsin/chymotrypsin digestion by reverse-phase HPLC (C18 column) using a 60-min linear gradient from 0 to 60% of solvent B and a flow rate of 1 mL/min with eluant peaks monitored by UV at 214 nm. Solvents are A: 90:10 water/acetonitrile containing 0.1% TFA, and B: 10/90 water/acetonitrile containing 0.1% TFA. Collect 2-mL fractions and freeze-dry them.
7. Analyze an aliquot (10%) of each collected fraction for sugar composition (*see* **Subheading 3.2.2.**) and select the fractions containing oligosaccharides.
8. Analyze the HPLC fractions containing the glycopeptides by MALDI-TOF to obtain the glycopeptide masses.
9. Digest the glycopeptides with PNGase A as previously described (*see* **Subheading 3.2.3.2.**). Separate the peptides from their oligosaccharide moiety on a C18 column as described above (*see* **Subheading 3.2.3.2.**).
10. Analyze the resulting deglycosylated peptides by MALDI-TOF mass spectrometry to determine their masses. These masses will allow the experimenter to identify the occupied glycosylation site(s) of the studied protein. The mass difference observed between glycosylated and deglycosylated peptides will provide structural information about the glycans borne by the protein (**Table 2**).

This procedure can also be performed for proteins with unknown sequences. In this case, the deglycosylated peptide analysis will be done by LC-MS/MS, to identify its amino acids sequence. Alternatively, the released glycans can be purified and analyzed by MALDI-TOF using specific procedures described before ***(18,19)***.

3.4. How Can the Whole Glycoproteome Be Analyzed?

The identification of glycoproteins contained within a plant extract implies the determination of (1) which genes encode the glycoproteins; (2) which sites, among their potential glycosylation sites, are actually glycosylated; and (3) what is the nature or the structure of the glycans borne by these glycoproteins. All these steps are referred now to as *glycoproteomics*. Such identifications have already been realized in animal cells, but nothing has been published yet for plant glycoproteome identifications.

The current strategy developed in our laboratory to gain access to plant glycoproteomes relies on their purification on immobilized lectins, their separation by 1D or 2D electrophoresis, and their identification by mass spectrometry. Glycoproteome isolation will be illustrated through two protocols: the purification of secreted glycoproteins bearing high-mannose type N-glycans and the purification of glycoproteins bearing *O*-GlcNAc. The protocols described here will only focus on the selection of the glycoproteins and will not detail their identification by electrophoresis and mass spectrometry. For more information on these techniques, see other chapters of this book.

3.4.1. Identification of Glycoproteins with High-Mannose N-Glycans

This protocol was developed in our laboratory for rapeseed, but it can easily be adapted to other plant material or species.

1. Extract the proteins by grinding 6 g of plant material in 50 mL of cold TBS buffer with a mortar and a pestle kept at 4°C (*see* **Note 13**). Centrifuge the extract successively at 10,000*g* for 30 min, to remove the insoluble material, and then at 150,000*g* for 1 h, to pellet the membranes. Estimate the protein concentration of the supernatant, representing the soluble protein fraction, by a Bradford assay. Divide the sample into fractions containing 20 mg of proteins and freeze them until use.
2. Thaw a fraction of 20 mg proteins by placing it at 4°C. Centrifuge it at 10,000*g* for 30 min and filter on a 0.20-μm membrane to remove all insoluble material. Adjust the final volume to 35 mL with cold TBS, and add $CaCl_2$ and $MgCl_2$ to a final concentration of 1 m*M*.
3. Resuspend the equivalent of 1 mL of Concanavalin A-Sepharose resin in 50 mL of TBS*. Wash the resin twice with 50 mL of TBS*. Recover the resin after each washing step by settlement or by centrifugation (3500*g* for 10 min).
4. Incubate the washed resin with the 35 mL sample for 2 h at 4°C on a rotary shaker. Eliminate the nonretained proteins by pouring the resin into a Poly-Prep® column and discarding the flowthrough.
5. Resuspend the resin in 50 mL of TTBS*, before washing it five times with 50 mL of TTBS* and five times with 50 mL of TBS* successively, as described above in **step 3**.

6. Pour the resin in a Poly-Prep® column and eliminate the remaining TBS*. Resuspend then the resin within the column with 10 mL of 0.3 *M* α-methyl-mannose in TBS*. Keep the preparation at 4°C on a rotary shaker for 1 h (*see* **Note 14**).
7. Elute the released glycoproteins and wash the resin with 15 vol of 0.3 *M* α-methyl-mannose in TBS*. Pool the two fractions. The resulting mixture represents the purified glycoproteins bearing high-mannose *N*-linked glycans.
8. After an estimation of protein concentration by a Bradford Assay, precipitate the glycoprotein sample overnight in the presence of 12.5% trichloroacetic acid (final concentration) at 4°C. Pellet the proteins by centrifugation at 10,000*g* for 15 min, wash the pellet with acetone/water (9:1 [v/v]), and centrifuge again. Repeat the washing step two more times, and solubilize the proteins in a suitable sample buffer for 2D electrophoresis analysis.
9. Separate the glycoprotein sample in a 2D gel, and stain it by colloidal blue staining. Collect each observable spot individually (*see* **Note 15**). Digest the collected spots by trypsin. Analyze these digests by LC-MS/MS. Submit the obtained peptide sequences to a blast search for "short, nearly exact matches" against the nonredundant database of NCBI, with a limitation to the organism used as a possible option (http://www.ncbi.nlm.nih.gov/BLAST/).

Once identified, the glycoproteins should be checked for two points: (1) are they potentially secreted, i.e., do they present a potential signal peptide? and (2) do they bear at least one potential *N*-glycosylation site? The putative subcellular location of the glycoproteins can be estimated in silico using softwares such as Predotar (http://genoplante-info.infobiogen.fr/predotar/), Target P (http://www.cbs.dtu.dk/services/TargetP/), or iPSORT (http://hc.ims.u-tokyo.ac.jp/iPSORT/). The occurence of *N*-glycosylation sites within a protein can be determined by matching this protein against Prosite (http://npsa-pbil.ibcp.fr/cgi-bin/npsa_automat.pl?page=npsa_prosite.html or http://www.expasy.org/tools/scanprosite/).

3.4.2. Identification O-GlcNAcylated Glycoproteins

The strategy involved in the identification of the *O*-GlcNAc glycoproteome requires glycoprotein purification by affinity chromatography on immobilized WGA. Previous experiments have shown that WGA is able to bind both *O*-GlcNAcylated proteins *(27)* and proteins with *N*-linked glycans *(28)*. Therefore, a preliminary elimination of *N*-linked glycans is an absolute necessity to purify *O*-GlcNAcylated glycoproteins selectively on the affinity column. This deglycosylation step is performed by PNGase F, an enzyme able to remove high-mannose and complex *N*-glycans. However, its effect is limited on plant complex *N*-linked glycans, as explained above (*see* **Subheading 3.2.3.2.**), owing to the presence of an α1-3 fucose linked to the proximal GlcNAc of the *N*-glycan core. To overcome this restriction, it is necessary to work on plant material in which *N*-glycans are kept in a high-mannose form, sensitive to PNGase F. This protocol has been developed in our laboratory using cultured cells of the *A. thaliana cgl* mutant, in which *N*-glycans are matured up to the $Man_5GlcNAc_2$ structure only ***(29)***. This protocol should be applicable to other plant materials, as long as their *N*-glycosylation is restricted to high-mannose *N*-linked glycans.

1. Filter 150 mL of cultured cells on Miracloth before incubating them in 150 mL of 0.5 *M* NaCl for 30 min at 4°C, to solubilize the proteins ionically bound to the cell wall (*see* **Notes 16** and **17**).
2. Filter the cells again on Miracloth and discard the supernatant. Grind the cells in 30 mL of 20 m*M* Tris-HCl, pH 7.8, containing 150 m*M* KCl, 2 m*M* $CaCl_2$, 10 m*M* $MgCl_2$, 1 m*M* DTT and protease inhibitors (WGA buffer).
3. Eliminate the insoluble material by centrifuging the extract at 5500*g* for 15 min at 4°C and at 25,000*g* for 30 min at 4°C, successively. Aliquot in 5-mL fractions.
4. Concentrate a 5-mL fraction using Centricon (Amicon Bioseparation YM-10) to a 0.5 mL final volume.
5. After concentration, adjust the sample to 0.1% SDS and incubate at 100°C for 5 min to denature the glycoproteins. Add Nonidet P40 to a final concentration of 0.5% to complex the SDS (*see* **Note 12**). After denaturation, deglycosylate the proteins by adding PNGase F (5 U) and incubate for 24 h at 37°C under agitation. During this step, only the *N*-linked glycans will be eliminated.
6. Dilute the deglycosylated sample 20 times in WGA buffer and incubate it with 100 μL of WGA-agarose for 4 h at 4°C on a rotary shaker. Discard the non-retained proteins and wash the resin with 20 column volumes of WGA buffer.
7. To elute the bound material, incubate the resin with 5 to 10 vol of WGA buffer containing 0.5 *M* GlcNAc for 30 min at 4°C (*see* **Note 18**). After collection of the eluate, repeat the same treatment once more. Pool the two fractions eluted from the column. This glycoprotein fraction only contains *O*-GlcNAcylated proteins.
8. Precipitate the glycoproteins in the presence of 12.5% trichloroacetic acid, final concentration, overnight at 4°C. Pellet the proteins by centrifugation for 15 min at 10,000*g*. Wash the protein pellet three times with acetone/water (9:1 [v/v]) and resuspend it in a suitable buffer for 1D electrophoresis.
9. Separate the *O*-GlcNAcylated proteins in 1D SDS-PAGE gel. After Coomassie blue staining, collect all the bands vizualized in the gel. After digestion by trypsin, identify the proteins by MALDI-TOF or LC-MS/MS.

3.5. Conclusions and Perspectives

Using the methods outlined in this chapter, information about the presence, the linkage, and the composition of glycans that are constitutive of a plant glycoprotein can be obtained. The strategies discussed here are already efficient to identify *N*-linked glycans, whereas analysis of *O*-linked glycans calls for further efforts.

Even when the physicochemical characterization of a plant glycan is deciphered, it is difficult to predict a priori the function(s) mediated by a given oligosaccharide or a given glycoprotein. Indeed, the same oligosaccharide sequence may present different functions when it is carried by different glycoproteins, or found at different locations within a plant, or at different times of the plant life cycle. Therefore, the still emerging era of glycoproteomics opens new avenues that will shed lights on the roles of *N*- and *O*-linked glycans at an increasing pace. Closing gaps in this field will expand our understanding of plant physiology and also offer wider perspectives in biotechnology and medicine.

4. Notes

1. Use fresh ampules of 3 *N* methanolic-HCl and silylation reagent for each experiment.
2. The experimenter should be careful in the blot interpretation, as WGA recognizes GlcNAc in *N*-linked glycans, as well as *O*-linked GlcNAc.
3. The solution used for blocking binding sites on the nitrocellulose needs to be devoid of glycoproteins. This is why we recommend using Tween-20 to coat the nitrocellulose ***(30)*** in this procedure.
4. Controls for specificity:
   a. It might be advisable to blot the proteins cited in **Table 1** on the membrane ("Suggested positive controls"), to get positive controls for the affinodetection.
   b. Lectin binding specificity must also be checked by running affinodetection in the presence of 0.3 *M* inhibitory sugar (**Table 1**).
5. The available anti-*O*-linked glycan antibodies are: anti-AGP antibodies: LM2 ***(31,32)***, JIM4, JIM13, and JIM15 ***(32)***, JIM8 ***(33)***, JIM14, JIM15, and JIM16 ***(34)***; anti-extensin antibodies: LM1 ***(35)***, JIM11, JIM12, and JIM20 ***(36)***, JIM 19 ***(37)***.
6. Control for *N*-glycan specificity of the immunodetection: it might be wise to check for the specificity of sera toward *N*-glycans attached to the protein studied. This can be done by performing a mild periodate oxidation on the blot prior to immunodetection. Mild periodate treatment oxidizes glycans and abolishes any recognition of the glycoprotein by the anti-glycan antibodies. Any remaining signal will be the consequence of a protein backbone antibody/recognition (38).
   a. After saturation with gelatin, incubate the blot in a 100 m*M* sodium acetate buffer, pH 4.5, containing 100 m*M* sodium metaperiodate for 1 h in the dark at room temperature, changing the incubation solution after 30 min.
   b. Incubate the blot in PBS containing 50 m*M* sodium borohydride for 30 min at room temperature.
   c. Rinse the blot with TBS, saturate for 15 min with TBS containing 1% gelatin, and perform the immunodetection, as described in **Subheading 3.2.1.1.**

7. Controls for fucose or xylose specificity: some proteins can be used as positive controls for the *N*-glycan immunodetection. Phospholipase A2 from honeybee venom contains the α1-3 fucose residue, and is devoid of β1-2xylose ***(39)***. PHA-L and recombinant avidin produced in maize are glycoproteins containing both β1-2 xylose and α1-3 fucose ***(18,40 ,41)***.
8. The protocol described calls for 1 mg of protein but can be adapted to a smaller amount.
9. Inositol is used as an internal standard.
10. Because reductive amination is a drastic treatment, some modifications of the protein may occur. For this reason, it is sometimes better to eliminate *O*-linked glycans by enzymatic digestion.
11. In general in our lab, we do not use chemical treatments to release *N*-linked glycans from glycoproteins.
12. The role of Nonidet P40 is to complex free SDS.
13. Sixty to 80 mg of soluble rapeseed proteins were necessary, as a starting material, to prepare enough glycoproteins for the loading of one 2D gel.
14. α-Methyl mannose is a ligand of concanavalin A and will displace the bound glycoproteins from the immobilized lectin (**Table 1**).
15. We observed an important release of concanavalin A while eluting the affinity chromatography column. This release polluted the glycoprotein preparation and obliged us to run, in parallel to the analytical 2D gel, another 2D gel loaded with concanavalin A alone. After staining, we selected the spots present only in the analytical gel and discarded the ones resolved simultaneously in the two 2D gels.
16. One hundred and fifty milliliters of 6-day-old *cgl A. thaliana* cultured cells roughly represent 10 g of plant material.
17. The elimination of cell wall-bound proteins represents a first purification step, as these proteins do not bear *O*-GlcNAc. This purification step should be included in the protocol whenever possible.
18. Free GlcNAc is a ligand of WGA and will displace the bound glycoproteins from the immobilized lectin (**Table 1**).

Acknowledgments

We thank present and former colleagues from our lab who contributed to the development and optimization of the strategies described in this manuscript. Work on glycobiology at the University of Rouen is supported by the French Ministère de la Recherche and the Centre National de la Recherche Scientifique (CNRS). Studies on plant glycoproteome were initiated thanks to funding from Genoplante II (grant AF 2001069) and FEDER (grant 1013). V. Gomord is aknowledged for coordinating the Genoplante II project "Glycoproteome from *A. thaliana* seeds."

References

1. Lerouge, P., Cabanes-Macheteau, M., Rayon, C., Fischette-Lainé, A.-C., Gomord, V., and Faye, L. (1998) *N*-glycosylation biosynthesis in plants: recent developments and future trends. *Plant Mol. Biol.* **3,** 31–48.
2. Cho, Y. P. and Chrispeels, M. J. (1976) Serine-*O*-galactosyl linkages in glycopeptides from carrot cell wall. *Phytochemistry* **15,** 165–169.
3. Matsuoka, K., Watanabe N., and Nakamura K. (1995) *O*-glycosylation of a precursor to a sweet potato protein, sporamin, expressed in tobacco cells. *Plant J.* **8,** 877–889.
4. Faye, L., Boulaflous, A., Benchabane, M., Gomord, V., and Michaud, D. (2005) Protein modifications in the plant secretory pathway: current states and practical implications in molecular pharming. *Vaccine* **23,** 1770–1778.
5. Schowalter, A. M. (1993) Structure and function of plant cell wall proteins. *Plant Cell* **5,** 9–23.
6. Moore, P. J., Swords, K. M. M., Lynch, M. A., and Staehelin, L. A. (1991) Spatial organization of the assembly pathways of glycoproteins and complex polysaccharides in the Golgi apparatus of plants. *J. Cell Biol.* **112,** 467–479.
7. Robinson, D., Andreae, M., and Sauer, A. (1985) Hydroxyproline-rich glycoprotein biosynthesis: a comparison with that of collagen, in *Biochemistry of Plant Cell Walls* (Brett, C. T. and Hillman, J. R., eds.), Cambridge University Press, Cambridge, UK, pp. 155–176.
8. Schowalter, A. M. (2001) Arabinogalactan-proteins: structure, expression and function. *Cell Mol. Sci.* **58,** 1399–1417.
9. Hart, G. (1999) The *O*-GlcNAc modification, in *Essentials of Glycobiology* (Varki, A., Cummings, R., Esko, J., Freeze, H., Hart, G., and Marth, J., eds.) Cold Spring Harbor Laboratory Press, Cold Spring Harbor, NY, pp. 183–194.
10. Slawson, C. and Hart, G. (2003) Dynamic interplay between *O*-GlcNAc and *O*-phosphate: the sweet side of protein regulation. *Curr. Opin. Struct. Biol.* **13,** 631–636.
11. Jacobsen, S. E., Binkowski, K. A., and Olszewski, N. E. (1996) SPINDLY, a tetratricopeptide repeat protein involved in gibberellin signal transduction in *Arabidopsis*. *Proc. Natl. Acad. Sci. USA* **93,** 9292–9296.
12. Thornton, T. M., Swain, S. M., and Olszewski, N. E. (1999) Gibberellin signal transduction presents ellipsis the SPY who *O*-GlcNAc'd me. *Trends Plant Sci.* **4,** 424–428.
13. Hartweck, L. M., Scott C. L., and Olsewski, N. E. (2002) Two *O*-linked *N*-acetylglucosamine transferase genes of *Arabidopsis thaliana* L. Heynh. have overlapping functions necessary for gamete and seed development. *Genetics* **161,** 1279–1291.
14. Heese Peck, A., Cole, R. N., Borkhsenious, O. N., Hart, G. W., and Raikhel, N. V. (1995) Plant nuclear pore complex proteins are modified by novel oligosaccharides with terminal *N*-acetylglucosamine. *Plant Cell* **7,** 1459–1471.

15. Heese Peck, A. and Raikhel, N. V. (1998) A glycoprotein modified with terminal N-acetylglucosamine and localized at the nuclear rim shows sequence similarity to aldose-1-epimerases. *Plant Cell* **10,** 599–612.
16. Holst, B., Bruun, A. W., Kielland-Brandt, M. C., and Winther, J. R. (1996) Competition between folding and glycosylation in the endoplasmic reticulum. *EMBO J.* **15,** 3538–3546.
17. Faye, L., Sturm, A., Bollini, R., Vitale, A., and Chrispeels, M. J. (1986) The position of the oligosaccharide side-chains of phytohemagglutinin and their accessibility to glycosidases determines their subsequent processing in the Golgi. *Eur J. Biochem.* **158,** 655–661.
18. Bardor, M., Loutelier-Bourhis, C., Marvin, L., et al. (1999) Analysis of plant glycoproteins by matrix-assisted laser desorption ionisation mass spectrometry: application to the *N*-glycosylation of bean phytohemagglutinin. *Plant Physiol. Biochem.* **37,** 319–325.
19. Bardor, M., Faye, L., and Lerouge, P. (1999) Analysis of the *N*-glycosylation of recombinant glycoproteins produced in transgenic plants. *Trends Plant Sci.* **4,** 376–380.
20. Faye, L. and Chrispeels, M. J. (1985) Characterization of *N*-linked oligosaccharides by affinoblotting with concanavalin A-peroxidase and treatment of the blots with glycosidases. *Anal. Biochem.* **149,** 218–224.
21. Faye, L., Gomord, V., Fitchette-Lainé, A.-C., and Chrispeels, M. J. (1993) Affinity purification of antibodies specific for Asn-linked glycans containing α1->3 fucose or β1->2 xylose. *Anal. Biochem.* **209,** 104–108.
22. Fitchette-Lainé, A.-C., Gomord V., Cabanes M., et al. (1997) *N*-glycans harboring the Lewis a epitope are expressed at the surface of plant cells. *Plant J.* **12,** 1411–1417.
23. Tomiya, N., Kurono, M., Ishihara, H., et al. (1987) Structural analysis of *N*-linked oligosaccharides by a combination of glycopeptidase, exoglycosidases, and high-performance liquid chromatography, *Anal. Biochem.* **163,** 489–499.
24. Ogawa, H., Hijikata, A., Amano, M., et al. (1996) Structures and contribution to the antigenicity of oligosaccharides of Japanese cedar (*Cryptomeria japonica)* pollen allergen *Cry j* I: relationship between the structures and antigenic epitopes of plant *N*-linked complex-type glycans. *Glycoconjugate J.* **13,** 555–566.
25. Linsley, K. B., Chan, S. Y., Chan, S., Reinhold, B. B., Lisi, P. J., and Reinhold, V. N. (1994) Applications of electrospray mass spectrometry to erythropoietin *N*- and *O*-linked glycans. *Anal. Biochem.* **219,** 207–217.
26. Jesperson, S., Koedam, J. A., Hoogerbrugge, C. M., Tjadn, U. R., Van der Greef, J., and Van den Brande, J. L. (1996) Characterization of *O*-glycosylated precursors of insuline-like growth factor II by matrix-assisted laser desorption/ionization mass spectrometry. *J. Mass Spectrom.* **31,** 893–900.
27. Hanover, J. A., Cohan, C. K., Willingham, M. C., and Park, M. K. (1987) *O*-linked *N*-acetylglucosamine is attached to proteins of the nuclear pore. *J. Biol. Chem.* **262,** 9887–9894.

28. Roquemore, E. P., Chou, T.-Y., and Hart, G. W. (1994) Detection of *O*-linked *N*-acetylglucosamine (*O*-GlcNAc) on cytoplasmic and nuclear proteins. *Methods Enzymol.* **230,** 443–460.
29. von Schaewen, A., Sturm, A., O'Neill, J., and Chrispeels, M. J. (1993) Isolation of a mutant *Arabidopsis* plant that lacks *N*-acetyl glucosaminyl transferase I and is unable to synthesize Golgi-modified complex *N*-linked glycans. *Plant Physiol.* **102,** 1109–1118.
30. Bird, C. R., Gearing, A. J. H., and Thorpe, R. (1988) The use of Tween-20 alone as a blocking agent for the immunoblotting can cause artefactual results. *J. Immunol. Methods* **106,** 175–179.
31. Smallwood, M., Yates, E. A., Willats, W. G. T., Martin, H., and Knox, J. P. (1996) Immunochemical comparison of membrane-associated and secreted arabinogalactan-proteins in rice and carrot. *Planta* **198,** 452–459.
32. Yates, E. A., Valdor, J.-F., Haslam, S. M., et al. (1996) Characterization of carbohydrate structural features recognized by anti-arabinogalactan-protein monoclonal antibodies. *Glycobiology* **6,** 131–139.
33. Pennell, R. I., Janniche L., Kjellbom, P., Scofield, G. N., Peart, J. M., and Roberts, K. (1991) Developmental regulation of a plasma membrane arabinogalactan protein epitope in oilseed rape flowers. *Plant Cell* **3,** 1317–1326.
34. Knox, J. P., Linstead, P. J., Peart, J., Cooper, C., and Roberts, K. (1991) Developmentally-regulated epitopes of cell surface arabinogalactan proteins and their relation to root tissue pattern formation. *Plant J.* **1,** 317–326.
35. Smallwood, M., Martin, H., and Knox, J. P. (1995) An epitope of rice threonine- and hydroxyproline-rich glycoprotein is common to cell wall and hydrophobic plasma-membrane glycoproteins. *Planta* **196,** 510–522.
36. Smallwood, M., Beven, A. Donovan, N., et al. (1994) Localization of cell wall proteins in relation to the developmental anatomy of the carrot root apex. *Plant J.* **5,** 237–246.
37. Knox, J. P., Peart, J., and Neill, S. J. (1995) Identification of novel cell surface epitopes using a leaf epidermal-strip assay system. *Planta* **196,** 266–270.
38. Lainé, A.-C. and Faye, L. (1988) Significant immunological cross-reactivity of plant glycoproteins. *Electrophoresis* **9,** 841–844.
39. Kubelka, V., Altmann, F., Staudacher, E., et al. (1993) Primary structures of the *N*-linked carbohydrate chains from honeybee venom phospholipase A_2. *Eur. J. Biochem.* **213,** 1193–1204.
40. Vitale, A., Warner, T. G., and Chrispeels, M. J. (1984) *Phaseolus vulgaris* phytohemagglutinin contains high-mannose and modified oligoasaccharide chains. *Planta* **160,** 256–263.
41. Bardor, M., Loutelier-Bourhis, C., Paccalet, T., et al. (2003) Monoclonal C5-1 antibody produced in transgenic alfalfa plants exhibits a *N*-glycosylation that is homogeneous and suitable for glyco-engineering into a human-compatible structure, *Plant Biotech. J.* **1,** 451–462.

42. Shibuya, N., Goldstein, I. J., van Damme, E. J. M., and Peumans, W. J. (1988) Binding properties of a mannose-specific lectin from the snowdrop (*Galanthus nivealis*) bulb. *J. Biol. Chem.* **263,** 728–734.
43. Animashaun, T., Mahmood, N, Hay, A. J., and Hughes, R. C. (1993) Inhibitory effect of novel mannose-binding lectins on HIV-infectivity and syncitium formation. *Antiv. Chem. Chemother.* **4,** 145–153.
44. Nagata, Y. and Burger, M. M. (1974) Wheat germ agglutinin. Molecular characteristics and specificity for sugar binding. *J. Biol. Chem.* **249,** 3116–3122.

26

Blue-Native Gel Electrophoresis for the Characterization of Protein Complexes in Plants

Jesco Heinemeyer, Dagmar Lewejohann, and Hans-Peter Braun

Summary

High-resolution protein separation procedures are an important prerequisite for proteome analyses. Classically, protein separations are based on 2D IEF/SDS-PAGE. Unfortunately, this technique only poorly recovers hydrophobic proteins, and it is not compatible with analyses of proteins in native state. Blue-native PAGE represents a powerful alternative. It is based on the careful integration of negative charge into proteins and protein complexes by the anionic wool dye Coomassie blue, and it allows protein analyses under native ("blue-native") conditions. Upon combination with SDS-PAGE for a second gel dimension, protein complexes separated by the blue-native dimension are dissected into their subunits, which form vertical rows of spots on the resulting 2D gels. 2D blue-native/SDS-PAGE ideally complements 2D IEF/SDS-PAGE in proteomics.

Key Words: Blue-native PAGE; membrane proteins; protein complexes; mitochondria; plastids; respiratory chain; photosynthesis.

1. Introduction

Blue-native polyacrylamide gel electrophoresis (BN-PAGE) was developed by Hermann Schägger and was originally optimized to characterize membrane-bound protein complexes of mitochondria and bacteria *(1)*. The basic idea of this procedure is to incubate protein mixtures with the protein stain Coomassie blue *before* an electrophoresis is carried out. At neutral pH values, Coomassie is a negatively charged molecule, which allows the introduction of negative charges into proteins. In contrast to sodium dodecyl sulfate (SDS), introduction of charges is not coupled to unfolding of proteins or disassembly of protein complexes. If Coomassie-pretreated protein fractions are subsequently separated on low-percentage polyacrylamide gels, protein complexes are resolved.

From: *Methods in Molecular Biology, vol. 335: Plant Proteomics: Methods and Protocols*
Edited by: H. Thiellement, M. Zivy, C. Damerval, and V. Méchin © Humana Press Inc., Totowa, NJ

They can be dissected into their subunits upon combination of the BN gel dimension with a second gel dimension, which is carried out in the presence of SDS. On the resulting 2D gels, protein spots are ordered in vertical rows, allowing their assignment to protein complexes. The first gel dimension indeed allows a *native* separation of proteins and protein complexes, as revealed by various in-gel activity stains ***(2–4)***. BN-PAGE is especially useful for the separation of membrane-bound protein complexes, which often lack the internal charges required for native protein separation by electrophoresis. Sample preparation of membrane fractions requires addition of a detergent for membrane solubilization prior to Coomassie treatment, which usually should be a mild nonionic detergent compatible with native protein conditions. However, BN-PAGE was also found to be very useful for the native separation of water-soluble proteins and protein complexes.

Today, BN-PAGE is a biochemical procedure of basic importance for a variety of protein fractions of interest that is widely used in plant proteomics. Applications include plant mitochondria ***(5–10)***, plastids ***(11–15)***, the plasma membrane ***(16)***, the apoplast ***(4)***, and various biochemically defined fractions ***(17–21)***. Scientific aims addressed with BN-PAGE range from proteomic approaches to the systematic characterization of protein complexes of cellular fractions and subcellular compartments on the one hand (**Fig. 1**), to specific characterization of individual protein complexes on the other. In the latter case, BN-PAGE represents an alternative to other biochemical procedures to separate protein complexes, like analytical ultracentrifugation or gel filtration chromatography. However, compared with these methods, BN-PAGE has an enhanced resolution capacity.

2. Materials

2.1. Sample Preparation for BN-PAGE

1. Dodecylmaltoside solubilization solution: 1.0% dodecylmaltoside (Roche, Basel, Switzerland), 50 m*M* BisTris, 750 m*M* amino caproic acid, 0.5 m*M* EDTA, pH 7.0 (store at 4°C). Add phenylmethylsulfonyl fluoride (PMSF) directly before use (final concentration: 2 m*M*; stock solution: 200 m*M* PMSF [w/v] in EtOH).
2. Triton X-100 solubilization solution: 1.0% Triton X-100 (Sigma, St. Louis, MO), 50 m*M* imidazole-HCl, 50 m*M* NaCl, 2 m*M* amino caproic acid, 0.5 m*M* EDTA, 10% (v/v) glycerol, pH 7.4 (store at 4°C). Add PMSF directly before use (final concentration: 2 m*M*; stock solution: 200 m*M* PMSF [w/v] in EtOH).
3. Digitonin solubilization solution: 5.0% digitonin (Fluka, Buchs, Switzerland), 30 m*M* HEPES, 150 m*M* potassium acetate, 10% (v/v) glycerol, pH 7.4 (store at 4°C). This buffer should be freshly prepared and briefly heated to 98°C to dissolve the detergent. Add PMSF directly before use (final concentration: 2 m*M*; stock solution: 200 m*M* PMSF [w/v] in EtOH).

4. Coomassie blue solution: 5 % Coomassie G 250 (e.g., Merck, Darmstadt, Germany), 750 m*M* amino caproic acid.

2.2. BN-PAGE

1. Acrylamide solution: 49.5%, acryl/bisacryl 32:1 (AppliChem, Darmstadt, Germany).
2. Gel buffer BN (6X): 1.5 *M* amino caproic acid, 150 m*M* BisTris, pH 7.0 (store at 4°C).
3. Cathode buffer BN (5X): 250 m*M* Tricine, 75 m*M* BisTris, 0.1% (w/v) Coomassie G 250 (e.g., Merck, Darmstadt, Germany), pH 7.0 (store at 4°C).
4. Anode buffer BN (6X): 300 m*M* BisTris, pH 7.0 (store at 4°C).

2.3. SDS PAGE for Second Gel Dimension

1. Acrylamide solution: 49.5%, acryl/bisacryl 32:1 (AppliChem).
2. Gel buffer sodium dodecyl sulfate (SDS): 3 *M* Tris-HCl, 0.3% (w/v) SDS, pH 8.45.
3. Gel buffer BN (6X): *see* **Subheading 2.2.**
4. Cathode buffer SDS: 0.1 *M* Tris-HCl, 0.1 *M* Tricine, 0.1% (w/v) SDS, pH 8.25.
5. Anode buffer SDS: 0.2 *M* Tris-HCl, pH 8.9.
6. Overlay solution: 1 *M* Tris-HCl, 0.1% (w/v) SDS, pH 8.45.
7. Denaturation solution: 1.0% (w/v) SDS, 1.0% (v/v) β-mercaptoethanol.
8. SDS solution: 10% (w/v) SDS.

2.4. Staining Procedures

1. Solution A: 2% (w/v) ortho-phosphoric acid, 10% ammonium sulfate.
2. Solution B: 5% (w/v) Coomassie G 250 (Merck).
3. Fixing solution: 40% (v/v) methanol, 10% (v/v) acetic acid.
4. Staining solution: 98% (v/v) solution A, 2% (v/v) solution B. Prepare this solution freshly the day before use, and shake for several hours, preferably over night.

3. Methods

3.1. Sample Preparation for BN-PAGE

All steps of the sample preparation should be carried out at 4°C. The BN gel (*see* **Subheading 3.2.**) should be prepared before sample preparation is started (*see* **Subheadings 3.1.1.** and **3.1.2.**).

3.1.1. Membrane Fractions

1. Prepare a membrane fraction of interest, e.g., plasma membranes, thylakoid membranes, or mitochondrial membranes according to standard procedures ***(14,16,22)***. Alternatively, entire organelles can also be used as starting material for BN-PAGE, e.g., entire plastids, mitochondria or peroxisomes.
2. Determine the protein concentration, e.g., according to Lowry ***(23)***.

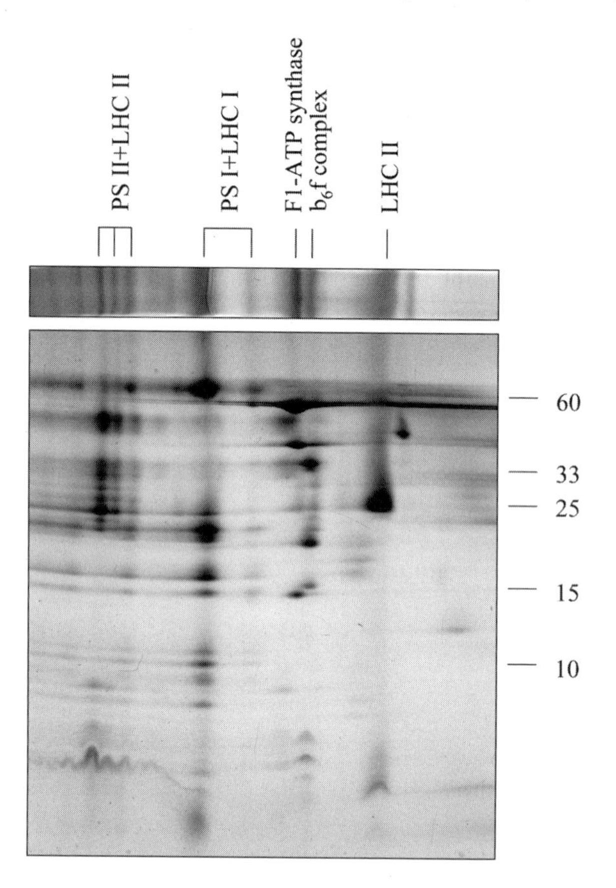

Fig. 1. Resolution of chloroplast protein complexes from *Arabidopsis* by 2D BN/SDS-PAGE. Proteins were solubilized in the digitonin solubilization solution. The gel was Coomassie stained. The numbers to the right refer to the molecular masses of standard proteins. PS II, photosystem II; PS I, photosystem I; LHC II, light harvesting complex II; LHC I, light harvesting complex I; F1-ATP synthase, F_1 part of the ATP synthase complex; b_6f complex, cytochrome b_6f complex.

3. Adjust the protein concentration to 10 µg/µL.
4. Centrifuge 100-µL fractions (about 1 mg protein) for 10 min at 15,000*g* to sediment membranes or organelles.
5. Resuspend pellets in 100 µL of dodecylmaltoside, Triton X-100, or digitonin solubilization solution (*see* **Note 1**).
6. Incubate the fractions for 15 min on ice.
7. Centrifuge the fractions for 20 min at 20,000*g* to remove insoluble material.
8. Supplement the fractions with 5 µL of Coomassie blue solution.
9. Load 50 to 100 µL of the supernatants (corresponding to 0.5–1.0 mg protein) directly into the wells of a BN gel. (Protein amounts are adjusted to allow stain-

ing of gels by Coomassie; if silver staining is used, protein amounts can be reduced by a factor of 10.)

3.1.2. Water-Soluble Fractions

1. Prepare soluble protein fractions of interest according to standard procedures. Buffers should not include high salt concentrations or ionic detergents. Protein concentrations should be adjusted to 10 μg/μL.
2. Supplement fractions of 100 μL (about 1 mg protein) with 2 μL of Coomassie blue solution.
3. Centrifuge the fractions for 20 min at 20,000*g* to remove insoluble material.
4. Directly load 50 to 100 μL of the supernatants (corresponding to 0.5–1.0 mg protein) into the wells of a BN gel. (Lower protein amounts are loaded if gels are stained by silver, *see* **Subheading 3.1.1., step 9**.)

3.2. BN-PAGE

The best resolution capacity of BN gels is achieved if the electrophoretic separation distance is greater than 12 cm. The following instructions refer to the Protean II electrophoresis unit (Bio-Rad, Richmond, CA; gel dimensions 0.15 × 16 × 20 cm). However, units from other manufacturers are of comparable suitability for BN-PAGE, e.g., the Hoefer SE-400 or SE-600 gel systems (GE Healthcare, Munich, Germany). Gradient gels are recommended, because molecular masses of protein complexes can vary between 50 kDa and several thousand of kDa. (*see* **Note 2**).

1. Prepare a 4.5% separation gel solution by mixing 1.8 mL of acrylamide solution with 3.3 mL of gel buffer BN and 14.9 mL ddH$_2$O.
2. Prepare a 16% separation gel solution by mixing 6.7 mL of acrylamide solution with 3.3 mL of gel buffer BN, 6.0 mL ddH$_2$O, and 4.0 mL glycerol.
3. Transfer the two gel solutions into the two chambers of a gradient former and connect the gradient former via a hose and a needle with the space in between two glass plates (preassembled in a gel casting stand). Gradient gels can be poured either from the top (16% gel solution has to enter the gel sandwich first) or from the bottom (4.5% gel solution has to enter first).
4. Add TEMED and APS to the two gel solutions (90 μL 10% APS/9 μL TEMED to the 4.5% gel solution; 65 μL APS/6.5 μL TEMED to the 16% gel solution).
5. Pour the gradient gel, leaving space for the stacking gel, and overlay with ddH$_2$O. The gel should polymerize in about 60 min.
6. Pour off the ddH$_2$O.
7. Prepare the stacking gel solution by mixing 1.2 mL of acrylamide solution, 2.5 mL of gel buffer BN, and 11.3 mL ddH$_2$O.
8. Add 65 μL APS and 6.5 μL TEMED, and pour the stacking gel around an inserted comb. The stacking gel should polymerize within 30 min.
9. Prepare 1X anode and 1X cathode buffers BN by diluting the corresponding stock solutions.

10. Once the polymerization of the stacking gel is finished, carefully remove the comb.
11. Add cathode and anode buffers BN to the upper and lower chambers of the gel unit, respectively. Cool down the unit to 4°C.
12. Load Coomassie blue-pretreated protein samples (*see* **Subheading 3.1.**) into the gel wells.
13. Connect the gel unit to a power supply. Start electrophoresis at constant voltage (100 V for 45 min) and continue at constant current (15 mA for about 8 h). Electrophoresis should be carried out at 4°C. Blue gel bands should already be visible during the electrophoresis run.

3.3. SDS PAGE for Second Gel Dimension

Protein complexes separated by 1D BN-PAGE can be either silver or Coomassie stained, their activity can be revealed by using in-gel enzyme assays, or they can be blotted for further analyses (*see* **Notes 3** and **4**). If protein complexes are further separated into their subunits, BN-PAGE can be combined with a second gel dimension, which is carried out in the presence of SDS. All published protocols for SDS-PAGE are suitable for combination with BN-PAGE, e.g., the system published by Laemmli ***(24)***. However, the Tricine-SDS PAGE system developed by Schägger and von Jagow ***(25)*** generally gives the best resolution. The following instructions refer to this gel system carried out in the Protean II gel unit (Bio-Rad). Gel electrophoresis units of other manufacturers are of equal suitability.

1. Cut out a strip of a BN gel and incubate it in the denaturation solution for 30 min at room temperature.
2. Wash the strip with ddH_2O for 30 to 60 s. (This step is important because β-mercaptoethanol inhibits polymerization of acrylamide.)
3. Place the strip on a glass plate of an electrophoresis unit at the position of the teeth of a normal gel comb.
4. Assemble the gel electrophoresis unit by adding 1-mm spacers, the second glass plate, and the clamps. Transfer all into the gel casting stand. (The reduced thickness of the gel of the second gel dimension [1 mm] in comparison with the strip of the first gel dimension [1.5 mm] keeps the gel strip from sliding down between the glass plates in the vertical position.)
5. Prepare a 16% separation gel solution by mixing 10 mL of acrylamide solution, 10 mL of gel buffer SDS, 10 mL ddH_2O, 100 µL APS, 10 µL TEMED, and pour the solution into the space in between the two glass plates below the BN gel strip. Leave space for the sample gel solution for embedding the BN gel strip. Overlay the separation gel with the overlay solution. The gel should polymerize in about 60 min (*see* **Note 5**).
6. Prepare a 10% sample gel solution by mixing 4.1 mL of acrylamide solution, 6.7 mL of gel buffer BN, 2 mL glycerol, 200 µL of SDS solution, 6.8 mL ddH_2O, 160 µL APS, and 16 µL TEMED.

7. Discard the overlay solution, and pour the sample gel embedding the BN gel lane. The casting stand should be held slightly diagonally to keep air bubbles from being captured underneath the BN gel lane (*see* **Note 6**).
8. Add the cathode and anode buffers SDS to the upper and lower chambers of the gel unit.
9. Connect the gel unit to a power supply. Carry out electrophoresis at 30 mA overnight.

As an alternative system, the second gel dimension also can be carried out under native conditions (*see* **Note 7**).

3.4. Staining Procedures

Visualization of proteins separated on 2D BN gels can be carried out using all standard protein staining procedures, e.g., Coomassie staining, silver staining, staining by fluorescence dyes, and so on. For subsequent protein identifications by mass spectrometry, Coomassie is mostly recommended. The following instructions are based on colloidal staining with Coomassie blue as developed by Neuhoff et al. *(26,27)*.

1. Incubate the gel in fixing solution for 1 h.
2. Supplement 80 mL of freshly prepared staining solution with 20 mL methanol.
3. Incubate the gel with this solution (overnight; *see* **Note 8**).
4. Destain the gel with ddH_2O. The water should be exchanged several times until the background is clear (20% methanol can be used if background staining is not sufficiently reduced by incubating with water).

4. Notes

1. Detergent concentrations for membrane protein solubilization should be optimized. Dodecylmaltoside and Triton X-100 concentrations are usually varied between 0.1 and 2.0% and digitonin concentrations between 1 and 10%.
2. If very large protein complexes (>3 MDa) have to be resolved, the acrylamide gradient gel of the BN gel dimension can be substituted by a 2.5% agarose gel prepared in gel buffer BN *(28)*.
3. BN gels can be blotted under native or denaturing conditions (presence of SDS). Short preblots should be carried out to destain gels electrophoretically from excess of Coomassie blue *(5)*. Alternatively, the cathode-buffer BN can be replaced by a cathode buffer BN without Coomassie blue after 50% completion of the electrophoretic run of the BN gel dimension. Protein complexes excised from 1D BN gels can also be electroeluted, precipitated, and analyzed by further gel dimensions, e.g., 2D IEF/SDS-PAGE. A corresponding 3D system (BN/IEF/SDS-PAGE) was found to be useful for the separation of isoforms of subunits of protein complexes in *Arabidopsis* *(29)*.
4. BN-PAGE is compatible with in-gel enzyme staining. For protocols, see, for instance **ref.** ***3***.

5. Concerning the Tricine-SDS-PAGE system for the second gel dimension, Schägger and von Jagow ***(25)*** originally proposed a two-step separation gel consisting of a large 16% phase and a smaller 10% phase (called a "spacer gel"). The advantage of this slightly more complicated gel system is better resolution of large protein subunits. The two gel solutions are poured one on the other. [Glycerol is added to the 16% gel solution to avoid mixing with the 10% gel solution; for details, see Schägger and von Jagow ***(25)***.]
6. The transfer of the strip of BN gel onto a second gel dimension can be carried out by embedding the gel lane by the sample gel of the second dimension. This procedure ensures optimal physical contact between the BN gel lane with the SDS gel of the second dimension. However, BN gel strips can also be fixed onto a prepoured SDS gel using agarose, as is usually carried out for 2D IEF/SDS-PAGE. Physical contact of the gels might not be optimal, but the time period between the end of the first and the start of the second electrophoresis is shortened.
7. BN-PAGE can also be combined with a second BN-PAGE ***(30)***. In this case, the two BN gel dimensions are usually carried out in the presence of different detergents. For instance, the first-dimension BN-PAGE can be carried out in the presence of digitonin and the second-dimension BN-PAGE in the presence of dodecylmaltoside. All protein complexes likewise stable in the presence of the two detergents are finally localized on a diagonal line on this gel system. In contrast, protein complexes specifically destabilized by the second detergent are dissected into subcomplexes of higher electrophoretic mobility. This 2D gel system nicely allows investigation of the substructures of protein complexes.
8. Extending staining duration for up to 3 d further increases the sensitivity of protein detection.

Acknowledgment

This work was supported by the Deutsche Forschungsgemeinschaft (grant Br 1829-7/1).

References

1. Schägger, H. and von Jagow, G. (1991) Blue native electrophoresis for isolation of membrane protein complexes in enzymatically active form. *Anal. Biochem.* **199,** 223–231.
2. Grandier-Vazeille, X. and Guerin, M. (1996) Separation by blue native and colorless native polyacrylamide gel electrophoresis of the oxidative phosphorylation complexes of yeast mitochondria solubilized by different detergents: specific staining of the different complexes. *Anal. Biochem.* **242,** 248–254.
3. Zerbetto, E., Vergani, L., and Dabbeni-Sala, F. (1997) Quantitation of muscle mitochondrial oxidative phosphorylation enzymes via histochemical staining of blue native polyacrylamide gels. *Electrophoresis* **18,** 2059–2064.
4. Fecht-Christoffers, M. M., Braun, H. P., Guillier, C., VanDorssealer, A., and Horst W. J. (2003) Effect of manganese toxicity on the proteome of the leaf apoplast in cowpea (*Vigna unguiculata*). *Plant Physiol.* **133,** 1935–1946.

5. Jänsch, L., Kruft, V., Schmitz, U. K., and Braun, H. P. (1996) New insights into the composition, molecular mass and stoichiometry of the protein complexes of plant mitochondria. *Plant J.* **9,** 357–368.
6. Eubel, H., Jänsch, L., and Braun, H. P. (2003) New insights into the respiratory chain of plant mitochondria: supercomplexes and a unique composition of complex II. *Plant Physiol.* **133,** 274–286.
7. Heazlewood, J. L., Howell, K. A., Whelan, J., and Millar, A. H. (2003) Towards an analysis of the rice mitochondrial proteome. *Plant Physiol.* **132,** 230–242.
8. Giegé, P., Sweetlove, L. J., and Leaver, C. (2003) Identification of mitochondrial protein complexes in *Arabidopsis* using two-dimensional blue-native polyacrylamide gel electrophoresis. *Plant Mol. Biol. Rep.* **21,** 133–144.
9. Sabar, M., Gagliardi, D., Balk. J., and Leaver, C. J. (2003) ORFB is a subunit of F(1)F(O)-ATP synthase: insight into the basis of cytoplasmic male sterility in sunflower. *EMBO Rep.* **4,** 1–6.
10. Dudkina, N. V., Eubel, H., Keegstra, W., Boekema, E. J., and Braun, H. P. (2005) Structure of a mitochondrial supercomplex formed by respiratory chain complexes I and III. *Proc. Natl. Acad. Sci. USA* **102,** 3225–3229.
11. Kügler, M., Jänsch, L., Kruft, V., Schmitz, U. K., and Braun, H. P. (1997) Analysis of the chloroplast protein complexes by blue-native polyacrylamide gelelectrophoresis. *Photosynthesis Res.* **53,** 35–44.
12. Rexroth, S., Meyer zu Tittingdorf, J. M., Krause, F., Dencher, N. A., and Seelert, H. (2003) Thylakoid membrane at altered metabolic state: challenging the forgotten realms of the proteome. *Electrophoresis* **24,** 2814–2823.
13. Ossenbühl, F., Gohre, V., Meurer, J., Krieger-Liszkay, A., Rochaix, J. D., and Eichacker, L. A. (2004) Efficient assembly of photosystem II in *Chlamydomonas reinhardtii* requires Alb3.1p, a homolog of *Arabidopsis* ALBINO3. *Plant Cell* **16,** 1790–1800.
14. Heinemeyer, J., Eubel, H., Wehmhöner, D., Jänsch, L., and Braun, H. P. (2004) Proteomic approach to characterize the supramolecular organization of photosystems in higher plants. *Phytochemistry* **65,** 1683–1692.
15. Ciambella, C., Roepstorff, P., Aro, E. M., and Zolla, L. (2005) A proteomic approach for investigation of photosynthetic apparatus in plants. *Proteomics* **5,** 746–757.
16. Kjell, J., Rasmusson, A. G., Larsson, H., and Widell, S. (2004) Protein complexes of the plant plasma membrane resolved by blue native PAGE. *Physiol. Plant* **121,** 546–555.
17. Rivas, S., Romeis, T., and Jones, J. D. (2002) The Cf-9 disease resistance protein is present in an approximately 420-kilodalton heteromultimeric membrane-associated complex at one molecule per complex. *Plant Cell* **14,** 689–702.
18. Piotrowski, M., Janowitz, T., and Kneifel, H. (2003) Plant C-N hydrolases and the identification of a plant N-carbamoylputrescine amidohydrolase involved in polyamine biosynthesis. *J. Biol. Chem.* **278,** 1708–1712.
19. Drykova, D., Cenklova, V., Sulimenko, V., Volc, J., Draber, P., and Binarova, P. (2003) Plant gamma-tubulin interacts with alphabeta-tubulin dimers and forms membrane-associated complexes. *Plant Cell* **15,** 465–480.

20. Salomon, M., Lempert, U., and Rüdiger, W. (2004) Dimerization of the plant photoreceptor phototropin is probably mediated by the LOV1 domain. *FEBS Lett.* **572,** 8–10.
21. Boldt, A., Fortunato, D., Conti, A., et al. (2005) Analysis of the composition of an immunoglobulin E reactive high molecular weight protein complex of peanut extract containing Ara h 1 and Ara h 3/4. *Proteomics* **5,** 675–686.
22. Werhahn, W. H., Niemeyer, A., Jänsch, L., Kruft, V., Schmitz, U. K., and Braun, H. P. (2001) Purification and characterization of the preprotein translocase of the outer mitochondrial membrane from *Arabidopsis thaliana*: identification of multiple forms of TOM20. *Plant Physiol.* **125,** 943–954.
23. Lowry, O. H., Rosebrough, N. J., Farr, A. L., and Randall, R. J. (1951) Protein measurement with the Folin phenol reagent. *J. Biol. Chem.* **193,** 265–275.
24. Laemmli, U. K. (1970) Cleavage of structural proteins during the assembly of the head of bacteriophage T4. *Nature* **227,** 680–685.
25. Schägger, H. and von Jagow, G. (1987) Tricine-sodium dodecyl sulfate-polyacrylamide gel electrophoresis for the separation of proteins in the range from 1 to 100 kDa. *Anal. Biochem.* **166,** 368–379.
26. Neuhoff, V., Stamm, R., and Eibl, H. (1985) Clear background and highly sensitive protein staining with Coomassie Blue dyes in polyacrylamide gels: a systematic analysis. *Electrophoresis* **6,** 427–448.
27. Neuhoff, V., Stamm, R., Pardowitz, I., Arold, N., Ehrhardt, W., and Taube, D. (1990) Essential problems in quantification of proteins following colloidal staining with Coomassie Brilliant Blue dyes in polyacrylamide gels, and their solution. *Electrophoresis* **11,** 101–117.
28. Henderson, N. S., Nijtmans, L. G., Lindsay, J. G., Lamantea, E., Zeviani, M., and Holt, I. J. (2000) Separation of intact pyruvate dehydrogenase complex using blue native agarose gel electrophoresis. *Electrophoresis* **21,** 2925–2931.
29. Werhahn, W. and Braun, H. P. (2002) Biochemical dissection of the mitochondrial proteome from *Arabidopsis thaliana* by three-dimensional gel electrophoresis. *Electrophoresis* **23,** 640–646.
30. Schägger, H. and Pfeiffer, K. (2000) Supercomplexes in the respiratory chain of yeasts and mammalian mitochondria. *EMBO J.* **19,** 1777–1783.

27

Electroelution of Intact Proteins from SDS-PAGE Gels and Their Subsequent MALDI-TOF MS Analysis

Zhentian Lei, Ajith Anand, Kirankumar S. Mysore, and Lloyd W. Sumner

Summary

A method has been developed for the extraction of intact proteins from SDS-PAGE gels and for subsequent MALDI-TOF MS analysis to determine precise molecular mass. The method consists of an electroelution step and a subsequent low-temperature matrix-analyte cocrystallization step prior to MALDI-TOF MS. Proteins were first extracted from polyacrylamide gels using electroelution (ProteoPLUS™). Following electroelution, salts, SDS, and dye were removed by dialysis using the same electroelution tubes. The proteins were then concentrated in a vacuum centrifuge. Using bovine serum albumin (BSA) and phosphorylase B (Phos B) as standards, it was found that electroelution was more effective than passive elution in extracting intact proteins from gels. Optimized protein recoveries were determined to be 89% for BSA and 58% for Phos B, supporting the hypothesis of an inverse relationship between recovery and protein size. The traditional practice of "fixing gels" using methanol and acetic acid was found to reduce protein recoveries significantly and should be avoided if possible. A subsequent sample preparation method involving the low-temperature cocrystallization of whole proteins with MALDI matrices (trans-3,5-dimethoxy-4-hydroxycinnamic acid, sinapinic acid, 10 mg/mL) was also developed to remove residual impurities further. Proteins and MALDI matrices were mixed, sealed, and stored at 4°C to allow for overnight cocrystallization of matrix and protein. The supernatant was removed and protein-matrix crystals redissolved using the same matrix solution. MALDI-TOF MS signal intensities were found to increase three- to fourfold using the low-temperature cocrystallization method relative to the more common method of immediate analysis following matrix-analyte mixing and "dried droplet" deposition. The reported method is illustrated using two β-1,3-endoglucanase isoforms from wheat apoplast extracts.

Key Words: Intact protein; recovery; SDS-PAGE, MALDI-TOF MS; electroelution; proteomics; cocrystallization.

From: *Methods in Molecular Biology, vol. 335: Plant Proteomics: Methods and Protocols*
Edited by: H. Thiellement, M. Zivy, C. Damerval, and V. Méchin © Humana Press Inc., Totowa, NJ

1. Introduction

Matrix-assisted laser desorption ionization time-of-flight mass spectrometry (MALDI-TOF MS) has been widely employed in the biological sciences. Coupled with sodium dodecyl sulphate-polyacrylamide gel electrophoresis (SDS-PAGE) and enzymatic digestion, it provides a rapid, sensitive, and accurate means for the identification of proteins through peptide mass fingerprinting (PMF) *(1,2)*. PMF typically yields sequence coverages of 10 to 40%, which is sufficient for confident protein identification; however, it may not be sufficient to differentiate highly similar protein isoforms. Protein isoforms/isozymes are widespread and originate through alternative splicing of mRNA or multiple gene families. Although they catalyze the same reactions in the cell, they are often differentially expressed either in various tissues or under different physiological conditions. Differentiation of protein isoforms by PMF can be challenging, as unique peptides that distinguish individual isoforms may not be observed in the 10 to 40% coverage. The differentiation of isoforms through PMF is further complicated by the lack of complete genomic, mRNA, and/or protein sequences for a large number of species of interest.

Accurate molecular weight determinations of intact proteins can provide additional information capable of differentiating isoforms. Mass determination of intact proteins using SDS-PAGE is characterized by its low accuracy (approximately 10%) and further complicated by the structural differences between analyte and protein molecular weight marker proteins, as well as post-translational modifications that could lead to anomalous migration behaviors of modified proteins during gel electrophoresis. However, modern MS offers high mass accuracy (0.1% or better) and is the method of choice for accurate measurement of intact protein masses.

Accurate mass measurement of proteins separated by SDS-PAGE depends, to a great extent, on the extraction of proteins from SDS-PAGE gels. Recovering intact proteins from gels is, however, more difficult than retrieving peptides resulting from in-gel enzymatic digestion. Several methods have been reported for the extraction of intact proteins from gels. They can be broadly classified into three categories: passive elution *(3–5)*, electroelution *(6–8)*, and electrotransfer (or electroblotting) onto membranes *(9–12)*.

Cohen and co-workers *(3)* developed a passive elution technique that utilized non-fixative negative stains (Cu or Zn staining), and organic solvents to extract proteins. After destaining, bands were excised from the gel, crushed, and vigorously agitated for several hours in an extraction solution (method A, formic acid/water/ isopropanol, F/W/I; 3:2:1 [v/v/v]) or MALDI matrix solution (method B, such as 4-hydroxy-α-cyano-cinnamic acid solution [4HCCA]). Using this approach, they were able to extract proteins with molecular weights up to 70 kDa. The key to the success of this approach was the use of nonfixative negative metal staining, which avoided in-gel protein precipitation and hence better recoveries. Unfortunately, negative staining is relatively expensive. Coomassie Blue is a lower cost protein stain that is widely utilized in many laboratories. Thus, a method to recover intact proteins from Coomassie Blue-stained gels would be beneficial.

Ehring and co-workers ***(4)*** described a passive elution method for protein extraction from Coomassie-Blue stained gels. Instead of using formic acid/water/isopropanol (F/W/I), they used formic acid/acetonitrile/isopropanol/H_2O (F/A/I/W; 50/25/15/10 [v/v/v/v]). The addition of acetonitrile was believed to enhance the elution of relatively hydrophobic proteins. Although easy to perform, this method only recovered proteins with molecular masses less than 30 kDa from Coomassie Blue-stained gels. Moreover, formylation of serine and threonine residues has been observed following exposure to formic acid for extended periods ***(4)***. To circumvent this problem, Claverol and co-workers ***(5)*** reported the use of SDS/sodium acetate buffer to extract proteins from Coomassie Blue-stained gels. The protein bands were incubated with the extraction buffer overnight at 37°C to allow proteins to diffuse into the buffer. Coomassie Blue, SDS, and salts were later removed with hydrophilic Ziptips ($\text{Ziptip}_{\text{HPL}}$, Millipore, Bedford, MA). This method reportedly recovered proteins with molecular masses up to 45 kDa. Retrieving larger proteins using this method, however, is still very challenging.

Electrotransfer (or electroblotting) of proteins from acrylamide gels onto membranes such as nitrocellulose (NC) or polymer membranes is an alternative method for intact protein recovery. Most membranes are compatible with MALDI-TOFMS, allowing direct analyses of intact proteins after electroblotting. Systematic investigations ***(9,12)*** of several commercially available membranes revealed that polyvinylidene fluoride (PVDF) and polypropylene (PP) membranes yielded the best results. However, poor reproducibility of spectra, low signal-to-noise (S/N) ratio, and poor resolution were observed with UV-MALDI-TOF MS. The poor S/N ratio was thought to be related to the short penetration depth (normally approx 200–300 nm) of the UV laser (355 nm) typically used in MALDI-TOFMS. It was also believed that proteins were strongly adsorbed onto the membranes, making desorption less efficient. MALDI-TOF MS using an IR laser at 2.94 µm proved to be more effective. The penetration depth of the IR laser is in the range of 3 to 5 µm, thus desorbing a greater number of proteins from the membranes, which have a thickness of 135 µm.

However, it was still noted that "considerable improvement in the quality of the IR-MALDI-TOFMS spectra was required, especially with respect to the observed protein signal peak width limiting the achievable mass determination accuracy" ***(12)***. Methods that directly dissolve membrane-blotted proteins in matrix solution have also been reported ***(10,11)***. Typically, matrix solutions were prepared with acetone, which also dissolves NC membrane. The matrix-protein solutions were then spotted onto the sample plate for MALDI-TOF MS analyses, resulting in better reliability and reproducibility. However, reduced mass resolution and mass accuracy were observed relative to methods involving elution and purification of proteins ***(11)***. It was speculated that the presence of NC on the target or the interaction of proteins with NC was responsible for the reduced mass accuracy.

Electroelution using a differential voltage potential provides a rapid and convenient means of extracting proteins from 1D or 2D polyacrylamide gels. In essence, electroelution is the continuation of electrophoresis resulting in the migration of proteins out of gel slices into solution. Salts, dye, and SDS are then removed by dialysis from the electroeluted solution, and the purified protein solution is concentrated. Owing to the applied voltage differential, proteins actively migrate rather than passively diffuse out of the gels. It is therefore more effective, especially for the elution of larger proteins. Several variants of electroelution have been reported. Clarke et al. ***(6)*** reported a one-step microelectroelution method, and Buzas et al. ***(13)*** developed a direct vertical electroelution approach that did not require protein band excision from gels.

These methods yielded good recoveries of proteins (up to 90%) but required specialty apparatuses not readily available. Galvani and co-workers ***(14)*** described the use of a commercially available electroeluter, the Centrilutor (Amicon, Beverly, MA), to extract proteins. The proteins were first eluted into a Centricon spin-filter. Salts, dye, and SDS were removed and proteins concentrated through centrifugation. Comparison of this method with ultrasonic assisted passive elution revealed that electroelution could recover high molecular weight proteins such as bovine serum albumin (BSA) and phosphorylase B, whereas passive elution failed to extract sufficient amounts of these proteins for MALDI-TOF MS. The authors, however, noted that passive elution generally yielded mass spectral data of better quality for the small proteins tested relative to electroelution. This could be owing to the incomplete removal of SDS, dye, and salts present in the electrophoresis buffer used in electroelution. Indeed, in our work, an intense blue color was observed for concentrated protein samples that had previously been dialyzed overnight, supporting the possibility of incomplete removal of these components.

This report utilizes another commercially available electroelution (ProteoPLUS™) apparatus to recover proteins from gels and a subsequent low-temperature protein/matrix cocrystallization step that further removes residual impurities for the enhancement of MALDI-TOF MS signals. Proteins were first extracted from gels, dialyzed overnight using the same electroelution tubes, and then concentrated in a vacuum centrifuge (SpeedVac). Using phos-

Table 1
Comparison of Protein Recovery[a]

Protein	Eletroelution (%)		Passive elution (%)	
	Fixed	Nonfixed	Fixed	Nonfixed
Phos B	6	58	n.d	n.d.
BSA	66	89	n.d	39

[a]SDS-PAGE proteins were fixed with methanol and acetic acid. Nonfixed gels did not use a fixative solution. n.d., not detected (by means of Coomassie Blue stain). Gel loading level: 5 µg. Data show that enhanced recovery is achieved without fixation and with electroelution. Also note that protein recovery is inversely proportional to protein mass, i.e., recovery decreases as protein mass increases.

phorylase B (Phos B) and BSA as standards, it was found that this method was more effective than passive elution, yielding a recovery of 89% for BSA and 58% for Phos B at a 5 µg gel loading level. Passive elution (overnight incubation of crushed protein bands with 1X running buffer) failed to recover Phos B and only extracted 39% of the loaded BSA (**Table 1**). The concentrated proteins were cocrystallized with matrix prior to MALDI-TOF MS analyses for further removal of residual impurities. Protein solutions and matrix solution (sinapinic acid, 10 mg/mL in acetonitrile/water; 3:7) were mixed and placed at 4°C overnight. The lower solubility of the matrix at lower temperature leads to the co-crystallization of proteins and matrix. Supernatant containing salts and dye was then removed and the crystals recovered using the same matrix solution.

The method was applied to the analysis of two β-1,3-endoglucanase isoforms previously purified from a transgenic wheat aploplast extract ***(15)***. A significant increase in MALDI-TOF MS signal was achieved after cocrystallization as shown in **Fig. 1**. For example, MALDI-TOF MS signal intensity increases threefold for the β-1,3-endoglucanase isoform 1(G1) and seven fold for the β-1,3-endoglucanase isoform 2 (G2). This demonstrates that the crystallization of proteins and matrix at lower temperature (4°C) is effective in enhancing MALDI-TOFMS signals through the removal of residual salts, detergents, and Coomassie Blue dye.

2. Materials

1. 1X running buffer: 25 m*M* Tris-HCl, 192 m*M* glycine, 0.1% (w/v) SDS, pH 8.3. (diluted from 10X TGS [Tris/glycine/SDS] running buffer; Bio-Rad, cat. no. 161-0772).
2. Horizontal electrophoresis gel apparatus (Horizon 58, Life Technologies, NY).

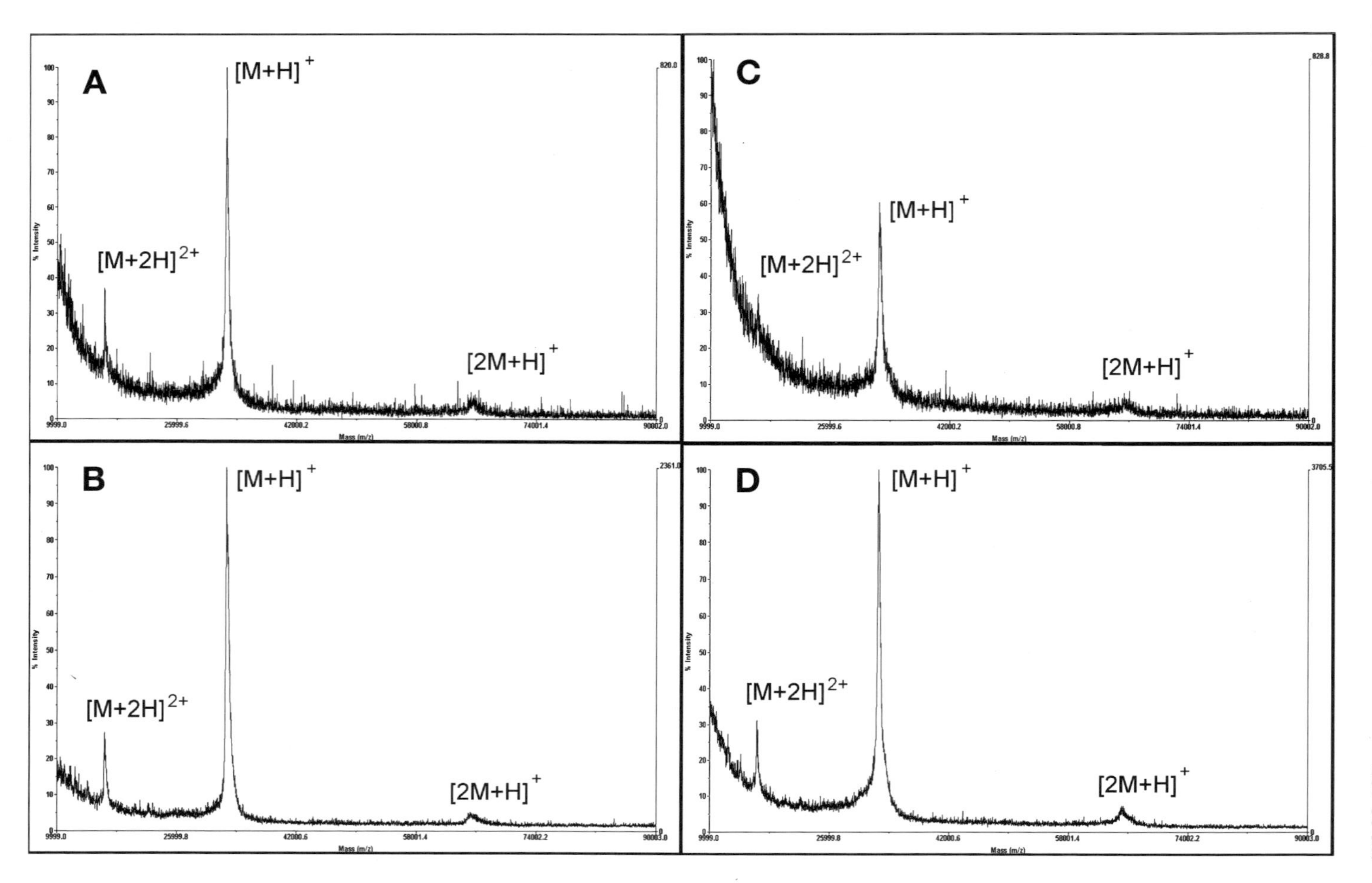
A
[M+H]+
[M+2H]2+
[2M+H]+
C
[M+H]+
[M+2H]2+
[2M+H]+
B
[M+H]+
[M+2H]2+
[2M+H]+
D
[M+H]+
[M+2H]2+
[2M+H]+

3. Electroelution kits (ProteoPLUS, cat. no. Prot0015, Qbiogene, Irvine, CA).
4. Protein staining solutions (ElectroBlue™, cat. no. Probs500, Qbiogene).
5. Vacuum centrifuge (SpeedVac).
6. 1.5-mL Microcentrifuge tubes.
7. 150-μL Microcentrifuge tubes.
8. Sinapinic acid (trans-3,5-dimethoxy-4-hydroxycinnamic acid) solution: 10 mg/mL in 69% water, 30% acetronitrile, and 1% formic acid.
9. Milli-Q water or equivalent high-purity distilled/dionized water.

3. Methods

3.1. Protein Detection

In general, gel staining and destaining are very straightforward. In this protocol, we use ElectroBlue dye rather than fixative Coomassie Blue stain solution to visualize proteins (*see* **Note 1**).

1. After SDS-PAGE, rinse gels (8 × 10 cm) with 300 mL of Milli-Q or distilled water three times for 10 min each (*see* **Note 2**).
2. Stain gels with ElectroBlue dye by incubating gels (8 × 10 cm) with approx 20 mL of stain solution and gently shaking for 2 h (*see* **Note 3**).
3. Destain gels with 300 mL of Milli-Q or distilled water until the gels reach the desired background. Change water every 30 min.

3.2. Electroelution

1. Set the electroelution tubes upright, fill with 0.8 mL of Milli-Q or distilled water, and close the caps. Wait 10 min and check for any leaks. Empty the tubes (*see* **Note 4**).
2. Use a clean, sharp, straight-edged razor to excise protein bands from the stained gels. Transfer the gel slices into electroelution tubes using flat-blade tweezers. Fill the tubes with 1X running buffer (0.8 mL) and gently close the cap. Carefully place and position the tubes in the electroelution support tray so the electric current can pass directly through the two membrane windows (*see* **Note 5**).
3. Fill the electrophoresis tank with 1X running buffer and place the electroelution support tray into the tank. Make sure that the tubes are fully immersed in the running buffer and the membranes of each tube are perpendicular to the electric field.
4. Electrophoresis at 120 V for 2 h (*see* **Note 6**).
5. After electrophoresis, reverse the polarity and electrophoresis for 30 to 60 s at the same voltage (*see* **Note 7**).

Fig. 1. MALDI-TOFMS spectra of two β-1,3-endoglucanase isoforms eluted from the SDS-PAGE *(15)*. **(A)** Spectrum of β-1,3-endoglucanase isoforms 1 (G1, MW: 32,620) before cocrystallization with matrix. **(B)** G1 spectrum after cocrystallization. **(C)** Spectrum of β-1,3-endoglucanase isoforms 2 (G2, MW: 32,541) before cocrystallization with matrix. **(D)** G2 after cocrystallization with matrix. These two proteins were previously purified from a wheat apoplast extract.

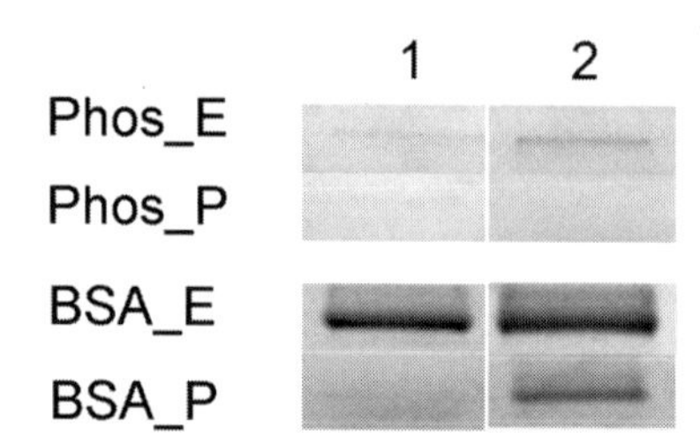

Fig. 2. SDS-PAGE analysis of Phos B and BSA recovered from 10 to 20% gels fixed with methanol/acetic acid (lane 1) and nonfixed (2) at a 5 μg gel loading level. Phos_E/BSA_E, proteins recovered using electroelution; Phos_P/BSA_P, proteins recovered using passive elution.

6. Remove gel slices from the electroelution tubes with tweezers. Gently close the cap and put the tubes back in the supporting tray.
7. Place the tray into a 2-L beaker containing desirable buffer or acidified water (1% formic acid). Dialyze overnight at 4°C. Change buffer once.
8. After dialysis, transfer protein solutions to 1.5-mL microcentrifuge tubes and concentrate in a vacuum centrifuge (SpeedVac) to about 5 μL (*see* **Note 8**).
9. Mix the concentrated protein solutions (1 μL) with matrix solution (1 μL, sinapinic acid, 10 mg/mL in 30% acetonitrile) in 150-μL microcentrifuge tubes. Cap tightly and place at 4°C overnight to allow for cocrystallization of proteins and matrix (*see* **Note 9**).
10. Gently open the lid of the microcentrifuge tubes and remove the supernatant using a 2-μL pipet. Add 1 μL of matrix solution to the tubes to resuspend the crystals, vortex, and deposit onto the MALDI sample plate (*see* **Note 10**).

4. Notes

1. Fixing proteins or staining gels in a fixative solution (mainly a mixture of methanol and acetic acid) is a common practice in SDS-PAGE analysis of proteins. The acetic acid and methanol mixture causes almost all proteins to precipitate, thus preventing protein loss through diffusion during the staining process. This, however, can significantly reduce the protein yield (**Table 1** and **Fig. 2**) during electroelution. For BSA, a 30% decrease in yield was observed in our work. The loss owing to fixing seems even more severe for larger proteins. For Phos B, the protein recovery dropped by 89% when the gels were fixed. Thus, avoiding "fixing" gels is beneficial for intact protein recovery from SDS-PAGE gels. Nonfixative Coomassie Blue stain should be used if possible. Other colloidal Coomassie-based staining solutions are commercially available and offer good sensitivity with little or no fixation of proteins. These include the Electro-Blue staining solution from Qbiogen (cat. no. PROBS500) and the PageBlue™ staining solution from Fermentas (cat. no. R0571). PageBlue does not contain fixing reagents like methanol or acetic acid, which is ideal for the recovery of protein

from gels. Electro-Blue staining solution slightly fixes proteins in the gel and is also suitable for the elution of intact proteins from gels.

2. It is necessary to rinse the gels thoroughly, as SDS interferes with the staining process.
3. Make sure that the gels are immersed completely in the staining solution. Adjust the staining solution volume for gels of different sizes. Gels can be stained overnight without increased background.
4. During the whole experiment, great care should be taken not to touch the electroelution membrane with hands, pipets, or tweezers.
5. When you are excising protein bands from gels, great care should be taken not to break the bands into several pieces. It is essential that one electroelution tube contain only one gel slice. Putting more than one gel slice into one electroelution tube may lower recovery, as protein eluting from one gel slice may enter another. If one band is broken into several pieces, use a sample tube for each piece. Alternatively, use 4% agrose (400 mg of agarose dissolved in 10 mL of 1X running buffer) to seal the gel pieces together.
6. The elution time depends on the protein size, the gel slice size, the applied voltage, the gel polyacrylamide percentage, and the fixing process. One needs to determine the optimal elution time for individual proteins. For reference, the minimum time needed to extract different-sized proteins from a 10% SDS-polyacrylamide gel (polyacrylamide/bisacrylamide; 29:1) at 100 V is given below (adapted from Qbiogen).

Protein (kD)	Time (min)
14	35–45
22	45–55
50	75–85
66	85–95
29	55–65
40	60–70
45	65–75
81	105–115
116	120–130
128	140–150

7. After electroelution, it is highly recommended that a reversed-polarity electrophoresis be conducted for 30 to 60 s. This releases proteins that adhere to the tube membranes during electroelution/electrophoresis back into the running buffer. It usually increases protein yield, especially when the protein loading is relatively low.
8. Do not dry the samples to complete dryness as it may be difficult to resolubilize the dried proteins completely. Check the samples frequently, especially when the samples are about to dry.
9. Matrix solution is prepared by adding 10 mg of sinapinic acid to a 1.5-mL microcentrifuge tube containing 1 mL of a mixture of acetonitrile and water (CAN/

H_2O, 3:7 [v/v]). Sonicate in a water bath for 10 min and heat at 50°C for 30 min. Add the matrix solution (1 μL) to the 150-μL microcentrifuge tube containing 1 μL of the recovered protein. After mixing the protein solution with the matrix solution, use the same pipet tip to scratch the inner wall and bottom of the microcentrifuge tube several times. The scratches will promote the growth of crystals.
10. Handle the tube gently after overnight cocrystallization of protein and matrix. Use a 2-μL pipet tip to remove the supernatant carefully while it is still cold. Avoid shaking, dropping, or touching the inner wall with the tip.

Acknowledgments

The authors thank Dr. Subbaratnam Muthukrishnan, Department of Biochemistry, Kansas State University for providing the apoplast extracts. This work was supported by The Samuel Roberts Noble Foundation.

References

1. Aebersold, R. and Mann, M. (2003) Mass spectrometry-based proteomics. *Nature* **422,** 198–207.
2. Newton, R. P., Brenton, A. G., Smith, C. J., and Dudley, E. (2004) Plant proteome analysis by mass spectrometry: principles, problems, pitfalls and recent developments. *Phytochemistry* **65,** 1449–1485.
3. Cohen, S. L. and Chait, B. T. (1997) Mass spectrometry of whole proteins eluted from sodium dodecyl sulfate-polyacrylamide gel electrophoresis gels. *Anal. Biochem.* **247,** 257–267.
4. Ehring, H., Stromberg, S., Tjernberg, A., and Noren, B. (1997) Matrix-assisted laser desorption/ionization mass spectrometry of proteins extracted directly from sodium dodecyl sulphate-polyacrylamide gels. *Rapid Commun. Mass Spectrom.* **11,** 1867–1873.
5. Claverol, S., Burlet-Schiltz, O., Gairin, J. E., and Monsarrat, B. (2003) Characterization of protein variants and post-translational modifications: ESI-MSn analyses of intact proteins eluted from polyacrylamide gels. *Mol. Cell. Proteomics* **2,** 483–493.
6. Clarke, N. J., Li, F., Tomlinson, A. J., and Naylor, S. (1998) One step microelectroelution concentration method for efficient coupling of sodium dodecylsulfate gel electrophoresis and matrix-assisted laser desorption time-of-flight mass spectrometry for protein analysis. *J. Am. Soc. Mass Spectrom.* **9,** 88–91.
7. Jacobs, E. and Clad, A. (1986) Electroelution of fixed and stained membrane proteins from preparative sodium dodecyl sulfate-polyacrylamide gels into a membrane trap. *Anal. Biochem.* **154,** 583–589.
8. Schuhmacher, M., Glocker, M. O., Wunderlin, M., and Przybylski, M. (1996) Direct isolation of proteins from sodium dodecyl sulfate-polyacrylamide gel electrophoresis and analysis by electrospray-ionization mass spectrometry. *Electrophoresis* **17,** 848–854.

9. Schleuder, D., Hillenkamp, F., and Strupat, K. (1999) IR-MALDI-mass analysis of electroblotted proteins directly from the membrane: comparison of different membranes, application to on-membrane digestion, and protein identification by database searching. *Anal. Chem.* **71,** 3238–3247.
10. Liang, X., Bai, J., Liu, Y. H., and Lubman, D. M. (1996) Characterization of SDS–PAGE-separated proteins by matrix-assisted laser desorption/ionization mass spectrometry. *Anal. Chem.* **68,** 1012–1018.
11. Dukan, S., Turlin, E., Biville, F., et al. (1998) Coupling 2D SDS-PAGE with CNBr cleavage and MALDI-TOFMS: a strategy applied to the identification of proteins induced by a hypochlorous acid stress in *Escherichia coli*. *Anal. Chem.* **70,** 4433–4440.
12. Strupat, K., Karas, M., Hillenkamp, F., Eckerskorn, C., and Lottspeich, F. (1994) Matrix-assisted laser desorption ionization mass spectrometry of proteins electroblotted after polyacrylamide gel electrophoresis. *Anal. Chem.* **66,** 464–470.
13. Buzas, Z., Chang, H. T., Vieira, N. E., Yergey, A. L., Stastna, M., and Chrambach, A. (2001) Direct vertical electroelution of protein from a PhastSystem band for mass spectrometric identification at the level of a few picomoles. *Proteomics* **1,** 691–698.
14. Galvani, M., Bordini, E., Piubelli, C., and Hamdan, M. (2000) Effect of experimental conditions on the analysis of sodium dodecyl sulphate polyacrylamide gel electrophoresis separated proteins by matrix-assisted laser desorption/ionisation mass spectrometry. *Rapid Commun. Mass Spectrom.* **14,** 18–25.
15. Anand, A., Lei, Z., Sumner, L. W., et al. (2004) Apoplastic extracts from a transgenic wheat line exhibiting lesion-mimic phenotype have multiple pathogenesis-related proteins that are antifungal. *Mol. Plant Microbe. Interact.* **17,** 1306–1317.

28

Generation of Plant Protein Microarrays and Investigation of Antigen–Antibody Interactions

Birgit Kersten and Tanja Feilner

Summary

The application of proteomics methods, such as the protein microarray technology, in plant science has been strongly supported by the completion of genome sequencing projects of *Arabidopsis thaliana* and rice. In this chapter we describe a method to generate plant protein microarrays and to use them for characterizing monoclonal antibodies or polyclonal sera with regard to their specificity and cross-reactivity. The method starts with characterized *E. coli* cDNA expression clones encoding His-tagged plant proteins. After expression and purification of these recombinant proteins in high throughput, protein microarrays are generated utilizing a contact printer. For the detection of the recombinant proteins on the microarrays, an anti-RGS-His_6 antibody is used. To characterize specific antibodies, the microarrays are incubated with the respective antibody solutions followed by fluorescently labeled secondary antibodies. Signal detection is performed by means of an arrayscanner system. Protein microarrays containing the whole proteome of a plant will represent the ideal format to test antibody specificity and cross-reactivity in the future.

Key Words: Protein microarray; antibody and serum profiling; interaction studies; *Arabidopsis thaliana*; high throughput; proteomics.

1. Introduction

Protein microarrays consist of hundreds or even thousands of addressable protein samples, which are arranged in a systematic order in high density on coated glass slides ***(1–4)***. They allow parallel, fast, and easy analysis of proteins for expression ***(5–7)*** and modification ***(8–11)*** as well as for their molecular interactions with antibodies ***(12–15)***, other proteins ***(16,17)***, DNA ***(18,19)***, or small molecules ***(20,21)***. Using protein antigen microarrays, several studies have been performed to characterize specific antibodies ***(12–15)*** or to screen sera from patients suffering from diverse diseases ***(15,22 ,23)***. As an ideal for-

From: *Methods in Molecular Biology, vol. 335: Plant Proteomics: Methods and Protocols*
Edited by: H. Thiellement, M. Zivy, C. Damerval, and V. Méchin © Humana Press Inc., Totowa, NJ

mat to characterize a specific antibody, proteome arrays are envisaged ***(13)***. To analyze several antibodies on a single microarray, different multiplexing approaches may be applied ***(24,25)***. Compared with Western blotting, the application of protein microarrays for antibody characterization has several advantages ***(13)*** including a higher sensitivity of protein detection ***(13,23)***.

Proteomics methods, e.g., 2D electrophoresis/mass spectrometry (2-DE/MS) methods or protein microarray technology, are finding increasing applications in the plant field ***(26–32)***. Here we describe a method to generate plant protein microarrays and to utilize them for the investigation of antigen-antibody interactions. We successfully applied the method to generate the first plant protein microarrays containing 96 *Arabidopsis* proteins, which were detectable with a limit of approx 2 to 4 fmol per spot with an anti-RGS-His_6 antibody on FAST™-slides ***(14)***. Utilizing these arrays, we were able to show that a monoclonal anti-TCP1 antibody and polyclonal anti-MYB6 and anti-DOF11 sera recognized their respective antigens on the microarrays and did not cross-react with the other immobilized proteins including other DOF and MYB transcription factors.

For this method, RGS-His_6-tagged plant proteins are expressed and purified in high throughput using characterized cDNA expression clones. Purified proteins are robotically arrayed onto FAST-slides. The generated protein microarrays are then incubated with the respective antibodies, followed by fluorescently labeled secondary antibodies. Signal detection and evaluation allows one to identify proteins interacting with the tested antibody.

2. Materials

2.1. High-Throughput Protein Expression and Purification

1. Media for culturing cells: 2 YT or LB medium containing 100 μg/mL ampicillin, 15 μg/mL kanamycin, and 2% glucose.
2. 96-Well microtiter plates (Greiner bio-one, Frickenhausen, Germany).
3. 384-Well microtiter plates (Genetix, Christchurch, UK).
4. 96-Deep-well plates: 96-well microtiter plates with 2-mL cavities (Qiagen, Hilden, Germany).
5. Replicator with 96 pins (Nunc, Wiesbaden, Germany).
6. 4X SB medium: 48 g/L bacto-tryptone, 96 g/L yeast extract, 0.8% (v/v) glycerol, sterilized by autoclaving at 120°C for 20 min.
7. 20X PP buffer (potassium phosphate buffer): 17 m*M* KH_2PO_4, 72 m*M* K_2HPO_4, sterilized by filtration through a 0.2-μm pore size filter.
8. Thiamin.
9. Isopropyl-β-D-thiogalactopyranosid (IPTG; MBI Fermentas, St. Leon-Rot, Germany).
10. $MultiScreen_{HTS}$ Vacuum Manifold (Millipore, Eschborn, Germany).
11. NiNTA-agarose (NTA: nickel-nitrilotriacetic acid; Qiagen).

12. Denaturing lysis buffer: 100 m*M* NaH_2PO_4, 10 m*M* Tris-HCl, 6 *M* GuHCl, pH 8.0.
13. Washing buffer: 100 m*M* NaH_2PO_4, 10 m*M* Tris-HCl, 8 *M* urea, pH 6.3.
14. Elution buffer: 100 m*M* NaH_2PO_4, 10 m*M* Tris-HCl, 8 *M* urea, pH 4.5.

For items 12 to **14**, *see* **Note 1**.

15. 96-Well filter plate with a non-protein-binding 0.65-µm pore size PVDF membrane (Millipore Multiscreen MADVN 6550).
16. Bradford reagent (Bio-Rad, Munich, Germany).

2.2. Generation of Plant Protein Microarrays

1. 384-Well microtiter plates (Genetix).
2. FAST-slides (Whatman Schleicher & Schuell, Dassel, Germany).
3. QArray system (Genetix, New Milton, UK).
4. Mouse, rat, and rabbit IgG antibodies (Santa Cruz, Biotechnology, Santa Cruz, CA).

2.3. Antibody Screening on Protein Microarrays

1. Arrayscanner: 428 Arrayscanner System (Affimetrix, Palo Alto, CA) or ScanArray 4000 (PerkinElmer Life Science, Cologne, Germany).
2. Cover slide (Carl Roth, Karlsruhe, Germany).
3. GenePixPro4.0 software.

2.3.1. Monoclonal Antibody Screening on FAST-Slides

1. TBS + Tween-20 (TBST): 10 m*M* Tris-HCl, pH 7.5, 150 m*M* NaCl, 0,1% (v/v) Tween-20.
2. Blocking solution: 2% bovine serum albumin (BSA; Sigma, St. Louis, MO) in TBST.
3. Mouse anti-RGS-His_6 antibody (Qiagen).
4. Specific monoclonal antibodies, e.g., rat anti-TCP1 antibody (Affinity Bioreagents, Cholden, USA).
5. Respective Cy3-labeled conjugates as secondary antibodies, e.g., rabbit antimouse IgG for anti-RGS-His_6 or rabbit antirat IgG for anti-TCP1 (Dianova, Hamburg, Germany).

2.3.2. Polyclonal Antibody Screening on FAST-Slides

1. TBS + Tween-20 (TBST): 10 m*M* Tris-HCl, pH 7.5, 150 m*M* NaCl, 0,1% (v/v) Tween-20.
2. Fish gelatine (gelatine from cold water fish skin; Sigma).
3. Polyclonal rabbit sera, e.g., anti-DOF11 or anti-MYB6 sera (Pineda, Berlin, Germany).
4. Goat antirabbit IgG-Cy3 conjugate (Dianova, Hamburg, Germany).

3. Methods

3.1. High-Throughput Protein Expression and Purification

For the production of recombinant plant proteins, we use characterized *E. coli* cDNA expression clones (in *E. coli* SCS1/pSE111; ***35***) constructed in pQE30-derived expression vectors, which allow for expression of recombinant plant proteins with an N-terminal RGS-His_6-tag. A detailed description of the generation of such expression clones is beyond the scope of this chapter. Therefore the authors provide only a short summary. Characterized plant expression clones in the described format may be obtained from the following ordered cDNA expression libraries: barley expression library ***(8)*** constructed in the *E. coli* expression vector pQE30NST (accession number: AF074376) and *Arabidopsis* library ***(33)*** in vector pQE30NASTattB (accession number: AY386205). These libraries have been constructed according to the protocol of Konrad Buessow's group using deoxythymidine oligonucleotides for priming to produce recombinant proteins with their complete N-terminus ***(34–36)***. As another source of characterized plant protein expression clones, we used full-length *Arabidopsis* clones generated by directional cloning of open reading frames (ORFs) via restriction-dependent cloning using gene-specific primers ***(14)*** or via recombination dependent cloning ***(33)*** utilizing the GATEWAY™ cloning technology ***(37,38)***.

To express and purify plant proteins from these resources in high throughput (96-well format) and small scale we apply the following protocols:

For expression:

1. 96-Deep-well plates with 2-mL cavities were filled with 100 µL 2YT medium supplemented with 2% glucose, 100 µg/mL ampicillin, and 15 µg/mL kanamycin.
2. Cultures of recombinant clones were inoculated from 384-well or from 96-well microtiter plates that had been stored at –80°C. For inoculation, replicators carrying 96 pins were used.
3. After 16 h of growth at 37°C with vigorous shaking (320 rpm), 900 µL of prewarmed medium (1X SB medium, 1X PP buffer supplemented with 100 µg/mL ampicillin, 15 µg/mL kanamycin, 20 µg/mL thiamin) were added, and the incubation was continued for 2 h.
4. To induce protein expression, IPTG was added to a final concentration of 1 m*M*, and incubation was continued for another 4 to 5 h.
5. Cells were harvested by centrifugation at 1500*g* for 10 min at 4°C, and the pellets were stored at –80°C.
6. Aliquots of the lysates of harvested cells can be used to control the efficiency of the expression on a 15% polyacrylamide gel (**Fig. 1**).

For protein purification:

1. First, 150 µL of denaturing lysis buffer were added to the thawed pellets. Pellets were resolved by extensive vortexing and afterward incubated for 30 min at room temperature on a shaker (approx 650 rpm) (*see* **Note 2**).
2. The lysated cells were centrifuged at 1900*g*. The supernatant was transferred to a 96-well filter plate (*see* **Note 3**) and immediately drawn into a fresh filter plate by using a vacuum manifold (*see* **Notes 4** and **5**).

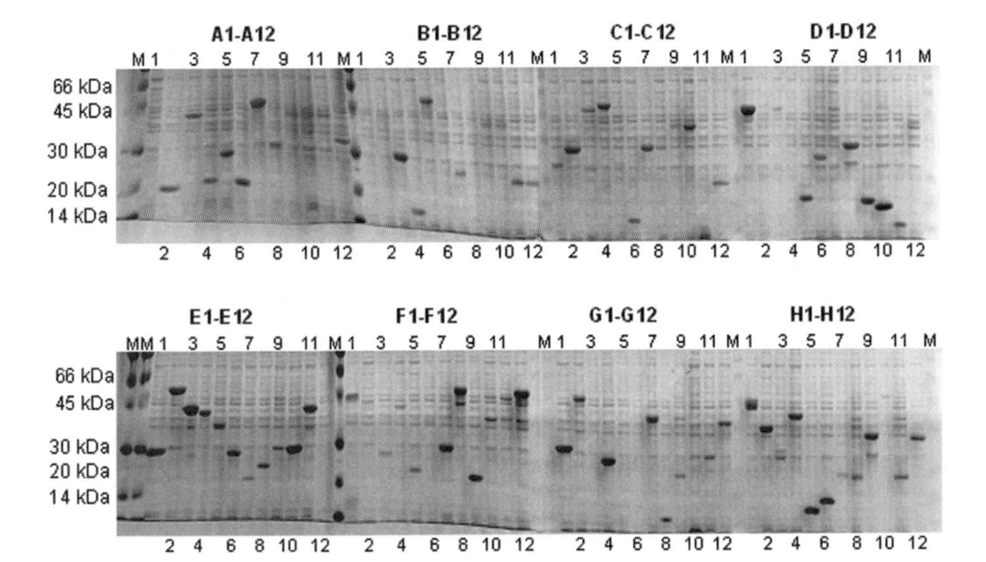

Fig. 1. SDS-Page (15%) of the lysates of 96 *Arabidopsis* expression clones after induction of expression. Expression clones were obtained from an *Arabidopsis* cDNA expression library ***(33)***. Lysates were separated using a 15% SDS-PAGE and then Coomassie stained. A1–H12, coordinates of the 96-well plate. M, marker. Approximate sizes of the protein markers are shown at the left in kDa.

3. Afterwards 30 μL NiNTA-agarose (1:2 diluted in lysis buffer) were added to each well, the plate was sealed with tape, and His-tagged proteins were bound by shaking for 1 h at 300 rpm at room temperature.
4. The agarose beads were washed three times by resuspending in 100 μL washing buffer, shaking for 5 min, and removing liquid on the vacuum manifold (*see* **Note 5**).
5. Finally, proteins were eluted from the agarose beads by incubation with 80 μL elution buffer without shaking for 20 min and subsequent filtration into a fresh 96-well microtiter plate (*see* **Note 6**).
6. Proteins were stored at 4°C.
7. Purified proteins can be separated, e.g., on a 15% polyacrylamide gel (**Fig. 2**), and protein concentrations can be determined by the Bradford assay ***(39)***.

Using the methods described, we obtained purified proteins with an average concentration of approximately 100 μg/mL (data not shown).

3.2. Generation of Plant Protein Microarrays

We used FAST-slides as the surface (*see* **Note 7**) to generate plant protein microarrays for antibody characterization ***(14)*** or for phosphorylation studies ***(8,33)*** (*see* Chapter 29 in this book).

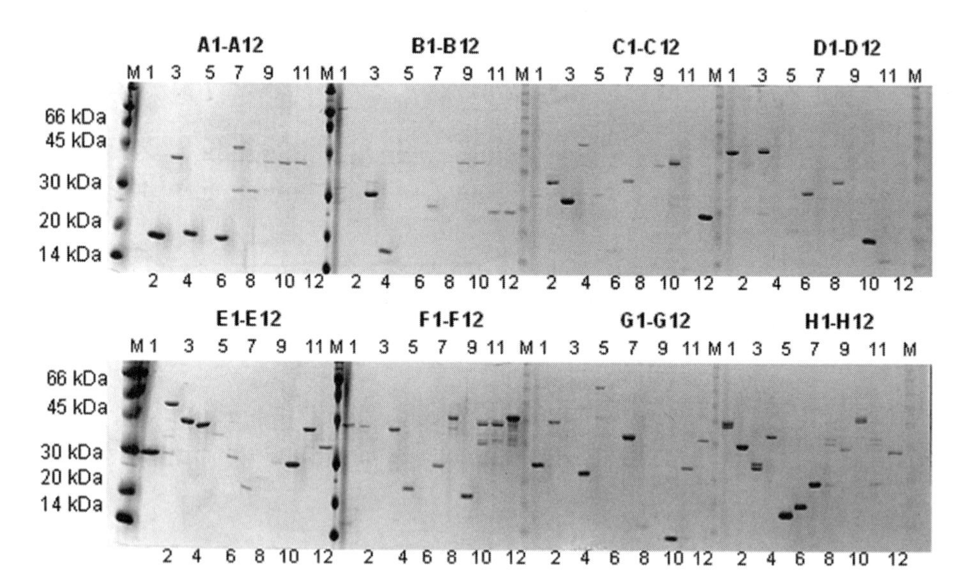

Fig. 2. SDS-Page (15%) of the 96 purified His-tagged *Arabidopsis* proteins. Proteins were expressed and purified from an *Arabidopsis* cDNA expression library ***(33)***. Proteins were separated using a 15% SDS-PAGE and then Coomassie stained. A1–H12, coordinates of the 96-well plate. M, Marker. Approximate sizes of the protein markers are shown at the left in kDa.

1. Prior to microarray production, 20 µL of the purified plant proteins and of the controls (*see* **Note 8**) were transferred to 384-well microtiter plates.
2. FAST-slides were placed in a QArray system equipped with humidity control (65 to 70%) and 16-blunt end stainless steel print pins with a tip diameter of 150 µm and a pin distance of 4.5 mm. We used a spotting depth of 120 µm.
3. Each sample was loaded once, transferring a volume of 0.6 nL per spot ***(40)***.
4. After each transfer the spotting gadget was washed in ddH_2O (6 s), dried with a heat fan (2 s), washed in 80% ethanol (6 s), and dried again (3 s) to prevent cross-contamination.
5. Protein microarrays were stored at 4°C in a closed slide holder.

Proteins can be spotted in different spotting patterns, e.g., in 4 × 4, 8 × 8, 10 × 10 patterns, and so on (*see* **Note 9**). Also, the protein can be spotted in duplicates, quadruplicates, and so forth. For antibody-antigen interaction studies, we recommend patterns with horizontal duplicates spotted in two identical fields (*see* **Note 10**).

Plant protein microarrays produced by this protocol can be utilized for antibody characterization, giving reproducible results for at least 3 wk when stored at 4°C. Furthermore, such plant protein microarrays may be applied for phos-

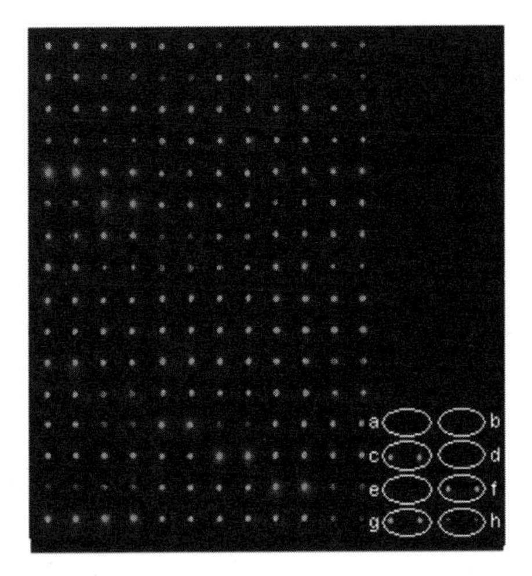

Fig. 3. Detection of 96 His-tagged *Arabidopsis* proteins on FAST-slides using an anti-RGS-His_6 antibody. Ninety six proteins were expressed and purified from an *Arabidopsis* cDNA expression library *(33)*. The proteins and several controls (a–h) were spotted as horizontal duplicates in a 4 × 4 spotting pattern on FAST-slides. The microarrays were screened with anti-RGS-His_6 antibody. Controls: a, elution buffer; b, PBS; c, rabbit antimouse IgG-Cy3 conjugate, diluted 1:25 in PBS; d, BSA, 20 pmol/μL in PBS; e, normal rabbit IgG, diluted 1:10 in PBS; f, normal rat IgG, diluted 1:10 in PBS; g, mouse anti-RGS-His_6 antibody, diluted 1:10 in PBS; h, normal mouse IgG, diluted 1:10 in PBS.

phorylation studies with protein kinases *(8,33)*, as also described in detail in this book (*see* Chapter 29).

For detection of the immobilized recombinant plant proteins, slides are incubated with an anti-RGS-His_6 antibody according to the following protocol (*see* **Subheading 3.3.1.**). As an example, **Fig. 3** shows a microarray containing 96 *Arabidopsis* proteins, which we expressed and purified from a recently described ordered cDNA expression library *(33)*. Proteins were spotted in a 4 × 4 spotting pattern with horizontally positioned duplicates in two identical fields. All the recombinant proteins gave a signal indicating that the efficiencies of protein expression, purification, and transfer were sufficient for detection of the tested recombinant proteins on the microarrays (**Fig. 3**).

3.3. Antibody Screening on Protein Microarrays

3.3.1. Monoclonal Antibody Screening on FAST-Slides

1. Plant protein microarrays were blocked for 1 h at room temperature with 2% BSA/TBST.

2. Mouse anti-RGS-His$_6$ (1:2000 dilution) or plant-specific antibodies (e.g., rat anti-TCP1 antibody; 1:1000 dilution) were diluted in blocking solution (*see* **Note 11**) and then applied onto the arrays for 1 h at room temperature (*see* **Note 12**).
3. Two 10-min wash steps with TBST were performed.
4. The slides were further incubated for 1 h at room temperature with the respective Cy3-labeled secondary antibody (rabbit antimouse IgG for anti-RGS-His$_6$ primary antibody; rabbit antirat IgG conjugate for anti-TCP1 antibody), which were applied at a 1:800 dilution in blocking solution (*see* **Notes 13** and **14**).
5. Then three wash steps of 30 min each were performed in TBST (**Note 15**).
6. Prior to scanning, we dried the microarrays using a microtiter plate centrifuge (e.g., Eppendorf, Hamburg, Germany, cat. no. 5810R) or by manual fanning.
7. Signal detection was performed by means of a 428 Arrayscanner System or a ScanArray 4,000.

All antibody incubation steps were carried out in a 200-μL volume underneath a cover slide.

To exclude nonspecific binding of the secondary antibody, we performed a parallel incubation of the protein microarrays with blocking solution without primary antibody followed by incubation with the respective secondary antibody, including all washing steps just described.

Figure 4 shows an example of the fluorescence image after incubation of an *Arabidopsis* protein microarray with a monoclonal anti-TCP1 antibody. Ninety six *Arabidopsis* proteins obtained from full-length cDNA expression clones (***14***) were immobilized on FAST-slides using a 4 × 4 spotting pattern with horizontally positioned duplicates in two identical fields. The image shows that the anti-TCP1 antibody specifically recognizes only the duplicate of TCP1 protein spots and does not cross-react with the other immobilized *Arabidopsis* proteins (**Fig. 4**).

3.3.2. Polyclonal Antibody Screening on FAST-Slides

1. The slides were blocked for 1 h at room temperature in fish gelatine (10% in TBST).
2. Rabbit sera were diluted in blocking solution (e.g., 1:1000 for anti-DOF11 or 1:500 for anti-MYB6 serum ***[14]***) and then applied to the arrays for 1 h at room temperature.
3. Two 10-min wash steps with TBST were performed.
4. The slides were further incubated for 1 h at room temperature with goat antirabbit IgG-Cy3 conjugate, which was applied at a 1:800 dilution in blocking solution (*see* **Note 13**).
5. Then three wash steps of 30 min each were performed in TBST.
6. Prior to scanning, we dried the microarrays using a microtiter plate centrifuge (e.g., Eppendorf, cat. no. 5810R) or by manual fanning.
7. Signal detection was performed by means of a 428 Arrayscanner System or a ScanArray 4000.

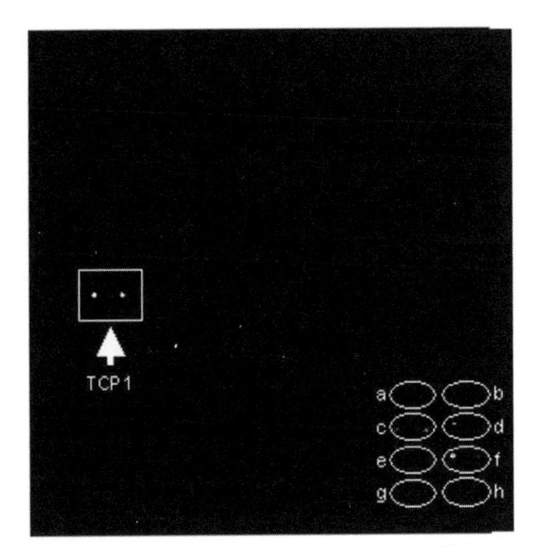

Fig. 4. Screening of an *Arabidopsis* protein microarray containing 96 proteins with an anti-TCP1 antibody. Proteins were expressed and purified from *Arabidopsis* full-length cDNA expression clones ***(14)***. The proteins and several controls (a–h) were spotted as horizontal duplicates in a 4 × 4 spotting pattern on FAST-slides. The microarrays were screened with anti-TCP1 antibody. For controls, see **Fig. 3** legend.

All antibody incubation steps were carried out in a 200-µL volume underneath a cover slide.

3.4. Image Analysis

The median spot intensity data (local background subtracted) were obtained with GenePixPro4.0; for further comparison, the average values of the intensities obtained from both spots of the duplicates were calculated.

4. Notes

1. pH values of the buffers have to be adjusted prior to every protein purification.
2. To yield extracts with higher content of specific recombinant proteins (e.g., for low-copy plasmids) for the subsequent purification, one can generate protein lysates from the pellets of two parallel 1-mL cultures using 100 µL lysis buffer each, which are then combined to 200 µL for the subsequent centrifugation ***(15)***.
3. Take off the supernatant very carefully. Only transfer clear supernatant to the filter plate to ensure that the plate will not occlude during the subsequent filtration.
4. Use a vacuum pump that is suitable for microtiter plates.
5. Ensure that no drops are retained underneath the filter plate after filtration; if so, they have to be transferred manually. To avoid this problem, the filtration should be carried out with enough pump power and as fast as possible. (The vacuum has

to build up quickly and should not be held up longer then necessary, because the filter plates become too leaky otherwise.)

6. Another possibility for performing high-throughput protein purification is to apply a Qiagen BioRobot 8,000 and the Ni-NTA Superflow 96 BioRobotKit (Qiagen) ***(14)*** using the same buffers for lysis, washing, and elution as in the case of manual purification (*see* **Subheading 2.1.**). Disadvantages of the robot method compared with the manual method are higher costs for consumables and the fact that the elution volume may not be adjusted to lower than 350 μL. Using manual purification, elution volumes from 30 to 80 μL may be used. Small elution volumes may be beneficial if rarely expressed proteins are purified.
7. FAST-slides are coated with a nitrocellulose-derived polymer. In terms of surface structure, they belong to the 3D microarrays, which have a higher immobilization capacity than 2D slides ***(2)***. For antigen-antibody interaction studies using monoclonal antibodies, we utilized self-made PAA slides in addition to FAST-slides ***(14)***. PAA slides have been prepared according to published protocols ***(15,41)***. Angenendt and co-workers compared different surfaces for antigen-antibody interaction studies ***(41,42)***. We reviewed different microarray surfaces for different microarray applications including antigen-antibody interaction studies ***(2)***.
8. Solutions without protein (elution buffer, PBS) as well as an unrelated protein without an RGS-His_6-tag (BSA) served as negative controls. As positive controls, we spotted the Cy3-labeled secondary antibodies (e.g., mouse IgG-Cy3 conjugate). To test the binding quality of the secondary antibodies we spotted respective monoclonal primary antibodies as well as IgG of the same species as the primary antibody (e.g., mouse IgG in the screening experiments with mouse anti-RGS-His_6 antibody or rabbit IgG in the screening experiments with polyclonal sera) as additional positive controls.
9. The spotting pattern used for antigen-antibody interaction studies on microarrays depends on the number of proteins and replicates to be analyzed. The maximum spot density at which individual spots can still be differentiated with fluorescence detection is highly dependent on the spotting method and the resolution of the microarray scanner. When we used our spotting protocol, a scanner with a resolution of 10 μm, and spotting patterns of 14 × 14 (spot distance of 321 μm) or 15 × 15 (spot distance of 300 μm), we obtained sufficient resolution to evaluate the fluorescence signals (data not shown). We have not tested higher spotting patterns up to now. Lueking and co-workers ***(15)*** successfully applied a 13 × 13 spotting pattern for antibody-antigen interaction studies with fluorescence detection using a similar spotting protocol on PAA slides but loading each sample five times to create one spot. (We loaded each sample once.) Michaud and colleagues ***(13)*** used yeast proteome arrays ***(17)*** spotted at even higher densities (16 × 18) to characterize antibodies utilizing fluorescence detection. We reviewed different antigen-antibody interaction studies using protein microarrays besides other applications ***(2)***.

10. In our proof-of-principle study to characterize plant antibodies, we used a 4 × 4 horizontal duplicate spotting pattern with a spot spacing of 1050 μm ***(14)***. Two identical fields were spotted on each slide.
11. The optimal antibody dilution, if not known from literature, may be determined prior to microarray studies by Western blotting experiments.
12. To prevent drying of the microarrays during antibody incubation, we put them on Whatman paper soaked with TBST.
13. Owing to fluorescence labeling of the secondary antibody, this incubation step and all subsequent steps have to be performed in the dark.
14. Fluorescently labeled conjugates are stored at 4°C or, after addition of glycerol (end concentration of 50%), at –20°C.
15. Alternatively to this protocol for screening of monoclonal antibodies on FAST-slides, we applied the following protocol ***(14)***: the FAST-slides were blocked with 1X PBS/0.5% (v/v) Tween-20 solution for 2 h at room temperature. A suitable dilution of monoclonal antibodies was then applied onto the microarrays for 2 h at room temperature, followed by two 10-min wash steps with blocking solution. The slides were further incubated with the respective secondary Cy3-labeled conjugate (species specific with regard to primary antibody) for 1 h at room temperature. The slides were washed two times with 1X PBS/0.5% (v/v) Tween-20 for 30 min and then two times with 1X PBS for 20 min at room temperature before the signal detection was performed.
    In the case of the monoclonal anti-PhyB antibody Pea-25 ***(43)***, only the application of this protocol led to the detection of PhyB on FAST-slides ***(14)***. Using the protocol described in **Subheading 3.3.1.** we were not able to detect PhyB.

Acknowledgments

We thank Alexandra Possling and Andrea König for reading the manuscript.

References

1. LaBaer, J. and Ramachandran, N. (2005) Protein microarrays as tools for functional proteomics. *Curr. Opin. Chem. Biol.* **9,** 14–19.
2. Feilner, T., Kreutzberger, J., Niemann, B., et al. (2004) Proteomic studies using microarrays. *Curr. Proteomics* **1,** 283–295.
3. Templin, M. F., Stoll, D., Schwenk, J. M., Potz, O., Kramer, S., and Joos, T. O. (2003) Protein microarrays: promising tools for proteomic research. *Proteomics* **3,** 2155–2166.
4. Zhu, H. and Snyder, M. (2003) Protein chip technology. *Curr. Opin. Chem. Biol.* **7,** 55–63.
5. Miller, J. C., Zhou, H., Kwekel, J., et al. (2003) Antibody microarray profiling of human prostate cancer sera: antibody screening and identification of potential biomarkers. *Proteomics* **3,** 56–63.
6. Nielsen, U. B. and Geierstanger, B. H. (2004) Multiplexed sandwich assays in microarray format. *J. Immunol. Methods* **290,** 107–120.

7. Shao, W., Zhou, Z., Laroche, I., et al. (2003) Optimization of Rolling-Circle Amplified Protein Microarrays for Multiplexed Protein Profiling. *J. Biomed. Biotechnol.* **2003,** 299–307.
8. Kramer, A., Feilner, T., Possling, A., et al. (2004) Identification of barley CK2alpha targets by using the protein microarray technology. *Phytochem.* **65,** 1777–1784.
9. Zhu, H., Klemic, J. F., Chang, S., et al. (2000) Analysis of yeast protein kinases using protein chips. *Nat. Genet.* **26,** 283–289.
10. Boutell, J. M., Hart, D. J., Godber, B. L., Kozlowski, R. Z., and Blackburn, J. M. (2004) Functional protein microarrays for parallel characterisation of p53 mutants. *Proteomics* **4,** 1950–1958.
11. Chan, S. M., Ermann, J., Su, L., Fathman, C. G., and Utz, P. J. (2004) Protein microarrays for multiplex analysis of signal transduction pathways. *Nat. Med.* **10,** 1390–1396.
12. Haab, B. B., Dunham, M. J., and Brown, P. O. (2001) Protein microarrays for highly parallel detection and quantitation of specific proteins and antibodies in complex solutions. *Genome Biol.* **2,** RESEARCH0004.
13. Michaud, G. A., Salcius, M., Zhou, F., et al. (2003) Analyzing antibody specificity with whole proteome microarrays. *Nat. Biotechnol.* **21,** 1509–1512.
14. Kersten, B., Feilner, T., Kramer, A., et al. (2003) Generation of *Arabidopsis* protein chips for antibody and serum screening. *Plant Mol. Biol.* **52,** 999–1010.
15. Lueking, A., Possling, A., Huber, O., et al. (2003) A nonredundant human protein chip for antibody screening and serum profiling. *Mol. Cell. Proteomics* **2,** 1342–1349.
16. Ramachandran, N., Hainsworth, E., Bhullar, B., et al. (2004) Self-assembling protein microarrays. *Science* **305,** 86–90.
17. Zhu, H., Bilgin, M., Bangham, R., et al. (2001) Global analysis of protein activities using proteome chips. *Science* **293,** 2101–2105.
18. Kersten, B., Possling, A., Blaesing, F., Mirgorodskaya, E., Gobom, J., and Seitz, H. (2004) Protein microarray technology and ultraviolet crosslinking combined with mass spectrometry for the analysis of protein-DNA interactions. *Anal. Biochem.* **331,** 303–313.
19. Snapyan, M., Lecocq, M., Guevel, L., Arnaud, M. C., Ghochikyan, A., and Sakanyan, V. (2003) Dissecting DNA-protein and protein-protein interactions involved in bacterial transcriptional regulation by a sensitive protein array method combining a near-infrared fluorescence detection. *Proteomics* **3,** 647–657.
20. MacBeath, G. and Schreiber, S. L. (2000) Printing proteins as microarrays for high-throughput function determination. *Science* **289,** 1760–1763.
21. Kim, S. H., Tamrazi, A., Carlson, K. E., and Katzenellenbogen, J. A. (2005) A proteomic microarray approach for exploring ligand-initiated nuclear hormone receptor pharmacology, receptor-selectivity,and heterodimer functionality. *Mol. Cell. Proteomics* **4,** 267–277.
22. Kim, T. E., Park, S. W., Cho, N. Y., et al. (2002) Quantitative measurement of serum allergen-specific IgE on protein chip. *Exp. Mol. Med.* **34,** 152–158.

23. Robinson, W. H., DiGennaro, C., Hueber, W., et al. (2002) Autoantigen microarrays for multiplex characterization of autoantibody responses. *Nat. Med.* **8,** 295–301.
24. Angenendt, P., Glokler, J., Konthur, Z., Lehrach, H., and Cahill, D. J. (2003) 3D protein microarrays: performing multiplex immunoassays on a single chip. *Anal. Chem.* **75,** 4368–4372.
25. Kersten, B., Wanker, E. E., Hoheisel, J. D., and Angenendt, P. (2005) Multiplexing approaches in protein microarray technology. *Expert Rev. Proteomics* **2,** 499–510.
26. Kersten, B., Bürkle, L., Kuhn, E. J., et al. (2002) Large-scale plant proteomics. *Plant Mol. Biol.* **48,** 133–141.
27. Kersten, B., Feilner, T., Angenendt, P., Giavalisco, P., Brenner, W., and Bürkle, L. (2004) Proteomic approaches in plant biology. *Curr. Proteomics* **1,** 131–144.
28. Agrawal, G. K., Yonekura, M., Iwahashi, Y., Iwahashi, H., and Rakwal, R. (2005) System, trends and perspectives of proteomics in dicot plants. Part III: Unraveling the proteomes influenced by the environment, and at the levels of function and genetic relationships. *J. Chromatogr. B Anal. Technol. Biomed. Life. Sci.* **815,** 137–145.
29. Agrawal, G. K., Yonekura, M., Iwahashi, Y., Iwahashi, H., and Rakwal, R. (2005) System, trends and perspectives of proteomics in dicot plants Part II: Proteomes of the complex developmental stages. *J. Chromatogr. B Anal. Technol. Biomed. Life. Sci.* **815,** 125–136.
30. Agrawal, G. K., Yonekura, M., Iwahashi, Y., Iwahashi, H., and Rakwal, R. (2005) System, trends and perspectives of proteomics in dicot plants Part I: Technologies in proteome establishment. *J. Chromatogr. B Anal. Technol. Biomed. Life. Sci.* **815,** 109–123.
31. Thiellement, H., Zivy, M., and Plomion, C. (2002) Combining proteomic and genetic studies in plants. *J. Chromatogr. B Analyt. Technol. Biomed. Life Sci.* **782,** 137–149.
32. Canovas, F. M., Dumas-Gaudot, E., Recorbet, G., Jorrin, J., Mock, H. P., and Rossignol, M. (2004) Plant proteome analysis. *Proteomics* **4,** 285–298.
33. Feilner, T., Hultschig, C., Lee, J., et al. (2005) High-throughput identification of potential *Arabidopsis* MAP kinases substrates. *Mol. Cell. Proteomics* **4,** 1558–1568.
34. Weiner, H., Faupel, T., and Bussow, K. (2004) Protein arrays from cDNA expression libraries, in *Methods in Molecular Biology* (Fung, E., ed.), Humana, Totowa, NJ, pp. 11–13.
35. Bussow, K., Cahill, D., Nietfeld, W., et al. (1998) A method for global protein expression and antibody screening on high-density filters of an arrayed cDNA library. *Nucleic Acids Res.* **26,** 5007–5008.
36. Clark, M. D., Panopoulou, G. D., Cahill, D. J., Bussow, K., and Lehrach, H. (1999) Construction and analysis of arrayed cDNA libraries [Review]. *Methods Enzymol.* **303,** 205–233.
37. Heyman, J. A., Cornthwaite, J., Foncerrada, L., et al. (1999) Genome-scale cloning and expression of individual open reading frames using topoisomerase I-mediated ligation. *Genome Res.* **9,** 383–392.

38. Walhout, A. J., Temple, G. F., Brasch, M. A., et al. (2000) GATEWAY recombinational cloning: application to the cloning of large numbers of open reading frames or ORFeomes. *Methods Enzymol.* **328,** 575–592.
39. Bradford, M. M. (1976) A rapid and sensitive method for the quantitation of microgram quantities of protein utilizing the principle of protein-dye binding. *Anal. Biochem.* **72,** 248–254.
40. Soldatov, A. V., Nabirochkina, E. N., Georgieva, S. G., and Eickhoff, H. (2001) Adjustment of transfer tools for the production of micro- and macroarrays. *Biotechniques* **31,** 848–854.
41. Angenendt, P., Glokler, J., Murphy, D., Lehrach, H., and Cahill, D. J. (2002) Toward optimized antibody microarrays: a comparison of current microarray support materials. *Anal. Biochem.* **309,** 253–260.
42. Angenendt, P., Glokler, J., Sobek, J., Lehrach, H., and Cahill, D. J. (2003) Next generation of protein microarray support materials: evaluation for protein and antibody microarray applications. *J. Chromatogr. A* **1009,** 97–104.
43. Cordonnier, M. M., Greppin, H., and Pratt, L. H. (1986) Identification of a highly conserved domain on phytochrome from angiosperms to algae. *Plant Physiol.* **80,** 982–987.

29

Phosphorylation Studies Using Plant Protein Microarrays

Tanja Feilner and Birgit Kersten

Summary

Identifying protein kinase substrates is one major focus of protein kinase research and supports the elucidation of signal transduction pathways and their complex regulation. In this chapter we describe a protein microarray-based in vitro method, which permits a systematic screening of immobilized proteins for their phosphorylation by specific protein kinases. This high-throughput method allows the identification of potential kinase substrates. The method starts with plant protein microarrays containing hundreds of purified recombinant His-tagged proteins. The microarrays have to be incubated with the soluble and active kinase in the presence of radioactive [γ^{33}-P]ATP. Only small volumes of active kinase are needed for one microarray experiment. Radioactive signals are then detectable by phosphor imager or X-ray film. The identified potential substrates have to be verified by other in vitro or in vivo methods, as both the kinase and the substrate may not interact with each other under *in vivo* conditions. This screening method could be generally applicable for direct identification of candidate substrates of various protein kinases.

Key Words: Protein microarrays; protein kinase; high throughput; phosphorylation screening; kinase substrates; targets.

1. Introduction

Posttranslational modification of proteins by phosphorylation is the most abundant type of cellular regulation affecting essentially every intracellular process of eukaryotes. Phosphorylation of a protein can cause changes in its structure, stability, enzymatic activity, the ability to interact with other molecules, or its subcellular localization. Protein kinases catalyze the reversible phosphorylation of protein substrates at serine, threonine or tyrosine residues. In *Arabidopsis*, e.g., serin/threonine kinases represent about 4% of the proteome *(1)*, but their biological function is not well understood. Therefore,

From: *Methods in Molecular Biology, vol. 335: Plant Proteomics: Methods and Protocols*
Edited by: H. Thiellement, M. Zivy, C. Damerval, and V. Méchin © Humana Press Inc., Totowa, NJ

high-throughput proteomics methods ***(2,3)*** for global analysis of protein kinase function are needed to identify downstream substrates.

Various techniques for determining consensus phosphorylation site sequences, including peptide libraries ***(4)*** and peptide arrays ***(5,6)*** led to the identification of kinase substrates. Furthermore, a solid-phase phosphorylation screening of λ phage cDNA expression libraries ***(7,8)*** as well as different protein-protein interaction screening methods, such as overlay methods ***(9–11)*** and the yeast two-hybrid system ***(12–14)*** have been applied in this respect. More recently, protein kinases were engineered to accept unnatural adenosine triphosphate (cyclopentyl ATP) analogs and have been used to identify specific substrates ***(15–17)***. Furthermore, preliminary studies have demonstrated the suitability of protein microarrays for the study of phosphorylation by kinases ***(18,19)***. To elucidate the sites of phosphorylation, antibodies against phosphorylated protein epitopes ***(6)*** may be used for detection on protein microarrays. Furthermore, peptide arrays ***(5,6)*** or mass spectrometry (MS)-based methods ***(20,21)*** may be applied in this respect.

We describe here a protein microarray-based method for phosphorylation screening of proteins and the identification of potential kinase substrates in a high-throughput manner. We successfully used this screening tool to identify novel targets for the barley casein kinase 2α (CK2α) ***(22)*** as well as for different *Arabidopsis thaliana* mitogen-activated protein (MAP) kinases ***(23)***. Our method utilizes plant protein microarrays generated as described in detail in the previous chapter in this book (*see* Chapter 28). Then microarrays have to be incubated with the soluble and active kinase in the presence of radioactive [γ^{33}-P]ATP. Radioactive signals detected by phosphor imager or X-ray film were then evaluated for the selection of potential substrates. We verified the potential substrates by on-blot phosphorylation in vitro. As a whole, our approach allows shortlisting candidate substrates of plant protein kinases for further analysis. Follow-up in vivo experiments are essential to evaluate their physiological relevance ***(13,24,25)***.

2. Materials

2.1. Purification of Recombinant Kinases Under Native Conditions

1. Medium for culturing cells: 2YT or LB containing 100 μg/mL ampicillin, 15 μg/mL kanamycin, and 2% glucose.
2. Isopropyl-β-D-thiogalactopyranosid (IPTG).
3. Native lysis buffer: 300 m*M* NaCl, 50 m*M* NaH_2PO_4, 10 m*M* imidazole, pH 8.0.
4. Native wash buffer: 300 m*M* NaCl, 50 m*M* NaH_2PO_4, 20 m*M* imidazole, pH 8.0.
5. Native elution buffer: 300 m*M* NaCl, 50 m*M* NaH_2PO_4, 250 m*M* imidazole, pH 8.0.
6. Lysozyme.
7. Phenylmethylsulfonyl fluoride (PMSF).

8. Ultrasonic homogenizer (Branson Ultrasonic, Danbury, CT).
9. NiNTA-agarose (NTA:nickel-nitrilotri-acetic acid, Qiagen).
10. 1 mL Polypropylene columns (Qiagen, Hilden, Germany).
11. Bradford reagent (Bio-Rad, Munich, Germany).

2.2. Kinase Assay on Protein Microarrays

1. TBS + Tween (TBST): 10 m*M* Tris-HCl, pH 7.5, 150 m*M* NaCl, 0.1% (v/v) Tween.
2. Blocking solution: 2% bovine serum albumin (BSA; Sigma, St. Louis, MO) in TBST.
3. [γ^{33}-P]ATP, 250 µCi/mL (Amersham Pharmacia Biotech Europe, Freiburg, Germany)
4. CK2α buffer: 25 m*M* Tris-HCl pH 8.5, 10 m*M* $MgCl_2$, 1 m*M* dithiothreitol (DTT).
5. FAST™-slides (Whatman Schleicher & Schuell, Dassel, Germany).
6. Cover slide (Carl Roth, Karlsruhe, Germany).
7. X-ray cassette (Hypercassette, Amersham Pharmacia, Freiburg, Germany).
8. X-ray film (Kodak, Stuttgart, Germany).
9. Imaging screen (BAS-SR 2025, Fujifilm, Japan).
10. Phosphor imager (BAS-Reader-5000, Fujifilm).

3. Methods

3.1. Purification of Recombinant Kinases Under Native Conditions

For phosphorylation screening, the kinase has to be soluble, active, and pure, which means that no other kinase should be present. To achieve this, large-scale production and purification of recombinant His-tagged protein kinases can be carried out from cDNA expression libraries, e.g., a barley expression library constructed in the *E. coli* vector pQE30NST (accession number: AF074376; ***22***) or an *Arabidopsis* expression library in vector pQE30NASTattB (accession number: AY386205; ***23***; see also Chapter 28 in this book, **Subheading 3.1.**). To express and purify protein kinases from these libraries, we apply the following protocol:

For expression:

1. Falcon tubes (50 mL) were filled with 10 mL of medium for culturing cells.
2. Cultures were inoculated with bacteria expressing the His-tagged protein kinase. They were taken from 384-well plates, which were stored at –80°C.
3. After overnight growth at 37°C with vigorous shaking, the cultures were transferred to 300-mL Erlenmeyer flasks, and 90 mL of prewarmed 2YT medium, supplemented with 100 µg/mL of ampicillin, and 15 µg/mL of kanamycin were added. The incubation was continued until an OD of 0.7 was reached.
4. To induce protein expression, IPTG was added to a final concentration of 1 m*M*, and incubation was continued for 4 h at 37°C.
5. Cells were transferred into 50-mL Falcon tubes (two for each culture), harvested by centrifugation at 1900*g* for 10 min, and stored immediately at –80°C.

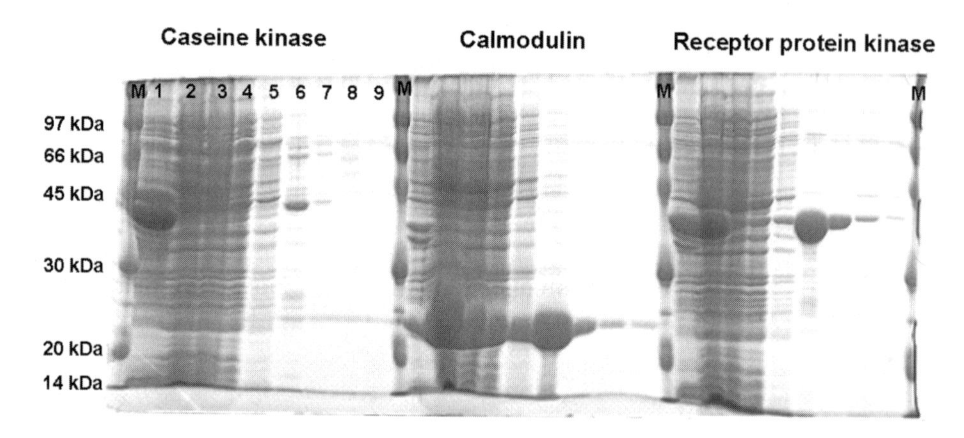

Fig. 1. SDS-PAGE (15%) of two native purified kinases and calmodulin (Coomassie-stained). Nine samples of each purification step were separated: 1, pellet (insoluble fraction after centrifugation); 2, supernatant (soluble fraction after centrifugation); 3, first column flowthrough; 4, 5, first and second washing fraction; 6–9, eluates. M, molecular weight marker. Approximate sizes of the protein markers are shown at the left in kDa.

For native protein purification (all the following steps were performed at 0–4°C):

1. The frozen pellets were thawed on ice.
2. Cells were lysed in 0.5 mL of native lysis buffer supplemented with 0.25 mg/mL of lysozyme and 0.1 m*M* of PMSF.
3. Lysates were pooled, and DNA was sheared with an ultrasonic homogenizer for 3X 1 min at 50% power on ice.
4. Lysates were transferred into 1.5-mL Eppendorf tubes and centrifuged at 20,000*g*, 4°C for 30 min.
5. Supernatants were transferred into fresh Eppendorf tubes.
6. Then 250 µL of NiNTA-agarose were added to bind the His-tagged proteins by shaking for 1 h at 300 rpm at 4°C.
7. The mixtures were transferred to 1-mL polypropylen columns.
8. The columns were washed with 10 bed vol native wash buffer.
9. Proteins were eluted with four elution steps using 0.5 mL of native elution buffer each.
10. Proteins were mixed with glycerol (20% [v/v] end concentration) and stored at –80°C (*see* **Note 1**).

We recommend collecting an aliquot from each centrifugation, lysis, wash, and elution step to control the efficiency of the purification steps using a polyacrylamide gel, e.g., 15% (*see* **Fig. 1**). Protein concentration was determined by the Bradford assay ***(26)***.

3.2. Kinase Assay on Protein Microarrays

We recommend checking the activity of the kinase prior to microarray experiments (*see* **Note 2**) using a known substrate as a positive control (*see* **Note 3**). Furthermore, prior to microarray experiments, one has to exclude the possibility that the kinase will phosphorylate 3' or 5' tags of the recombinant proteins, such as His-tags (*see* **Note 4**).

Here, we intended to perform a qualitative signal evaluation after the kinase assay. Therefore we spotted proteins and controls (*see* **Note 3**) in quadruplicates on FAST-slides (*see* **Note 5**) using a 10 × 10 spotting pattern (*see* **Note 6**), as demonstrated in **Fig. 2** *(22)*. Spotting patterns for following quantitative evaluation have been described recently *(23)*.

For kinase assay:

1. Microarrays (for generation *see* Chapter 28) which were spotted the day before and stored at 4°C in a closed slide holder, were washed for 1 h in TBST with vigorous shaking at room temperature (*see* **Note 7**).
2. Washed microarrays were blocked for 1 h at room temperature with 2% BSA/TBST.
3. Blocked microarrays were placed on Whatman paper soaked with TBST to avoid drying of the microarrays during the subsequent kinase reaction.
4. Microarrays were incubated with 250 µL kinase solution containing 13 µg/mL of CK2α and 25 µCi/mL of radioactive labeled [γ^{33}-P]ATP in CK2α buffer for 1 h at room temperature (*see* **Note 1**). This incubation was performed underneath a cover slide.
5. Six wash steps of 30 min each were performed in TBST.
6. Finally, the microarrays were dried using a microtiter plate centrifuge (e.g., Eppendorf, Hamburg, Germany, cat. no. 5810R) or by manual fanning and transferred to an X-ray cassette.
7. Signal detection was performed by means of an X-ray film. The film was laid on the microarrays, and the cassette was stored for a suitable time at –80°C (*see* **Note 8**). Afterwards, the film was developed in a dark chamber. Alternatively, the microarrays could be exposed to an imaging screen that would be screened afterward by a phosphor imager for signal detection (*see* **Note 8**).

Figure 2 (left) shows a typical result of a phosphorylation screening experiment with CK2α after exposure of the microarrays to X-ray film. Quadruples of several proteins gave distinct signals on the film.

To determine the immobilization of the recombinant His-tagged proteins on the microarrays we screened them with an anti-RGS-His_6 antibody, as described in detail in this book (*see* Chapter 28). Nearly all the spotted proteins were detectable (approx 95%), as shown in **Fig. 2** (right).

3.3. Selection of Potential Kinase Targets

For qualitative signal evaluation, the results of two independent experiments with different preparations of the kinase were used.

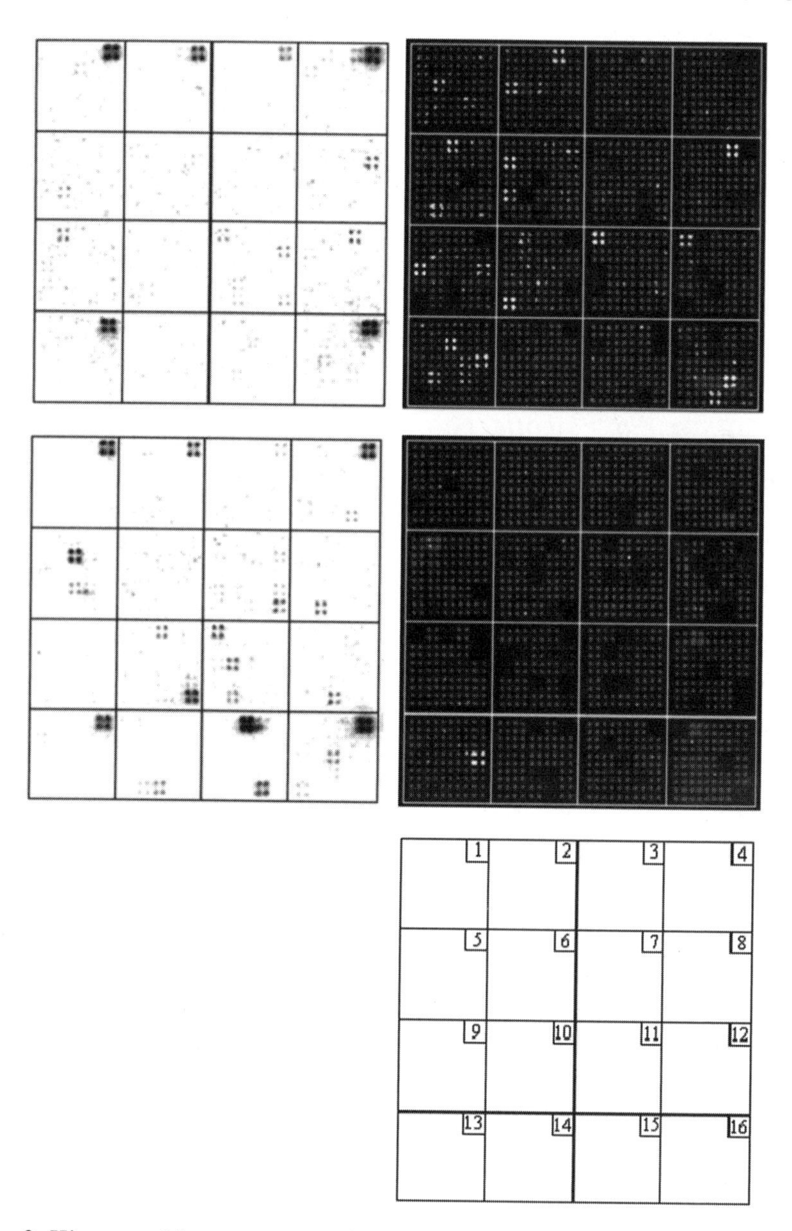

Fig. 2. Kinase and immunoassay with 768 different barley proteins, immobilized on FAST-slides (384 proteins in each field). Left: X-ray film image with the results of a representative CK2α assay. Right: equal microarrays screened with an anti-RGS-His_6 antibody. The map at the bottom shows the positions of the 16 controls. The purified proteins and the controls were spotted in quadruplicate. Controls: 1, 4, 13, and 16, high-mobility group (HMG)(1), 0.29 mg/mL; 2, HMG(2), 0.76 mg/mL; 3, Fig.

Criteria for selection of the potential kinase targets were:

1. In one experiment, a protein was considered positive if all four spots of the corresponding quadruplicate gave a distinct signal on the X-ray film or in the phosphor imager (**Fig. 2**).
2. A protein was regarded as a potential target protein if it attained positive results in both experiments.

Thus, for example, 21 potential CK2α target proteins were identified out of nearly 800 different barley proteins *(22)*. The selection of potential targets using a threshold-based quantification system after phosphorylation on microarrays has been described recently *(23)*.

3.4. Verification of Potential Phosphorylation Targets

As this microarray-based phosphorylation method is a screening tool to identify potential phosphorylation targets in vitro they have to be verified, by other methods. For such verifications, various in vitro or in vivo approaches may be applied. A few of them have already been mentioned in the introduction.

We performed an on-blot phosphorylation assay to verify potential substrates in vitro:

1. Proteins were separated using an SDS–PAGE, e.g., 15%, followed by blotting of the proteins on PVDF membranes. After blotting, gels were Coomassie stained to check the transfer efficiency of the proteins.
2. The phosphorylation reaction was carried out on the blot membrane using reaction conditions as described for the microarray-based assay in an appropriate volume.

Figure 3 shows a typical example of such an on-blot phosphorylation experiment. Forty-eight proteins, which were identified in both microarray experiments as potential substrates using a MAP kinase, were verified with this method. Nearly all were confirmed, as shown in **Fig. 3**. We used the same positive controls as in the microarray experiments: different concentrations of myelin basic protein (MBP; artificial substrate of MAP kinases). For negative controls proteins were used that had been tested in the microarray-based kinase assay and had not been identified as potential substrates (**Fig. 3**).

---

2. (*continued*) HMG(3), 0.28 mg/mL; 5,CK2α, 0.22 mg/mL; 6, other library kinase, 0.21 mg/mL; 7, human gapdh (His-tagged recombinant human glyceraldehyde-3-phosphate dehydrogenase), 0.36 mg/mL; 8, BSA, 20 pmol/μL in PBS; 9, native elution buffer; 10, denaturing elution buffer; 11, mouse anti-RGS-His_6 antibody, diluted 1:10 in PBS; 12, rabbit antimouse IgG-Cy3 conjugate, diluted 1:25 in PBS; 14, H_2O; 15, PBS. (Reprinted with permission from **ref. *22***. Copyright 2004 Elsevier Science.)

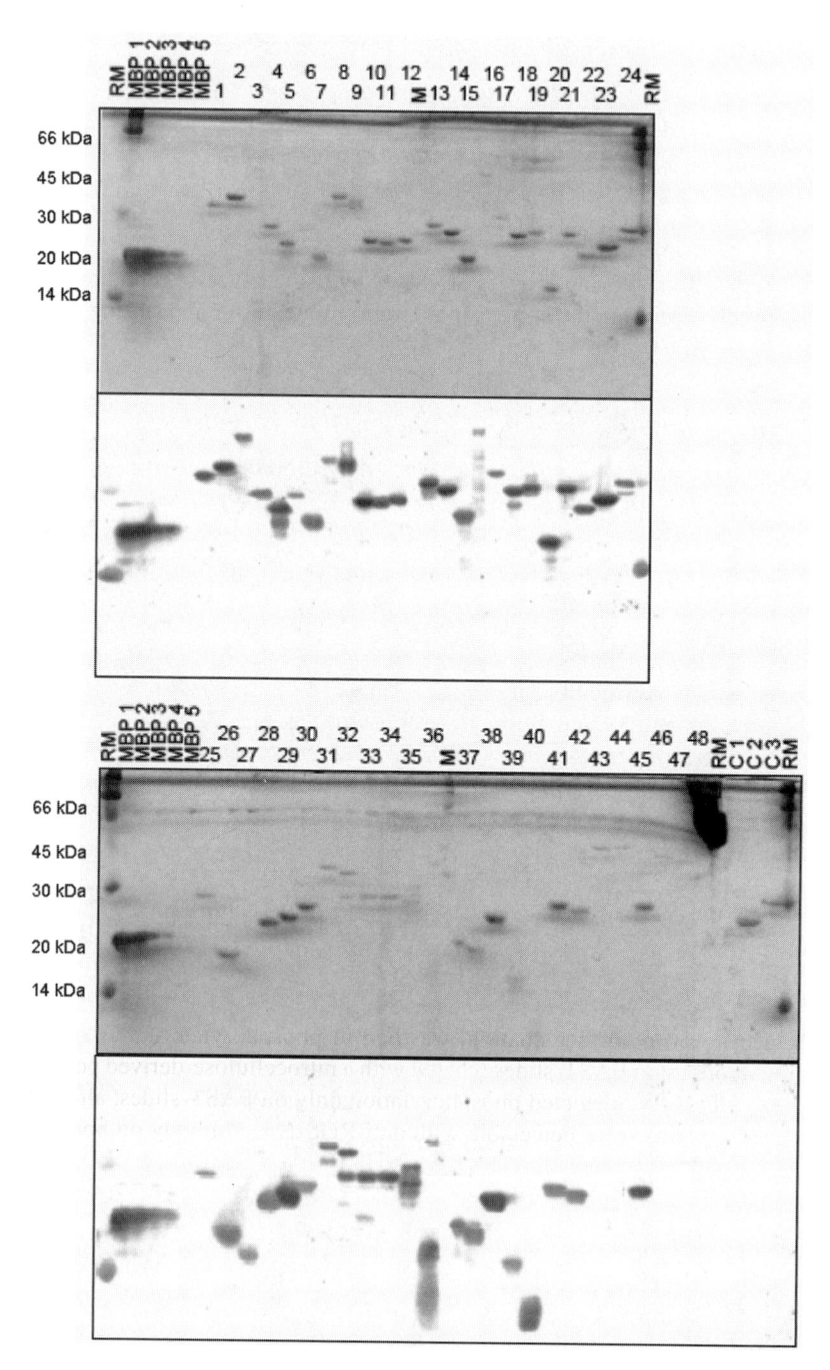

Fig. 3. Verification of kinase substrates. SDS-PAGE (15%) gels of potential kinase substrates (upper panel) and autoradiograms of these proteins after blotting and kinase reaction (lower panel). Gels were Coomassie stained after electrophoresis and blotting.

Other in vitro verification methods are phosphorylation assays performed in solution with subsequent detection of the phosphorylated proteins by SDS electrophoresis and autoradiography ***(8,23)***. Such an assay has been used by Fukunaga and Hunter to verify candidate kinase substrates identified by phosphorylation screening of an expression phage cDNA library ***(8)***.

4. Notes

1. We recommend storing kinases at –80°C by using glycerol (end concentration 20%) or lyophilizing the kinases in the appropriate buffer and storing them afterward at –80°C. The lyophilized kinases need to be dissolved in ddH_2O shortly before use.
2. Prior to performing microarray-based kinase experiments, we recommend testing the activity of the prepared kinase by in-gel assays ***(27)*** or kinase assays in solution ***(8)*** using a known substrate of the respective kinase (*see* **Note 3**). Optimal conditions for kinase activation (e.g., optimal buffer conditions, kinase concentration, and incubation time of the kinase) must be determined experimentally for each individual kinase, if they are not known from the literature.
3. The use of a positive control is recommended (e.g., a known substrate of the kinase). In studies with barley CK2α ***(22)***, we used as positive controls different barley proteins, which share a strong homology with different plant HMG (high mobility group) proteins. These proteins are well-known targets of CK2α ***(28,29)***. In another study, we analyzed different *Arabidopsis* MAP kinases ***(23)***. In this case a known artificial substrate, MBP, was used as a positive control.
4. Recombinant proteins often contain 3' or 5' tags, which are encoded by the cloning vector in addition to the coding sequence of the respective protein. Phosphorylation of these tags may result in false-positive results and should therefore be excluded prior to microarray experiments. One possibility is to perform test assays in solution with the active kinase using some recombinant proteins that are known not to be substrates of the respective kinases and that are expressed the same manner (same tags) as the recombinant proteins used for the microarray experiments. These selected proteins should yield negative results in the test.
5. With respect to surface coating, we tried to phosphorylate immobilized denatured proteins on FAST-slides (coated with a nitrocellulose-derived polymer) and epoxy slides. We detected phosphorylation only on FAST-slides, although His-tagged proteins were detectable with anti-RGS-His_6 antibody on both surfaces. In other studies, nanowells ***(19)***, BSA-NHS (BSA-*N*-hydroxysuccinimide) slides ***(18)***, or SAM2 (streptavidin-coated membrane) slides ***(30)*** have been used as

---

Fig. 3. (*continued*) MBP1–MBP5, positive controls. MBP in various dilution steps (MBP1: 2000 ng/μL; MBP2: 1000 ng/μL; MBP3: 500 ng/μL; MBP4: 250 ng/μL; MBP5: 125 ng/μL. 1–48, potential kinase substrates identified in the microarray experiments. C1–C3, negative controls proteins that were not identified as potential substrates using microarrays. RM, rainbow marker; M, molecular weight marker. Approximate sizes of the protein markers are shown at the left in kDa.

surfaces in microarray-based kinase assays. We reviewed different micorarray surfaces for different microarray applications including phosphorylation studies *(31)*.

6. The maximum spot density, which at these individual spots can still be differentiated in a phosphorylation assay, is highly dependent on the resolution of the phosphor imager used to scan the imaging plates. For determination of this density, we recommend the use of commercially available protein kinase (e.g., PKA from New England Biolabs). When using a scanner with a resolution of 10 μm (e.g., BioImager FLA 8,000, Fujifilm, Japan) in combination with spotting patterns up to 11 × 11 (spot distance: 410 μm), we received sufficient resolution to evaluate the radioactive signals. Applications of higher spotting patterns (14 × 14-pattern with a spot distance of 321 μm; 15 × 15 pattern with a spot distance of 300 μm) were not suitable for radioactive detection, as signals of intense spots interfered with neighboring spot signals. This led to faulty analysis of the signal intensity of neighboring proteins as well as faulty background corrections.
7. This washing step in TBST was performed to remove urea from the microarrays, as we recognized that urea reduces the activity of the kinase in the microarray assay.
8. For detection of radioactive signals with X-ray film, the film was laid onto the microarrays, which were covered with cling film. The exposure time depends on the signal intensity and can vary between 30 min to a few days. The sensitivity of the X-ray films can cause problems, as blackening of the films is only in a very limited area that is linear and proportional to signal intensity. For detection with a phosphor imager, the imaging plates are laid onto the covered microarrays and then scanned. Detection by phosphor imager is 10 to 100 times more sensitive than by X-ray films, and therefore it is possible to get faster results and/or detect poor signals, respectively. Furthermore, the system has a higher dynamic range compared with the X-ray film. Strong signals as well as poor signals are detected within one exposure and the linear signal-to-intensity range has a dimension over 5 (100,000:1). This shows that the phosphor imager is better for quantification of radioactive signals than X-ray films.

Acknowledgments

We thank Armin Kramer for providing **Fig. 1** and Jasmin Bastian for reading the manuscript.

References

1. Champion, A., Kreis, M., Mockaitis, K., Picaud, A., and Henry, Y. (2004) *Arabidopsis* kinome: after the casting. *Funct. Integr. Genomics* **4,** 163–187.
2. Phizicky, E., Bastiaens, P. I., Zhu, H., Snyder, M., and Fields, S. (2003) Protein analysis on a proteomic scale. *Nature* **422,** 208–215.
3. Kersten, B., Feilner, T., Angenendt, P., Giavalisco, P., Brenner, W., and Bürkle, L. (2004) Proteomic approaches in plant biology. *Curr. Proteomics* **1,** 131–144.
4. Hutti, J. E., Jarrell, E. T., Chang, J. D., et al. (2004) A rapid method for determining protein kinase phosphorylation specificity. *Nat. Methods* **1,** 27–29.

5. Houseman, B. T., Huh, J. H., Kron, S. J., and Mrksich, M. (2002) Peptide chips for the quantitative evaluation of protein kinase activity. *Nat. Biotechnol.* **20,** 270–274.
6. Lesaicherre, M. L., Uttamchandani, M., Chen, G. Y., and Yao, S. Q. (2002) Antibody-based fluorescence detection of kinase activity on a peptide array. *Bioorg. Med. Chem. Lett.* **12,** 2085–2088.
7. Fukunaga, R. and Hunter, T. (1997) MNK1, a new MAP kinase-activated protein kinase, isolated by a novel expression screening method for identifying protein kinase substrates. *EMBO J.* **16,** 1921–1933.
8. Fukunaga, R. and Hunter, T. (2004) Identification of MAPK substrates by expression screening with solid-phase phosphorylation, in *Methods in Molecular Biology* (Seger, R., ed), Humana Press, Totowa, NJ, pp. 211–236.
9. Zhao, J., Dynlacht, B., Imai, T., Hori, T., and Harlow, E. (1998) Expression of NPAT, a novel substrate of cyclin E-CDK2, promotes S-phase entry. *Genes. Dev.* **12,** 456–461.
10. Zhao, J., Kennedy, B. K., Lawrence, B. D., et al. (2000) NPAT links cyclin E-Cdk2 to the regulation of replication-dependent histone gene transcription. *Genes. Dev.* **14,** 2283–2297.
11. Gao, G., Bracken, A. P., Burkard, K., et al. (2003) NPAT expression is regulated by E2F and is essential for cell cycle progression. *Mol. Cell. Biol.* **23,** 2821–2833.
12. Waskiewicz, A. J., Flynn, A., Proud, C. G., and Cooper, J. A. (1997) Mitogen-activated protein kinases activate the serine/threonine kinases Mnk1 and Mnk2. *EMBO J.* **16,** 1909–1920.
13. Fujita, H., Fujita, H., Takemura, M., et al. (2003) An *Arabidopsis* MADS-box protein, AGL24, is specifically bound to and phosphorylated by meristematic receptor-like kinase (MRLK). *Plant Cell Physiol.* **44,** 735–742.
14. Anderson, G. H. and Hanson, M. R. (2005) The *Arabidopsis* Mei2 homologue AML1 binds AtRaptor1B, the plant homologue of a major regulator of eukaryotic cell growth. *BMC Plant Biol.* **5,** 2.
15. Shah, K., Liu, Y., Deirmengian, C., and Shokat, K. M. (1997) Engineering unnatural nucleotide specificity for Rous sarcoma virus tyrosine kinase to uniquely label its direct substrates. *Proc. Natl. Acad. Sci. USA* **94,** 3565–3570.
16. Liu, Y., Shah, K., Yang, F., Witucki, L., and Shokat, K. M. (1998) Engineering Src family protein kinases with unnatural nucleotide specificity. *Chem. Biol.* **5,** 91–101.
17. Eblen, S. T., Kumar, N. V., Shah, K., et al. (2003) Identification of novel ERK2 substrates through use of an engineered kinase and ATP analogs. *J. Biol. Chem.* **278,** 14926–14935.
18. MacBeath, G. and Schreiber, S. L. (2000) Printing proteins as microarrays for high-throughput function determination. *Science* **289,** 1760–1763.
19. Zhu, H., Klemic, J. F., Chang, S., et al. (2000) Analysis of yeast protein kinases using protein chips. *Nat. Genet.* **26,** 283–289.
20. Glinski, M., Romeis, T., Witte, C. P., Wienkoop, S., and Weckwerth, W. (2003) Stable isotope labeling of phosphopeptides for multiparallel kinase target analysis

and identification of phosphorylation sites. *Rapid Commun. Mass. Spectrom.* **17,** 1579–1584.
21. Ballif, B. A., Villen, J., Beausoleil, S. A., Schwartz, D., and Gygi, S. P. (2004) Phosphoproteomic analysis of the developing mouse brain. *Mol. Cell. Proteomics* **3,** 1093–1101.
22. Kramer, A., Feilner, T., Possling, A., et al. (2004) Identification of barley CK2alpha targets by using the protein microarray technology. *Phytochemistry* **65,** 1777–1784.
23. Feilner, T., Hultschig, C., Lee, J., et al. (2005) High throughput identification of potential *Arabidopsis* mitogen-activated protein kinase substrates. *Mol. Cell. Proteomics* **4,** 1558–1568.
24. Liu, Y. and Zhang, S. (2004) Phosphorylation of 1-aminocyclopropane-1-carboxylic acid synthase by MPK6, a stress-responsive mitogen-activated protein kinase, induces ethylene biosynthesis in *Arabidopsis*. *Plant Cell* **16,** 3386–3399.
25. Riera, M., Figueras, M., Lopez, C., Goday, A., and Pages, M. (2004) Protein kinase CK2 modulates developmental functions of the abscisic acid responsive protein Rab17 from maize. *Proc. Natl. Acad. Sci. USA* **101,** 9879–9884.
26. Bradford, M. M. (1976) A rapid and sensitive method for the quantitation of microgram quantities of protein utilizing the principle of protein-dye binding. *Anal. Biochem.* **72,** 248–254.
27. Usami, S., Banno, H., Ito, Y., Nishihama, R., and Machida, Y. (1995) Cutting activates a 46-kilodalton protein kinase in plants. *Proc. Natl. Acad. Sci. USA* **92,** 8660–8664.
28. Grasser, K. D., Maier, U. G., and Feix, G. (1989) A nuclear casein type II kinase from maize endosperm phosphorylating HMG proteins. *Biochem. Biophys. Res. Commun.* **162,** 456–463.
29. Stemmer, C., Schwander, A., Bauw, G., Fojan, P., and Grasser, K. D. (2002) Protein kinase CK2 differentially phosphorylates maize chromosomal high mobility group B (HMGB) proteins modulating their stability and DNA interactions. *J. Biol. Chem.* **277,** 1092–1098.
30. Boutell, J. M., Hart, D. J., Godber, B. L., Kozlowski, R. Z., and Blackburn, J. M. (2004) Functional protein microarrays for parallel characterisation of p53 mutants. *Proteomics* **4,** 1950–1958.
31. Feilner, T., Kreutzberger, J., Niemann, B., et al. (2004) Proteomic studies using microarrays. *Curr. Proteomics* **1,** 283–295.

Index

N

O, P

R,S